A Mathematical Approach to Engineering

Volume I

A Mathematical Approach to Engineering Volume I

Edited by **Matt Ferrier**

Clanrye International
New Jersey

Published by Clanrye International,
55 Van Reypen Street,
Jersey City, NJ 07306, USA
www.clanryeinternational.com

A Mathematical Approach to Engineering: Volume I
Edited by Matt Ferrier

International Standard Book Number: 978-1-63240-005-5 (Hardback)

Printed in the United States of America.

Contents

Preface

Mathematical problems in engineering usually refer to the branch of applied mathematics and techniques that is generally used in engineering fields. This can be seen as an interdisciplinary field with offshoots and applications in various branches of engineering. It is motivated and based on the foundation of an engineer's requirement for practical, theoretical and other considerations with their specialization and to deal with restrictions to be effective in their work. People with skills in the arena of mathematical problems in engineering are often required to use their knowledge to formulate a problem, translate the problem into mathematical terms, use their mathematical understanding to question and interpret results, to write computer programs and to use their knowledge of engineering to implement solution. Graduates of this discipline are students who are equipped with transferable and technical skills that enable them to control leading and creative roles as engineers and mathematicians in industry, academic research, and elsewhere. They are trained in the analysis, design and development of numerous real world and practical engineering problems. Thus this aptitude which is embedded into engineering mathematicians is beneficial to all disciplines in engineering.

This book is an attempt to compile, collate and understand the research being done in the field of engineering mathematics and the data found in the advancements in this field. I am grateful to those who put in their hard work and efforts into this field as well as those who supported us in this endeavor. I would also like to thank the publishing team for their outstanding technical assistance. Lastly, I wish to convey my regards to my friends and family who have always supported me in every endeavor of my life.

Editor

1

Automated Visual Inspection of Ship Hull Surfaces Using the Wavelet Transform

Carlos Fernández-Isla, Pedro J. Navarro, and Pedro María Alcover

Universidad Politécnica de Cartagena, Campus Muralla del Mar, 30202 Cartagena, Spain

Correspondence should be addressed to Carlos Fernández-Isla; carlos.fernandez@upct.es

Academic Editor: Wen-Jer Chang

A new online visual inspection technique is proposed, based on a wavelet reconstruction scheme over images obtained from the hull. This type of visual inspection to detect defects in hull surfaces is commonly carried out at shipyards by human inspectors before the hull repair task starts. We propose the use of Shannon entropy for automatic selection of the band for image reconstruction which provides a low decomposition level, thus avoiding excessive degradation of the image, allowing more precise defect segmentation. The proposed method here is capable of on-line assisting to a robotic system to perform grit blasting operations over damage areas of ship hulls. This solution allows a reliable and cost-effective operation for hull grit spot blasting. A prototype of the automated blasting system has been developed and tested in the Spanish NAVANTIA shipyards.

1. Introduction

Main ships' maintenance care consists of periodical (every 4-5 years) hull treatment which includes blasting works; blasting consists in projecting a high-pressure jet of abrasive matter (typically water or grit) onto a surface to remove adherences or rust traces. The object of this task is to maintain hull integrity, guarantee navigational safety conditions, and assure that the surface offers little resistance to the water in order to reduce fuel consumption. That object can be achieved by grit blasting [1] or ultra-high pressure water jetting [2]. In most cases these techniques are applied using manual or semiautomated procedures with the help of robotized devices [3]. In either case defects are detected by means of human operators; this is therefore a subjective task and hence vulnerable to cumulative operator fatigue and highly dependent on the experience of the personnel performing the task. Figure 1 shows a view of ship's hulls under repair at NAVANTIA's shipyards.

From an operational point of view, there are two working modes: full blasting and spot blasting. Full blasting consists of blasting the entire hull of the ship, while spot blasting consists of blasting numerous isolated areas where corrosion has been observed. Spot blasting is the most demanded operation due to cost saving reasons. This second working mode demands very precise information about position, size, and shape of damaged portions of the hull to make robotic devices [3–5] to achieve maximum efficiency.

This paper proposes a computer vision algorithm which equips a machine vision system (see Figure 2), capable for precisely detecting defects in ship hulls which is simple enough to be implemented in such a way as to meet the real-time requirements for the application.

Because of the textured appearance of the hull's surface under inspection (see Figures 1(c) and 1(d)), we have used the wavelet transform, and the developed computer vision algorithm includes an image reconstruction approach based on automatic selection of the optimal wavelet transform resolution level, using Shannon entropy.

2. Defect Detection in Textured Surfaces

Texture is a very important characteristic when identifying defects or flaws, as it provides important information for defect detection. In fact, the task of detecting defects has been largely seen as a texture analysis problem. Figure 3 shows several texture images from ship hull surfaces.

In his review Xie [6] classified texture analysis techniques for defect detection in four categories: statistical approaches,

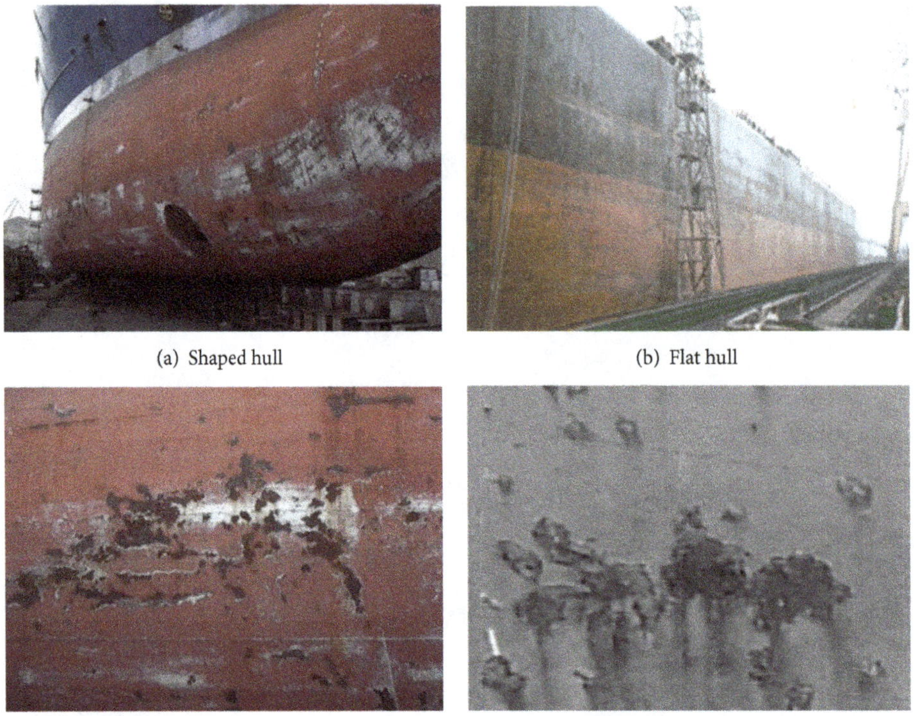

(a) Shaped hull

(b) Flat hull

(c) Damaged portions

(d) Detail visual appearance

FIGURE 1: Ship's hulls at a repair yard.

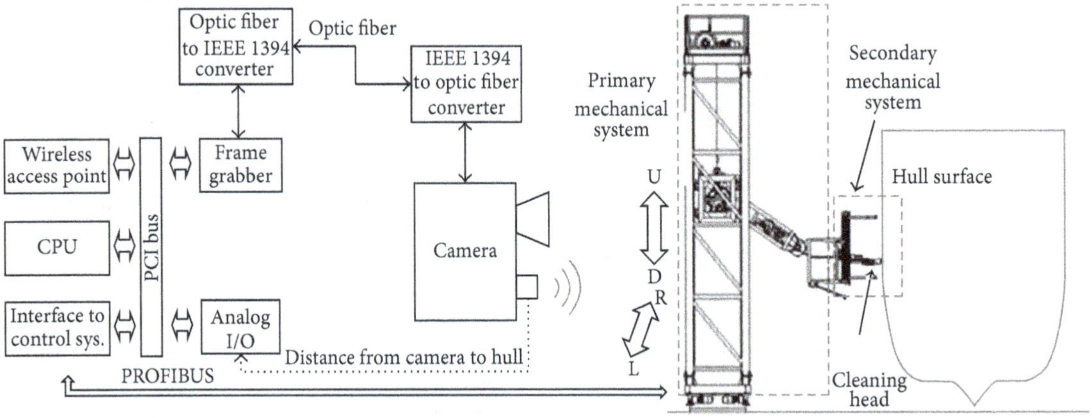

FIGURE 2: Machine vision system for hull blasting.

structural approaches, filter based approaches, and model based approaches. In his review of fabric defect detection similarly Kumar [7] classified the proposed solutions into three categories: statistical, spectral, and model based. In his review of automated defect detection in fabric, Ngan et al. [8] classified defect detection techniques in textured fabric into nonmotif-based and motif-based approaches. The motif-based approach [9] uses the symmetry property of motifs to calculate the energy of moving subtraction and its variance among different motifs. Many defect detection methods usually use clustering techniques which are mainly based on texture feature extraction and texture classifications. These features are collated using methods such as cooccurrence matrix [10], Fourier transform [11], Gabor transform [12], or the wavelet transform [13].

Spectral-approach methods for texture analysis characterize the frequency contents of a texture image—Fourier transform—or provide spatial-frequency analysis—Gabor filters, wavelets. A two-dimensional spectrum of a visual texture frequently contains information about the periodicity and directionality of the texture pattern. For example, a texture with a coarse appearance analysed from the spectral point of view shows high-frequency components, while a texture with a fine appearance shows low-frequency components. The analytical methods based on Fourier transform show good results in texture patterns with high regularity

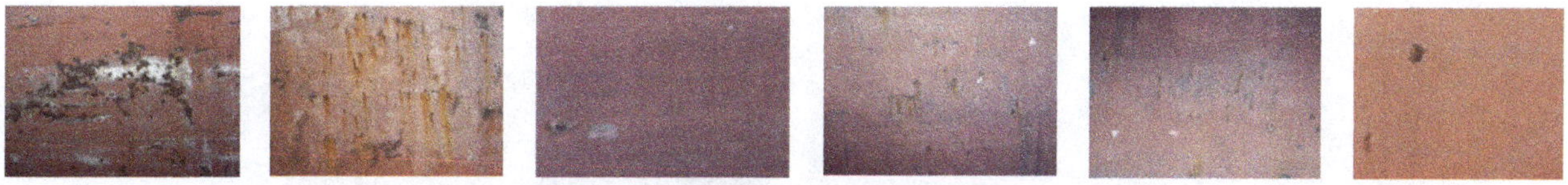

FIGURE 3: Texture images from ships' hull surface.

and/or directionality, but they are limited by a lack of spatial localization. In this field, Gabor filters provide better spatial localization, although their utility in natural textures is limited because there is no single filter resolution that can localize a structure. The wavelet transform has some advantages over the Gabor transform, such as the fact that the variation of the spatial resolution makes it possible to represent the textures in the appropriate scale, as well as to choose from a wide range of wavelet functions.

3. The Wavelet Transform for Defect Detection

The suitability of wavelet transforms for use in image analysis is well established: a representation in terms of the frequency content of local regions over a range of scales provides an ideal framework for the analysis of image features, which in general are of different size and can often be characterised by their frequency domain properties [14]. This makes the wavelet transform an attractive option when attempting defect detection in textured products, as reported by Truchetet and Laligant [15] in his review of industrial applications of wavelet-based image processing. He reported different uses of wavelet analysis in successful machine vision applications: detecting defects for manufacturing applications for the production of furniture, textiles, integrated circuits, and so forth, from their wavelet transformation and vector quantization-related properties of the associated wavelet coefficients; printing defect identification and classification (applied to printed decoration and tampon printed images) by analysing the fractal properties of a textured image; online inspection of the loom under construction using a specific class of the 2D discrete wavelet transform (DWT) called the multiscale wavelet representation with the objectives of attenuating the background texture and accentuating the defects; online fabric inspection device performing an independent component analysis on a subband decomposition provided by a 2-level DWT in order to increase the defect detection rate.

The review of the literature shows two categories of defect detection methods based on wavelet transform. The first category includes direct thresholding methods [10, 11], whose design is based on the fact that texture background can be attenuated by the wavelet decomposition. If we remove the texture pattern from real texture, it is feasible to use existing defect detecting techniques for nontexture images, such as thresholding techniques [16]. Textural features extracted from wavelet-decomposed images are another category which is widely used for defect detection [17, 18]. Features extracted from the texture patterns are used as feature vectors to feed a classifier (Bayer, Euclidean distance, Neural Networks, or Support Vector Machines), which has unavoidable drawbacks when dealing with the vast image data obtained during inspection tasks. For instance, proximity-based methods tend to be computationally expensive and there is no straightforward way of defining a meaningful stopping criterion for data fusion (or division). Often, the learning-based classifiers need to be trained by the nondefect features, which is a troublesome and usually time consuming procedure, thus limiting its real-time applications [10]. For this reason we have focused on direct thresholding methods. The use of direct thresholding presents a main challenge: how to select the decomposition level. On the other hand, direct thresholding presents two main drawbacks: (1) an excessive wavelet decomposition level produces a fusion of defects with the texture pattern and (2) a wrong reconstruction scheme produces false positives when defects are detected.

The work presented here is based on the authors' research on previous works [10, 19] and addresses abovementioned drawbacks by a new approach based on Shannon Entropy calculation. Its main contribution is the formulation of a novel use of the normalized Shannon Entropy, calculated on the different detail subimages, to determine the optimal decomposition level in textures with low directionality. For this purpose we propose to calculate the optimal decomposition level as the maximum of the ratio between the entropy of the approximation subimage and the total entropy, as the sum of entropies calculated for every subimage.

3.1. Wavelet Decomposition. For an image $f(x, y)$ of size $M \times N$ pixels, each level of wavelet decomposition is obtained by applying two filters: a low-pass filter (L) and a high-pass filter (H). The different combinations of these filters produce four images that are here denoted with the subscripts LL, LH, HL, and HH. In the first decomposition level four subimages or bands are produced: one smooth image, also called approximation, $f_{LL}^{(1)}(x, y)$, that represents an approximation of the original image $f(x, y)$ and three detail subimages $f_{LH}^{(1)}(x, y)$, $f_{HL}^{(1)}(x, y)$, and $f_{HH}^{(1)}(x, y)$, which represent the horizontal, vertical and diagonal details, respectively. With this notation, $f_{LL}^{(0)}(x, y)$ represents the original image, $f(x, y)$, and $f_{LL}^{(j)}(x, y)$ represent the approximation image in the decomposition level j. From each decomposition level $f_{LL}^{(j)}(x, y)$ we obtain four subimages, designated here as $f_{LL}^{(j+1)}(x, y)$, $f_{LH}^{(j+1)}(x, y)$, $f_{HL}^{(j+1)}(x, y)$, and $f_{HH}^{(j+1)}(x, y)$, which together form the decomposition level $j + 1$. The pyramid algorithm to obtain $f_{LL}^{(j)}(x, y)$, $f_{LH}^{(j)}(x, y)$, $f_{HL}^{(j)}(x, y)$, and $f_{HH}^{(j)}(x, y)$, as well as the calculation of the inverse transform, can be found in [20]. We will designate $F(x, y) = W^{-1}[f_{LL}^{(j)}]$ as the inverse wavelet transform of a subimage $f_{LL}^{(j)}$ from the resolution level j to level 0.

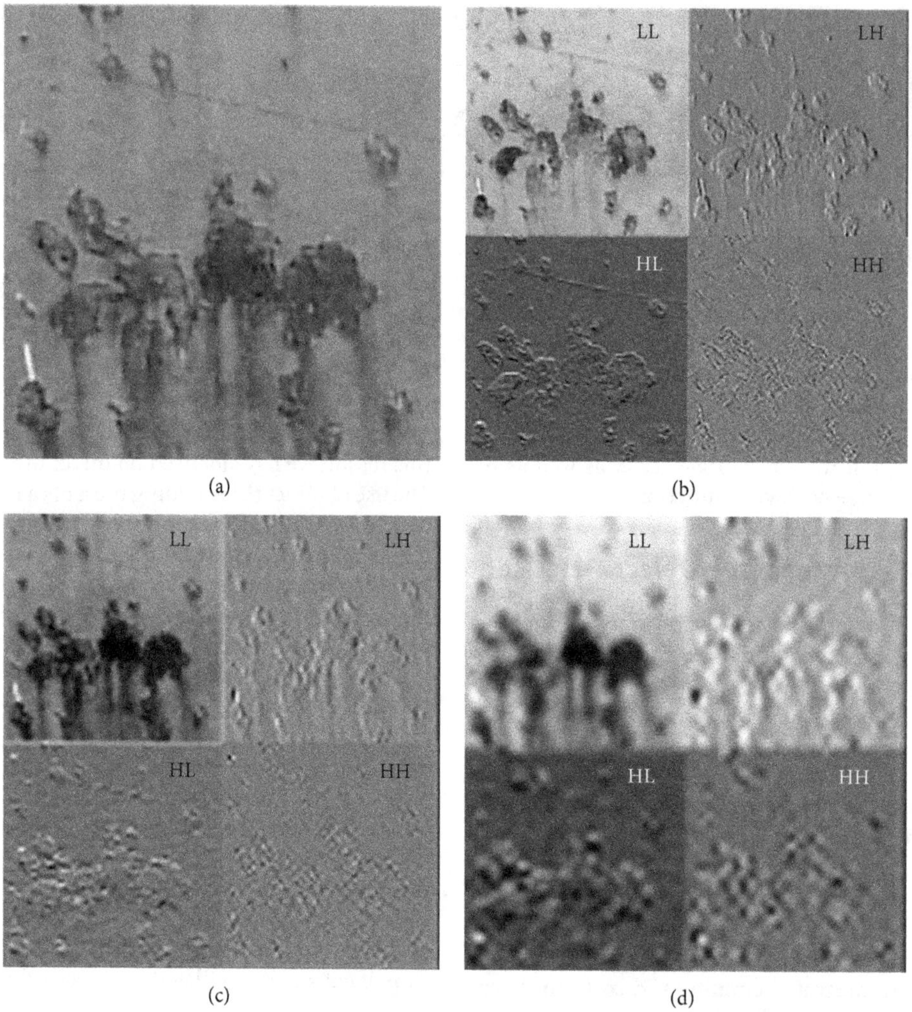

FIGURE 4: Decomposition of an image from a damaged hull, using Haar wavelet with two coefficients, at three decomposition levels: (a) original image, (b) first decomposition level, (c) second decomposition level, and (d) third decomposition level.

Figure 4 shows the wavelet decomposition for three conveniently scaled levels ($j = 3$) of a statistical texture pattern—painted surface—with corrosion defects; the different subimages or bands are shown (named LL, LH, HL, and HH). These were obtained after applying the different coefficients of the wavelet filters. More specifically, the image in Figure 4(a) was decomposed through the application of the Haar wavelet with two coefficients. At level j, images of size $(N/2^j) \times (M/2^j)$ pixels are obtained by iterative application of the pyramid algorithm. Note also that the subimages corresponding to the different decomposition levels are produced by successively applying the low-pass and high-pass filters and reducing the rows and columns by a factor of two.

4. Entropy-Based Method for the Automatic Selection of the Wavelet Decomposition Level

In image processing, entropy has been used by many authors as part of the algorithmic development procedure. There are examples of the use of entropy in the programming of thresholding algorithms [21] and image segmenting [22] as a descriptor for texture classification [23]; as one of the parameters selected by Haralick et al. for application to gray level concurrence matrixes and used for texture characterization [24]; as an element in characteristic vector groups used for classification by Bayesian techniques [25], neuronal networks [26], compact support vectors [27], and so forth.

4.1. Automatic Selection of the Appropriate Decomposition Level. In this work we propose a novel approach for the automatic selection of the appropriate decomposition level by means of Shannon entropy. The entropy function was used to identify the resolution level that provides the most information about defects in real textures. For this purpose, the intensity levels of the subimages of the wavelet transform were considered as random samples. The concept of information entropy—Shannon entropy—describes how much randomness (or uncertainty) there is in a signal or an image; in other words, how much information is provided

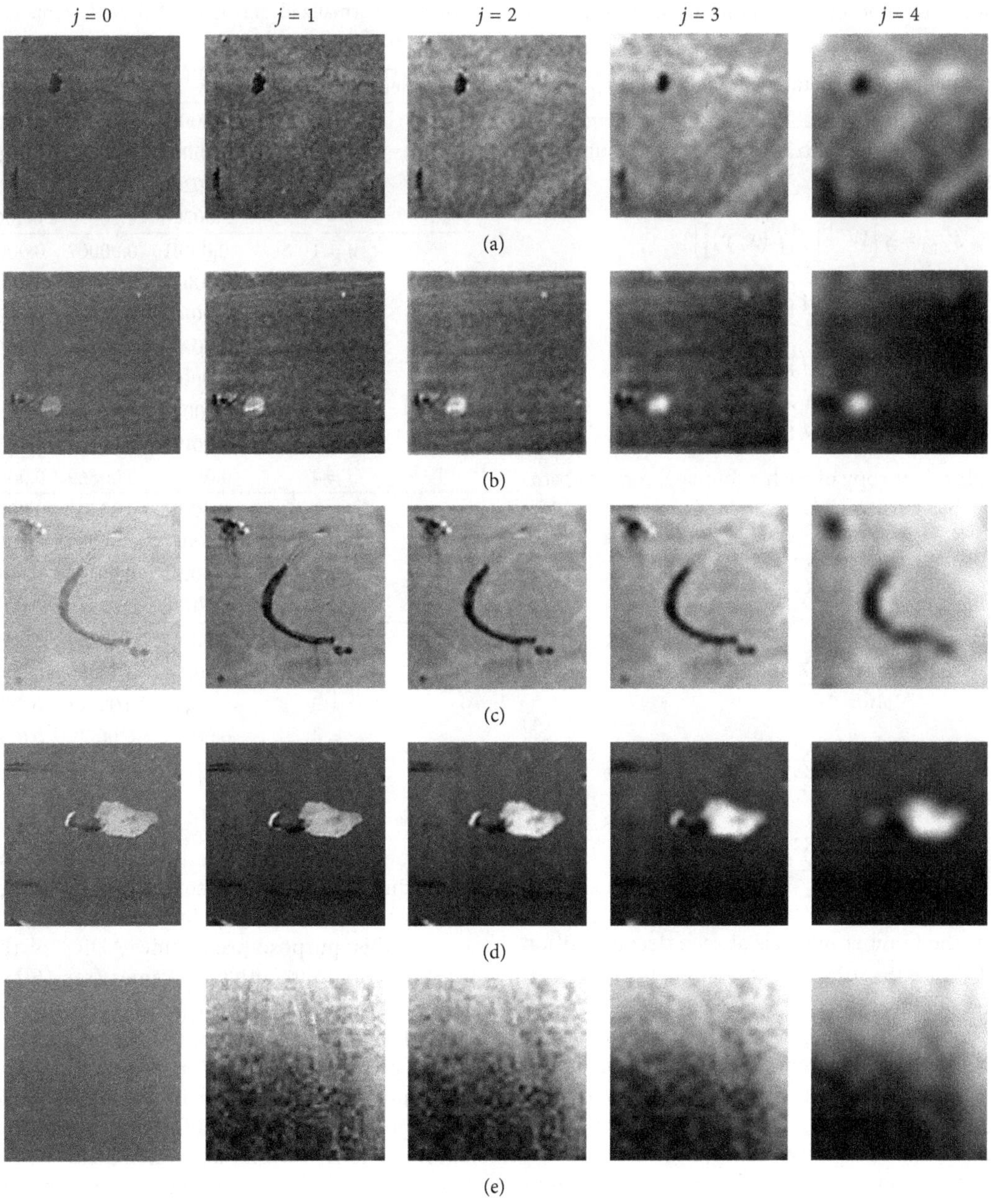

FIGURE 5: Approximation subimages ($f_{LL}^{(j)}$) of four wavelet decomposition levels for different images ((a) H225, (b) H34, (c) H241, (d) H137, and (e) H10) from portions of ship's hulls.

by the signal or image. In terms of physics, the greater the information entropy of the image is, the higher its quality will be [28].

Figure 5 shows how the texture pattern degrades as the decomposition level increases. This degradation is distributed among the different decomposition levels depending on the texture nature and can be quantified by means of the Shannon entropy.

The Shannon entropy function [28, 29] is calculated according to the expression

$$s(X) = -\sum_{i=1}^{T} p(x_i) \log p(x_i), \quad (1)$$

where $X = \{x_1, x_2, \ldots, x_T\}$ is a set of random variables with T outcomes and $p(x_i)$ is the probability of occurrence associated with x_i.

For a 256-gray-level image of size N_t pixels, we define a set of random variables $X = \{x_1, x_2, \ldots, x_i, \ldots, x_{256}\}$ as the number of pixels in the image that have gray level i. The probability of this random variable x_i is calculated as the number of occurrences, hist$[x_i]$, divided by the total number of pixels, N_t

$$p(x_i) = \frac{\text{hist}[x_i]}{N_t}. \quad (2)$$

To calculate the value of the Shannon entropy on the approximation subimage ($f^{(j)}_{\mathrm{LL}}(x,y)$) and on the horizontal, vertical and diagonal detail subimages ($f^{(j)}_{\mathrm{LH}}(x,y)$, $f^{(j)}_{\mathrm{HL}}(x,y)$, and $f^{(j)}_{\mathrm{HH}}(x,y)$) in each decomposition level j, we obtain first the inverse wavelet transform of every subimage and then we apply (3)

$$\begin{aligned} s^j_{\mathrm{LL}} &= s\left[W^{-1}\left[f^{(j)}_{\mathrm{LL}}(x,y)\right]\right], \\ s^j_{\mathrm{LH}} &= s\left[W^{-1}\left[f^{(j)}_{\mathrm{LH}}(x,y)\right]\right], \\ s^j_{\mathrm{HL}} &= s\left[W^{-1}\left[f^{(j)}_{\mathrm{HL}}(x,y)\right]\right], \\ s^j_{\mathrm{HH}} &= s\left[W^{-1}\left[f^{(j)}_{\mathrm{HH}}(x,y)\right]\right]. \end{aligned} \tag{3}$$

The normalized entropy of each subimage, for a decomposition level j, has been calculated as

$$\begin{aligned} S^j_s &= \frac{1}{N^j_{\mathrm{pixels}}}\sum_x\sum_y s^j_{\mathrm{LL}}(x,y), \\ S^j_h &= \frac{1}{N^j_{\mathrm{pixels}}}\sum_x\sum_y s^j_{\mathrm{LH}}(x,y), \\ S^j_v &= \frac{1}{N^j_{\mathrm{pixels}}}\sum_x\sum_y s^j_{\mathrm{HL}}(x,y), \\ S^j_d &= \frac{1}{N^j_{\mathrm{pixels}}}\sum_x\sum_y s^j_{\mathrm{HH}}(x,y), \end{aligned} \tag{4}$$

where N^j_{pixels} is the number of pixels at each decomposition level j. Table 1 shows the values for Shannon entropy calculated for images of Figure 5.

Shannon entropy brings us information about the amount of texture pattern that remains after every decomposition level. Considering (2), entropy provides a measurement of the histogram distribution; the higher the entropy the greater the histogram uniformity; that is, a greater amount of texture pattern is contained in the image. As the decomposition level increases, the texture pattern is being removed; that is, the information content decreases; so the histogram distribution gains uniformity. An optimal reconstruction scheme would eliminate the texture pattern, without loss of defect information. To determine this optimal decomposition level we use a ratio R_j (see (5)) between the entropy of the approximation subimage and the sum of the entropies for all detail subimages, so R_j indicates how much information about the texture pattern is contained in decomposition level j. Variations in this ratio allow detecting changes in the amount of information about the texture pattern between two consecutive decomposition levels

$$R_j = \frac{S^j_s}{S^j_s + S^j_h + S^j_v + S^j_d}, \quad j = 1, 2, \ldots. \tag{5}$$

The goal is to find the optimal decomposition level which provides the maximum variation among two consecutive R_j values because this indicates that, in decomposition level j, the texture pattern still present in level $j-1$ has been removed, keeping useful information (defects).

For this purpose we define ADR_j as the difference between two consecutive R_j values (see (6)). The optimal decomposition level J^* is calculated as the value of j for which ADR_j takes a maximum value. This maximum value points out the greatest variation of information content among two consecutive decomposition levels, which means that both decomposition levels are sufficiently separated in terms of texture pattern information content, and the decomposition process should end. For decomposition levels $j < J^*$, ADR_j indicates that significant texture pattern information still remains in the approximation subimage, and the decomposition process should continue. For decomposition levels $j > J^*$; ADR_j indicates that the approximation subimage is oversmoothed, and the reconstruction result from such smooth approximation subimage will cause defect loss

$$\begin{aligned} \mathrm{ADR}_j &= \left\{\begin{matrix} 0 & j = 1 \\ R_j - R_{j-1} & j = 2, \ldots \end{matrix}\right\}, \\ J^* &= \arg\left\{\max_j\left(\left|\mathrm{ADR}_j\right|\right)\right\}. \end{aligned} \tag{6}$$

Table 2 shows values for R_j coefficients at every image decomposition level (j) for the different textures shown in Figure 5, together with the ADR_j values.

TABLE 1: Normalized entropies of four decomposition levels for textures of Figure 5.

Decomposition level, j		S_s^j	S_h^j	S_v^j	S_d^j
H225	$j = 1$	0.00011	0.00008	0.00007	0.00004
	$j = 2$	0.00048	0.00042	0.00042	0.00037
	$j = 3$	0.00200	0.00172	0.00168	0.00179
	$j = 4$	0.00796	0.00750	0.00681	0.00716
H34	$j = 1$	0.00011	0.00007	0.00008	0.00006
	$j = 2$	0.00046	0.00038	0.00039	0.00038
	$j = 3$	0.00185	0.00179	0.00169	0.00186
	$j = 4$	0.00732	0.00750	0.00662	0.00698
H241	$j = 1$	0.00011	0.00007	0.00007	0.00005
	$j = 2$	0.00046	0.00038	0.00037	0.00036
	$j = 3$	0.00194	0.00159	0.00151	0.00164
	$j = 4$	0.00763	0.00669	0.00640	0.00643
H137	$j = 1$	0.00011	0.00007	0.00006	0.00004
	$j = 2$	0.00047	0.00028	0.00027	0.00031
	$j = 3$	0.00191	0.00127	0.00132	0.00118
	$j = 4$	0.00726	0.00593	0.00581	0.00524
H10	$j = 1$	0.00013	0.00004	0.00004	0.00002
	$j = 2$	0.00057	0.00035	0.00037	0.00025
	$j = 3$	0.00214	0.00190	0.00182	0.00140
	$j = 4$	0.00841	0.00711	0.00737	0.00665

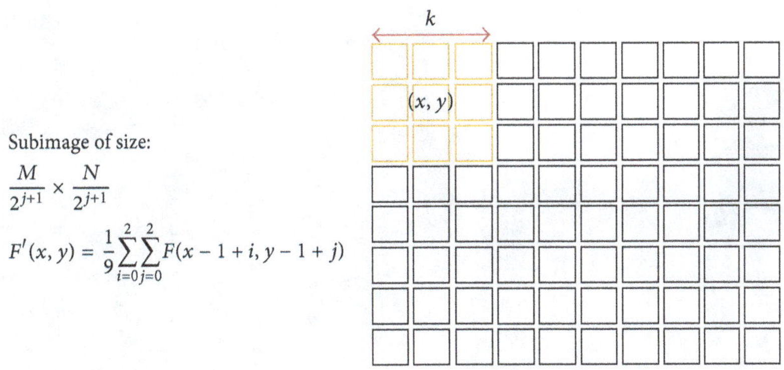

FIGURE 6: Smoothing mask ($k = 3$) for the wavelet coefficients.

Step 1. Compute Shannon Entropy: S_S^j, S_h^j, S_v^j and S_d^j
Step 2. Compute $R_j = S_S^j/(S_S^j + S_h^j + S_v^j + S_d^j)$ for $j = 1, 2, 3, \ldots, J$
Step 3. Compute optimal decomposition level:
$J^* = \arg\left\{\max_j\{\text{ADR}_j\}\right\}$, $j = 1, 2, 3, \ldots, J$
Step 4. Compute $F = W^{-1}[f_{LL}^{(J^*)}]$
Step 5. Compute $F' = m_{k\times k}[F]$
Step 6. Binarize F'

PSEUDOCODE 1: Pseudocode to implement the developed algorithm.

TABLE 2: R_j, ADR_j, and optimal decomposition level obtained for texture images of Figure 5.

Image	Level (j)	1	2	3	4	J^*
H225	R_j	0.3785	0.2812	0.2778	0.2704	2
	ADR_j	0	**0.0973**	0.0035	0.0074	
H34	R_j	0.3460	0.2852	0.2576	0.2575	2
	ADR_j	0	**0.0608**	0.0276	0.0002	
H241	R_j	0.3626	0.2931	0.2899	0.2811	2
	ADR_j	0	**0.0695**	0.0032	0.0088	
H137	R_j	0.388067	0.353254	0.336133	0.299523	4
	ADR_j	0	0.034813	0.017121	**0.036610**	
H10	R_j	0.545018	0.369480	0.294868	0.284585	2
	ADR_j	0	**0.175538**	0.074612	0.010283	

Once the optimal decomposition level is obtained, the process ends with the production of the reconstructed image using (7)

$$F(x, y) = W^{-1}\left[f_{\text{LL}}^{(j)}(x, y)\right]. \tag{7}$$

4.2. Smoothing Mask. To remove the noise running through the successive decomposition levels, we applied average-based smoothing over image $F(x, y)$ to obtain $F'(x, y)$ as shown in (8)

$$F'(x, y) = \frac{1}{k^2}\sum_{i=0}^{k-1}\sum_{j=0}^{k-1} F\left(x - \left\lfloor\frac{k}{2}\right\rfloor + i, y - \left\lfloor\frac{k}{2}\right\rfloor + j\right), \tag{8}$$

where k is the size of the smoothing mask (see Figure 6).

5. Results

5.1. Algorithm Implementation. The proposed computer vision algorithm was implemented as shown in the Pseudocode 1, using the C++ programming language. The mother wavelet used for decomposition was the Haar base function with two coefficients, applied up to a fourth decomposition level. A decomposition level higher than four produced the fusion between defects and background, thus reducing the probability of defect detection.

5.2. Implementation of the Computer Vision System. The computer vision system for visual inspection of ship hull surfaces (Figure 2) has been implemented on a Pentium computer with a Meteor II/1394 card. This card is connected to the microprocessor via a PCI bus and is used as a frame-grabber. For that purpose the card had a processing node based on the TMS320C80 DSP from Texas Instruments and the Matrox NOA ASIC. In addition, the card had a firewire input/output bus (IEEE 1394) which enables it to control a half-inch digital colour camera (15 fps, 1024 × 768 square pixel) equipped with a wide-angle lens (f 4,2 mm).

The software development environment used to implement the system software modules was the Visual C++ programming language powered by the Matrox Imaging Library v9.0. The system also had a Siemens CP5611 card which acted as a PROFIBUS-DP interface for connection with the corresponding robotized blasting system. A Honeywell sensor was used to measure the distance to the ship by ultrasound, with a range of 200–2000 mm and an output of 4–20 mA. User access to the computer vision system was by

FIGURE 7: Robotized blasting system.

means of an industrial PDS (Mobic T8 from Siemens) and a wireless access point. Among other functions, the software that has been developed allows the operator to (1) enter the system configuration parameters, (2) visualize the detected areas to blast for validation by the operator before blasting commences, and (3) calibrate the computer vision system.

5.3. Validation Environment. The proposed computer vision algorithm was assessed at the NAVANTIA shipyard in Ferrol (Spain) on a robotized system used for automatic spot blasting. This operation accounts for 70% of all cleaning work carried out at that shipyard. The robotized system (Figure 7) consists of a mechanical structure divided into two parts: primary and secondary. The primary structure holds the secondary structure (XYZ table), which supports the cleaning head and the computer vision system. More information regarding this system can be found in [5].

With the help of this platform, 260 images of ship hulls' surfaces (with and without defects) were taken, similar to those shown in Figure 3. In this way a catalogue was compiled of typical surface defects as they appear before grit blasting.

5.4. Metrics. To conduct a quantitative analysis of the quality of the proposed segmentation method, we need to use the best suited metrics to that purpose. The performance of image segmentation methods has been assessed by such authors as Zhang [30] and Sezgin and Sankur [16]. They proposed various different metrics for measurement of the quality of the segmentation in a given method, using parameters like position of the pixels, area, edges, and so forth. Out of these, one of the quantitative appraisal methods proposed by Sezgin was selected and examined: Misclassification Error (ME).

ME represents the percentage of the background pixels that are incorrectly allocated to the object (i.e., to the foreground) or vice versa

$$\mathrm{ME} = 1 - \frac{|B_P \cap B_T| + |O_P \cap O_T|}{|B_P| + |O_P|}. \quad (9)$$

The error can be calculated by means of (9), where B_P (background pattern) and O_P (object pattern) represent the pattern image of the background and of the object taken as reference, and B_T (background test) and O_T (object test) represent the image to be assessed. In the event that the test image coincides with the pattern image, the classification error will be zero and therefore the performance of the segmentation will be the maximum.

The performance of the implemented algorithms is assessed according to the equation:

$$\eta = 100 \cdot (1 - \mathrm{ME}). \quad (10)$$

5.5. Algorithm Appraisal. The proposed visual inspection algorithm (see Pseudocode 1) was applied to the above mentioned catalogue that had been taken at the shipyard (some samples are shown in column (a) of Figure 8). The Shannon entropy was calculated and normalized for four wavelet decomposition levels and the optimal J^* level was calculated (6). Images were also processed applying algorithms proposed by Han and Shi [10] and Tsai and Chiang [19]. The result was 3 sets of 260 reconstructed images in which the defects have been isolated from texture. To check the quality of the defect detection algorithms we have concluded with a binarization stage. For that purpose we have selected Kapur's method [21] which belongs to the group of entropy-based methods, as classified by Sezgin and Sankur [16] in his review of thresholding methods; this has resulted in 3 sets of 260 images (column (b) of Figure 8 shows some results obtained with the proposed algorithm; column (c) of Figure 8 shows some results obtained with the Tsai algorithm and column (d) of Figure 8 shows some results obtained with Han algorithm).

To apply the metrics described above, human inspectors were needed to segment each of the catalogue images manually (samples of these are shown in column (v) of Figure 8). Table 3 shows the performance (η) when sample texture images of Figure 8 were segmented using the three algorithms.

As can be observed from above results, the proposed entropy-based algorithm achieved better results than Tsai algorithm and significantly better results than Han algorithm.

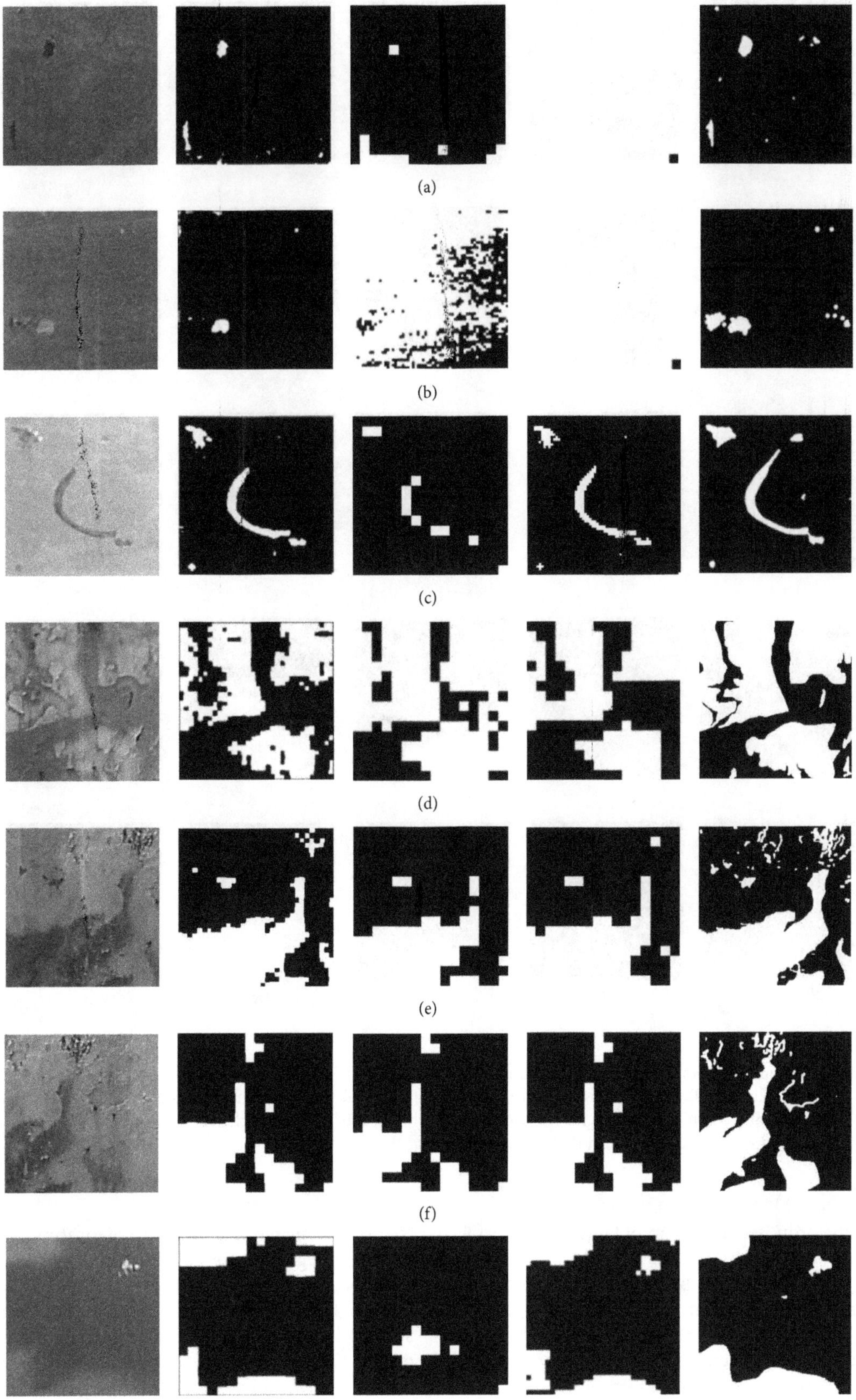

(a)

(b)

(c)

(d)

(e)

(f)

(i) (ii) (iii) (iv) (v)

(g)

FIGURE 8: Continued.

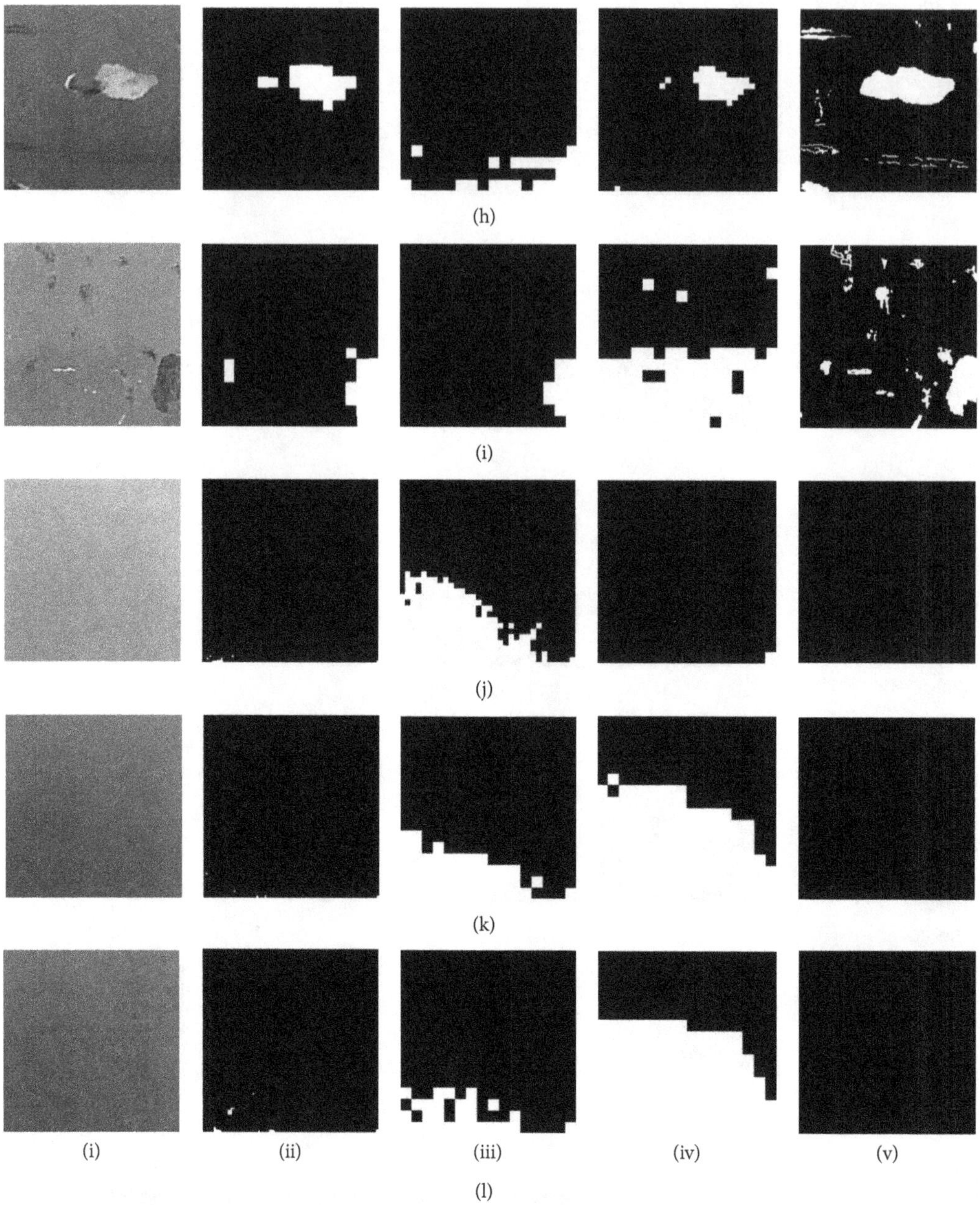

FIGURE 8: Columns: column (i) shows texture images of portions of hulls; column (ii) shows reconstructed images resulting from the proposed reconstruction scheme; column (iii) shows reconstructed images resulting from Tsai algorithm; column (iv) shows reconstructed images resulting from Han algorithm; column (v) shows defects segmented by hand, the "ground truth." Image Rows: (a) H225, (b) H34, (c) H241, (d) H1, (e) H120, (f) H121, (g) H99, (h) H137, (i) H48, (j) H9, (k) H10, and (l) H11.

In both cases the proposed algorithm obtains higher performance with low decomposition level.

We have also analysed the behaviour of the proposed algorithm as misclassification rates. A set of 120 images were processed by the proposed algorithm and also by Han and Tsai algorithms. Results were then analysed by a skilled blasting operator, who assessed what portions of the shown hull surface would be blasted in real conditions at the repair yard. Table 4 shows the average number of defect points classified as Type I and Type II errors for 120 samples of the 260-image set indicated above.

As we can see, the proposed algorithm produced better results as regards false positives—that is, points marked as defective when they are not (Type I error). This is essentially because the operator tends to blast larger areas than necessary, and moreover he is less able to control the cut-off of the grit jet. On the other hand, the proposed algorithm identified similar false negatives (Type II error). This difference was not very significant and is quite acceptable in view of the clear advantage offered by the computer vision system equipped with the proposed inspection algorithm as regards Type I errors.

TABLE 3: η in defect segmentation of texture images of Figure 8.

	Entropy based	J^*	Tsai	J^*	Han	J^*
Defect samples						
H225	97.09	2	94.51	4	2.14	4
H34	97.41	2	74.52	2	3.43	4
H241	96.04	2	95.17	4	97.41	2
H1	86.50	3	82.63	4	84.69	4
H20	91.40	3	89.22	4	90.08	4
H121	89.36	4	89.91	4	90.23	4
H99	89.30	4	73.76	4	91.46	3
H137	93.94	4	82.87	4	94.26	3
H48	93.06	4	93.97	4	64.23	4
Average on defect samples	*92.68*	*3.11*	*86.28*	*3.78*	*68.66*	*3.56*
Nondefect samples						
H9	98.30	2	74.04	3	99.61	4
H10	98.33	2	80.85	4	48.05	4
H11	98.22	2	89.06	4	48.65	4
Average on nondefect samples	*98.28*	*2.00*	*81.32*	*3.67*	*65.44*	*4.00*
Total Average	**95.48**	**2.56**	**83.80**	**3.72**	**67.05**	**3.78**

TABLE 4: Automated inspection examined by a skilled blasting operator.

	Entropy-based algorithm	Han algorithm	Tsai algorithm
Type I error	6.8%	9.2%	11.1%
Type II error	0.9%	1.1%	0.7%

6. Conclusions

This paper has presented a computer vision algorithm based on the wavelet transform which brings a robust method for detecting defects in ship hull surfaces. To achieve this, we used an image reconstruction approach based on automatic selection of the optimal wavelet transform resolution level by means of a novel use of the Shannon entropy, calculated on the different detail subimages.

The algorithm has been incorporated to a computer vision system that masters a robotized system for blasting ship hulls, making it possible to fully automate grit blasting operation. The results as regards reliability were very similar to those achieved with human workers, while faster inspection was provided (among 8% for flat surfaces in oil tankers and 15% for shaped hulls like frigates) and the consequences of operator fatigue minimized.

Acknowledgments

The work submitted here was carried out as a part of the projects ViSel-TR (ref. TIN2012-39279) and EXPLORE (ref. TIN2009-08572) funded by the Spanish National R&D&I Plan.

References

[1] A. Momber, *Blast Cleaning Technologies*, Springer, Berlin, Germany, 2008.

[2] P. Le Calve, "Qualification of paint systems after UHP water jetting and understanding the phenonemon of "blocking" of flash rusting," *Journal of Protective Coatings and Linings*, vol. 24, no. 8, pp. 13–26, 2007.

[3] P. J. Navarro, J. Suardíaz, C. Fernández, P. Alcover, and R. Borraz, "Teleoperated service robot for high quality ship maintenance," in *Proceedings of the 8th IFAC International Workshop on Intelligent Manufacturing Systems*, vol. 8, pp. 152–157, Alicante, Spain, May 2007.

[4] C. Fernández, A. Iborra, B. Álvarez, J. A. Pastor, and J. M. Fernández, "Ship shape in Europe: co-operative robots in the ship repair industry," *IEEE Robotics and Automation Magazine*, vol. 12, no. 3, pp. 65–77, 2005.

[5] A. Iborra, J. A. Pastor, P. J. Navarro et al., "A cost-effective robotic solution for the cleaning of ship hulls," *Robotica*, vol. 28, pp. 453–464, 2010.

[6] X. Xie, "A review of recent advances in surface defect detection using texture analysis techniques," *Electronic Letters on Computer Vision and Image Analysis*, vol. 7, no. 3, pp. 1–22, 2008.

[7] A. Kumar, "Computer-vision-based fabric defect detection: a survey," *IEEE Transactions on Industrial Electronics*, vol. 55, no. 1, pp. 348–363, 2008.

[8] H. Y. T. Ngan, G. K. H. Pang, and N. H. C. Yung, "Automated fabric detect detection-A review," *Image and Vision Computing*, vol. 29, pp. 442–458, 2011.

[9] H. Y. T. Ngan, G. K. H. Pang, and N. H. C. Yung, "Performance evaluation for motif-based patterned texture defect detection," *IEEE Transactions on Automation Science and Engineering*, vol. 7, no. 1, pp. 58–72, 2010.

[10] Y. Han and P. Shi, "An adaptive level-selecting wavelet transform for texture defect detection," *Image and Vision Computing*, vol. 25, no. 8, pp. 1239–1248, 2007.

[11] D. M. Tsai and T. Huang, "Automated surface inspection for statistical textures," *Image and Vision Computing*, vol. 21, pp. 307–323, 2003.

[12] A. Kumar and G. Pang, "Defect detection in textured materials using Gabor filters," *IEEE Transactions on Industry Applications*, vol. 38, no. 2, pp. 425–440, 2002.

[13] H. Y. T. Ngan, G. K. H. Pang, S. P. Yung, and M. K. Ng, "Wavelet based methods on patterned fabric defect detection," *Pattern Recognition*, vol. 38, no. 4, pp. 559–576, 2005.

[14] S. Mallat, "Multifrequency channel decompositions of images and wavelet models," *IEEE Transactions on Acoustics Speech and Signal Processing*, vol. 37, no. 12, pp. 2091–2110, 1989.

[15] F. Truchetet and O. Laligant, "Review of industrial applications of wavelet and multiresolution-based signal and image processing," *Journal of Electronic Imaging*, vol. 17, no. 3, Article ID 031102, pp. 3–11, 2008.

[16] M. Sezgin and B. Sankur, "Survey over image thresholding techniques and quantitative performance evaluation," *Journal of Electronic Imaging*, vol. 13, no. 1, pp. 146–168, 2004.

[17] W. K. Wong, C. W. M. Yuen, D. D. Fan, L. K. Chan, and E. H. K. Fung, "Stitching defect detection and classification using

wavelet transform and BP neural network," *Expert Systems With Applications*, vol. 36, no. 2, pp. 3845–3856, 2009.

[18] H.-D. Lin, "Automated visual inspection of ripple defects using wavelet characteristic based multivariate statistical approach," *Image and Vision Computing*, vol. 25, pp. 1785–1801, 2007.

[19] D. M. Tsai and C. H. Chiang, "Automatic band selection for wavelet reconstruction in the application of defect detection," *Image and Vision Computing*, vol. 21, no. 5, pp. 413–431, 2003.

[20] S. G. Mallat, "Theory for multiresolution signal decomposition: the wavelet representation," *IEEE Transactions on Pattern Analysis and Machine Intelligence*, vol. 11, no. 7, pp. 674–693, 1989.

[21] J. N. Kapur, P. K. Sahoo, and A. K. C. Wong, "A new method for gray-level picture thresholding using the entropy of the histogram," *Computer Vision, Graphics, & Image Processing*, vol. 29, no. 3, pp. 273–285, 1985.

[22] C. Yan, "Local entropy-based transition region extraction and thresholding," *Pattern Recognition Letters*, vol. 24, no. 16, pp. 2935–2941, 2003.

[23] L. Setia and H. Burkhardt, "Feature selection for automatic image annotation," in *Proceedings of the 28th DAGM Symposium*, pp. 294–303, Berlin, Germany, September 2006.

[24] R. M. Haralick, K. Shanmugam, and I. Dinstein, "Texture features for image classification," *IEEE Transactions on Systems, Man, and Cybernetics*, vol. 3, pp. 610–621, 1973.

[25] M. Toews and T. Arbel, "Entropy-of-likelihood feature selection for image correspondence," in *Proceedings of the 9th IEEE International Conference on Computer Vision*, vol. 2, pp. 1041–1047, Nice, France, October 2003.

[26] E. Avci, "An expert system based on Wavelet Neural Network-Adaptive Norm Entropy for scale invariant texture classification," *Expert Systems with Applications*, vol. 32, no. 3, pp. 919–926, 2007.

[27] F. Harirchi, P. Radparvar, H. Abrishami Moghaddam, F. Dehghan, and M. Giti, "Two-level algorithm for MCs detection in mammograms using Diverse-Adaboost-SVM," in *Proceedings of the 20th International Conference on Pattern Recognition (ICPR '10)*, pp. 269–272, Istanbul, Turkey, August 2010.

[28] C. E. Shannon, "A mathematical theory of communication," *The Bell System Technical Journal*, vol. 27, p. 379–423, 623–656, 1948.

[29] R. R. Coifman and M. V. Wickerhauser, "Entropy-based Algorithms for best basis selection," *IEEE Transactions on Information Theory*, vol. 38, no. 2, pp. 713–718, 1992.

[30] Y. J. Zhang, "A review of recent evaluation methods for image segmentation," in *Proceedings of the 6th International Symposium on Signal Processing and its Applications*, pp. 148–151, Kuala Lumpur, Malaysia, August 2001.

2

A Study of Wavelet Analysis and Data Extraction from Second-Order Self-Similar Time Series

Leopoldo Estrada Vargas,[1] Deni Torres Roman,[1] and Homero Toral Cruz[2]

[1] *Department of Electrical Engineering, Center of Research and Advanced Studies (CINVESTAV), 45019 Guadalajara, JAL, Mexico*
[2] *Department of Sciences and Engineering, University of Quintana Roo (UQROO), 77019 Chetumal, QROO, Mexico*

Correspondence should be addressed to Leopoldo Estrada Vargas; lestrada@gdl.cinvestav.mx

Academic Editor: Marcelo Moreira Cavalcanti

Statistical analysis and synthesis of self-similar discrete time signals are presented. The analysis equation is formally defined through a special family of basis functions of which the simplest case matches the Haar wavelet. The original discrete time series is synthesized without loss by a linear combination of the basis functions after some scaling, displacement, and phase shift. The decomposition is then used to synthesize a new second-order self-similar signal with a different Hurst index than the original. The components are also used to describe the behavior of the estimated mean and variance of self-similar discrete time series. It is shown that the sample mean, although it is unbiased, provides less information about the process mean as its Hurst index is higher. It is also demonstrated that the classical variance estimator is biased and that the widely accepted aggregated variance-based estimator of the Hurst index results biased not due to its nature (which is being unbiased and has minimal variance) but to flaws in its implementation. Using the proposed decomposition, the correct estimation of the *Variance Plot* is described, as well as its close association with the popular *Logscale Diagram*.

1. Introduction

In the past decades, the self-similar processes and long-range dependence (LRD or long memory) have been applied to the study and modeling of many natural and man-made complex phenomena. These kinds of processes have been particularly attractive in the pursuit of optimal design and configuration of network communications.

The published work of Leland et al. in 1993 and 1994 [1, 2] demonstrated that Ethernet traffic is statistically self-similar and that the commonly used models are unable to capture that fractal behavior, highlighting that a *burstiness* and LRD are present when $H > 0.5$. Since then, researchers have been studying extensively long memory processes and their impact on network performance, for example, Karagiannis et al. stated that the identification of LRD is not trivial and that not all scenarios in modern networks present LRD characteristics, for example, traffic in the Internet backbone is more likely to be Poisson type instead of LRD [3].

Many researchers have also addressed their studies to determine if network traffic is sufficiently modeled by self-similar processes or a more general model is needed, for example, one that considers *multiscaling* or *multifractality* [4–6]. The advantage of the capability to model complex systems with self-similar processes is that the correlation structure is defined by a single parameter: the Hurst index (H).

Unlike other statistics, the Hurst index, although it is mathematically well defined, cannot be estimated unambiguously from real world samples. Several methods have been developed then in order to estimate it. Examples of classical estimators are those based on R/S statistic [7] (and its unbiased version [8]), detrended fluctuation analysis (DFA) [8, 9], maximum likelihood (ML) [10], aggregated variance (VAR) [7], wavelet analysis [11, 12], and so forth. In [13], Clegg developed an empirical comparison of estimators for data in raw form and corrupted. An important observation is that the estimation of the Hurst index may differ from one estimator to another, and the selection of the most adequate estimator is a difficult task. This selection depends greatly on how well the data sample meets the assumptions the estimator is based on. However, through analytical and empirical studies, it has been discovered that the estimators that have the best performance in bias and standard deviation, and, consequently, in mean squared error (MSE) are Whittle ML and the wavelet-based estimator proposed by Veitch and Abry in [11].

From these two estimators, the wavelet based is computationally simpler and faster [7, 11].

In addition to the Hurst index, other statistical characteristics are needed to describe the phenomenon under study. The most common are the first and second-order statistics, that is, mean, variance, and correlation. The classical estimators of these characteristics have been proposed decades ago, for example, Kenney and Keeping demonstrated in 1939 and 1951 that the classical variance estimator is unbiased for independent and identically distributed Gaussian observations [14, 15]. Confidence interval is also given for these estimations, for example, $P[\overline{X} \in (-3/\sqrt{N}, 3/\sqrt{N})] \approx 99\%$ (where $\overline{X}$ is the sample mean estimated from a sample of size N) for a standardized white noise process. This confidence interval is narrower as the sample size increases.

It has been claimed that most processes satisfy those common assumptions [16]. As it has been expressed, however, other authors (Leland et al. [1], Taqqu et al. [5], Tsybakov and Georganas [17], Veitch and Abry [11], and many others) conclude that traffic characteristics present correlation and that the estimation of these statistics (including confidence interval) with the classical estimators (which do not consider correlation) may lead to estimation errors and, consequently, to wrong decisions or inaccurate models, especially when data presents accentuated LRD.

Several estimators of the Hurst index have been proposed, but many of them do not consider the effect of correlation on the estimation of first- and second-order statistics, thus applying incorrectly the classical formulae. Particularly, it has been claimed that the aggregated variance method can only be used as a heuristic method, and that the *Variance Plot* (also named *Variance-time Plot*; see Section 2.3) can only be used to check whether the time series is self-similar or not and, if so, to obtain a crude guess for the Hurst index [22, page 44]. This work clarifies this point, demonstrating that the *Variance Plot* can be estimated efficiently and that the estimation of the Hurst index from it is actually unbiased and has minimum variance (similarly to the wavelet-based estimator).

This work is motivated by the mentioned importance of the self-similar processes in many areas, especially in the analysis and modeling of Internet traffic, and by the fact that there are still some misunderstandings and bad practices that must be overcome.

This document is organized as follows. Section 2 defines the discrete time self-similarity and some of its statistical properties. Section 3 describes the proposed set of basis functions and the analysis and synthesis equations, which are used to generate self-similar samples whose Hurst index matches the estimator proposed by Veitch et al. and the corrected version of the variance-based estimator. Section 4 defines statistics for the sample mean and variance of self-similar discrete processes; Section 5 explains how the variance-based estimator has been misunderstood and defines the correct estimator, which coincides with the Haar wavelet based in a particular case. Section 6 presents some simulations and measurements and, finally, Section 7 concludes the work.

2. Self-Similarity

Self-similarity describes the phenomenon where certain properties are preserved irrespective of scaling in space or time. Deterministic self-similarity is clearly exemplified by popular figures as Sierpinski's triangle or Koch's snow flake. This form of self-similarity is named scale invariance and makes different scales of the same object undistinguishable. Stochastic self-similarity is not that obvious, it refers to how statistical properties of a stochastic process are preserved under time expansion. Stochastic self-similarity is defined for continuous and discrete time stochastic processes.

Self-similarity (either continuous or discrete time) is tightly related to short- and long-range dependencies (SRD and LRD, resp.). The degree of self-similarity is defined and measured through the so named Hurst index H ($0 < H < 1$). It is known that processes with $H < 0.5$ are SRD, and processes with $H > 0.5$ are LRD. If $H = 0.5$, neither SRD nor LRD are present. For example, the commonly used white Gaussian noise (WGN) has always $H = 0.5$ and does not present any time dependency.

Processes with LRD are also named long memory, as current and future realizations of these are strongly correlated. The dividing line between SRD and LRD processes is not ambiguous; for LRD processes, the autocovariance function is not absolutely convergent (i.e., the sum is not finite), while it is for SRD processes. This work refers only to stochastic discrete time self-similarity.

2.1. Discrete Time Self-Similarity. The definition of discrete time stochastic self-similarity is given in terms of the aggregated processes. Let $\{X_t; t \in \mathbb{N}\}$ be a discrete time series derived from a self-similar process with stationary increments and Hurst index H (H-SSSI). The aggregated time series, derived from X_t is the sequence given by [17]

$$X(m) = \left\{X_k^{(m)};\ k \in \mathbb{N}\right\}, \tag{1}$$

where each term $X_k^{(m)}$ is defined as

$$X_k^{(m)} = \frac{1}{m} \sum_{i=(k-1)m+1}^{km} X_i, \quad k \in \mathbb{N}, \tag{2}$$

where m represents the aggregation level. That is, each new time series is obtained by partitioning the original time series into nonoverlapping blocks of size m and then averaging each block to obtain its respective values.

Let X_t be a covariance stationary discrete time series with mean $\mu_X = 0$, variance σ_X^2 and autocovariance function (ACvF) $\gamma_X(k)$ and $X_k^{(m)}$ its aggregated series. Then it is said that X_t is self-similar (H-SS), if the following holds [18]:

$$X_k^{(m)} \sim m^{H-1} X_t, \tag{3}$$

where $\sim$ means equality in distribution.

The many methods for generating artificial discrete time self-similar sequences are classified in *sequential* and *fixed length*. All these have particular advantages and shortcomings associated to accuracy, generation time, memory or processing resources, and so forth. Sequential generators are sometimes more appropriate for long duration simulations, but the level of approximation and several parameters are needed, while fixed length only depends on the desired Hurst index but may need to store part or all of the sequence in memory. For the purpose of the simulations described in Section 6, a fixed-length Davies-Harte *fractional Gaussian noise* (FGN) generator is used. FGN is a special type of noise where the autocovariance function has a special shape (expression (8)). Section 3.2 describes a theoretical method to synthesize self-similar discrete sequences using Haar wavelet-based decomposition. Although Daubechies wavelet- (DW-) based approximations are better in mean and variance, the Haar-based method generates a self-similar sequence of specified Hurst index from practically any given sequence, regardless of whether it is self-similar or not if some conditions are met [19]. For other applications, DW are preferred; see [20] as an example.

2.2. Properties of Self-Similar Discrete Time Series. The definition of discrete stochastic self-similarity (3) has important implications about the stochastic process X_t; these implications include the following properties [21].

(i) Zero-mean:

$$E(X_t) = E\left(X_k^{(m)}\right) = 0. \tag{4}$$

(ii) Power law of the qth order moments:

$$E\left[\left(X_k^{(m)}\right)^q\right] = m^{q(H-1)} E\left[(X_t)^q\right]. \tag{5}$$

(iii) Power law of the qth order absolute moments:

$$E\left(\left|X_k^{(m)}\right|^q\right) = m^{q(H-1)} E\left(\left|X_t\right|^q\right). \tag{6}$$

2.3. Second-Order Discrete Self-Similarity. The second-order definition of self-similarity is derived from (5) for $q = 2$. The variance of the aggregated time series is defined by the following [17]:

$$\operatorname{var}\left(X_k^{(m)}\right) = m^{2H-2} \operatorname{var}(X_t). \tag{7}$$

An equivalent definition is

$$\gamma_X^{(m)}(k) = \frac{\sigma_X^2}{2}\left[(k+1)^{2H} - 2k^{2H} + (k-1)^{2H}\right], \quad k \geq 0, \tag{8}$$

where $\gamma_X^{(m)}(k)$ is the autocovariance function of $X_k^{(m)}$.

If a discrete time series X_t satisfies these conditions, it is called second-order self-similar with Hurst index H (H-SOSS). Note that the mean of an H-SOSS process is not necessarily zero.

The plot $\log[\operatorname{var}(X_k^{(m)})]$ versus $\log(m)$ is known as *Variance Plot*. It is a straight line of slope $2H - 2$ for self-similar processes. This plot is the basis of the variance-based estimator of the Hurst index. It has been "shown" in the literature that the variance-based estimator underestimates the Hurst index and that the variance-based estimator throws a coarse estimation of the true Hurst index. Section 5 demonstrates that this is a consequence of inadequate implementations of this estimator, that is, the aggregated variance is estimated with the classical formula (36), which is not correct if any correlation exists. An apparent solution is to use the proposed unbiased estimator, but that leads to an ill-conditioned problem: the Hurst index is needed to estimate the variance and vice versa. The solution for this situation is also described (see Section 5.1).

2.4. Wavelet Decomposition and the Logscale Diagram. The wavelet decomposition transforms a signal X_t into a sum of orthogonal components as follows:

$$X_t = \sum_{j=i}^{J} \sum_{k=1}^{2^j} d_X(j,k)\, \psi_{j,k}(t), \tag{9}$$

where each function $\psi_{j,k}(t)$ is derived from a basis function $\psi_0(t)$, namely, the mother wavelet, by scaling and displacement, that is,

$$\psi_{j,k}(t) = 2^{-j/2}\psi_0\left(2^{-j}t - k\right), \tag{10}$$

and coefficients $d_X(j,k)$ is the value at time k of scale j, computed as an inner product between the signal X_t and the wavelet function $\psi_{j,k}(t)$:

$$d_X(j,k) = \left\langle X(t), \psi_{j,k}(t)\right\rangle. \tag{11}$$

The statistic $s_2(j)$ is then defined from these coefficients as

$$S_2(j) = E\left|d_X(j,\cdot)\right|^2, \tag{12}$$

which, for an H-SOSS process, is related to the Hurst index as

$$S_2(j) = c_f C 2^{j(2H-1)}, \tag{13}$$

where the quantity $c_f C$, related to the power of the process, is considered a constant.

The plot $\log_2 S_2(j)$ versus j forms the widely known *Logscale Diagram* described by Veitch and Abry [11]. The *Logscale Diagram* of an H-SOSS process is a straight line of slope $2H - 1$. To obtain an unbiased estimation of the Hurst index based on the *Logscale Diagram*, it is also necessary to subtract the bias that results of averaging the logarithms of the respective variance of real world time series, which is estimated as [11]

$$g_j = \frac{\Psi\left(n_j/2\right)}{\ln(2)} - \log_2\left(\frac{n_j}{2}\right), \tag{14}$$

where n_j is the number of coefficients available at octave, that is

$$E\left\{\log_2\left[\operatorname{var}\left(\widehat{C}_{X,t}^{n,i}\right)\right]\right\} \approx E\left\{\log_2\left[\operatorname{var}\left(\widetilde{C}_{X,t}^{n,i}\right)\right]\right\} - g_j. \tag{15}$$

If the *Logscale Diagram* of a time series cannot be adequately modeled with a linear model, possibly the scaling behavior needs to be described with more than one scaling parameter, that is, the Hurst parameter is not adequate (or insufficient) [22, 23]. However, even if the time series under study is not self-similar, the *Logscale Diagram* can show whether or not it presents LRD.

3. An Orthogonal Decomposition Performed by Subtracting Aggregated Series

The time series X_t can be decomposed into a set of time series, each one defined as

$$C_{X,t}^{n,j} = X_t^{(n^{j-1}E)} - X_t^{(n^jE)}, \quad n, j \in \mathbb{N}, \tag{16}$$

where $X_t^{(n^jE)}$ is the time series X_t after two operations, which are as follows.

(1) Aggregation at level n^j, as defined by (1) and (2), that is, $m = n^j$.

(2) Expansion of level n^j, which consists of "repeat" each element of a time series n^j times, that is, $X_t^{(n^jE)} = X_k^{(n^j)}$ for $k = 1 + \lfloor (t-1)/n^j \rfloor$ and $j \in \mathbb{N}$.

These zero-mean components $C_{X,t}^{n,j}$ have three important properties.

(i) They synthesize the original time series without loss (assuming zero-mean), that is,

$$X_t = \sum_j C_{X,t}^{n,j}. \tag{17}$$

(ii) They are pair-wise orthogonal:

$$\left\langle C_{X,t}^{n,j_1}, C_{X,t}^{n,j_2} \right\rangle = 0, \quad j_1 \neq j_2. \tag{18}$$

(iii) If X_t is exactly or at least second-order self-similar, then the variance of its components satisfies

$$\operatorname{var}\left(C_{X,t}^{n,j}\right) = r \cdot \operatorname{var}\left(C_{X,t}^{n,j-1}\right), \tag{19}$$

where

$$r = n^{2H-2}. \tag{20}$$

Another useful property relates the variance of the component to the variances of the aggregated series, that is,

$$\operatorname{var}\left(C_{X,t}^{n,j}\right) = \operatorname{var}\left[X_t^{(n^{j-1}E)}\right] - \operatorname{var}\left[X_t^{(n^jE)}\right]. \tag{21}$$

It is easy to proof (21): from (16), it turns out that $\operatorname{var}[C_{X,t}^{n,j}] = \operatorname{var}[X_t^{(n^{j-1}E)}] + \operatorname{var}[X_t^{(n^jE)}] - 2\operatorname{cov}[X_t^{(n^{j-1}E)}, X_t^{(n^jE)}]$, but as $X_t^{(n^jE)}$ is itself an aggregation of $X_t^{(n^{j-1}E)}$ it turns out that $\operatorname{cov}[X_t^{(n^{j-1}E)}, X_t^{(n^jE)}] = \operatorname{var}[X_t^{(n^jE)}]$, and (21) comes after a substitution.

Properties (i), (ii), and (iii) imply that

$$\sigma_X^2 = \sum_i \operatorname{var}\left(C_{X,t}^{n,j}\right), \qquad \sigma_X^2 = \frac{1}{1-r}\operatorname{var}\left(C_{X,t}^{n,1}\right). \tag{22}$$

Then, the variance of the jth component is related to the variance of X_t as

$$\operatorname{var}\left(C_{X,t}^{n,j}\right) = (1-r)\, r^{j-1}\sigma_X^2. \tag{23}$$

It is easy to prove the following relation:

$$\operatorname{var}\left(C_{X,t}^{n,j}\right) = n^{-j} S_2(j). \tag{24}$$

An immediate consequence of (24) is that the plot $j + \log_n[\operatorname{var}(C_{X,t}^{n,j})]$ versus j is equivalent to the *Logscale Diagram*. It is a straight line for H-SOSS time series, and the slope is related to the Hurst index so that $s = 2H - 1$.

For example, for $n = 2$ the statistics $\operatorname{var}(C_{X,t}^{2,j})$ and $S_2(j)$ are related as

$$\operatorname{var}\left(C_{X,t}^{2,j}\right) = 2^{-j} S_2(j), \tag{25}$$

and in this case the basis function is the Haar wavelet, which is defined as [24, 25]

$$\psi_0(t) = \begin{cases} +1, & 0 \le t < \dfrac{1}{2}, \\ -1, & \dfrac{1}{2} \le t < 1, \\ 0, & \text{otherwise}. \end{cases} \tag{26}$$

Figure 1 shows the components obtained from an H-SOSS sample of size 32 and $H = 0.9$. The squared form of the components is due to the expansion of the aggregated series and disappears after the downsampling.

The authors of [26] described the aggregation as an inner product with the signal and the Haar "father" wavelet, and then, the relation between wavelet coefficients and aggregation levels is obvious. At this point there are similarities between this section and that previous work, the most important is that the relation between aggregation levels and Haar wavelet is described by Abry et al. However, two differences must be highlighted: (1) the decomposition presented in this work only coincides with that definition for $n = 2$ in (16); for higher values, the Haar wavelet is not sufficient to describe the components of (16); and (2) authors of [26] discard anyway the estimation of the Hurst index based on the so named "a-aggregation." In this work, it is clarified that the "a-aggregation" is misunderstood leading to incorrect implementations (i.e., the "classical" variance estimator), and a generalization of the "d-aggregation" is proposed.

3.1. General Waveform of the Basis Functions. The orthogonal decomposition defined by (16) can be expressed in terms of an inner product between the signal and a set of orthogonal wavelet-type functions. Let us describe the waveform for the general case.

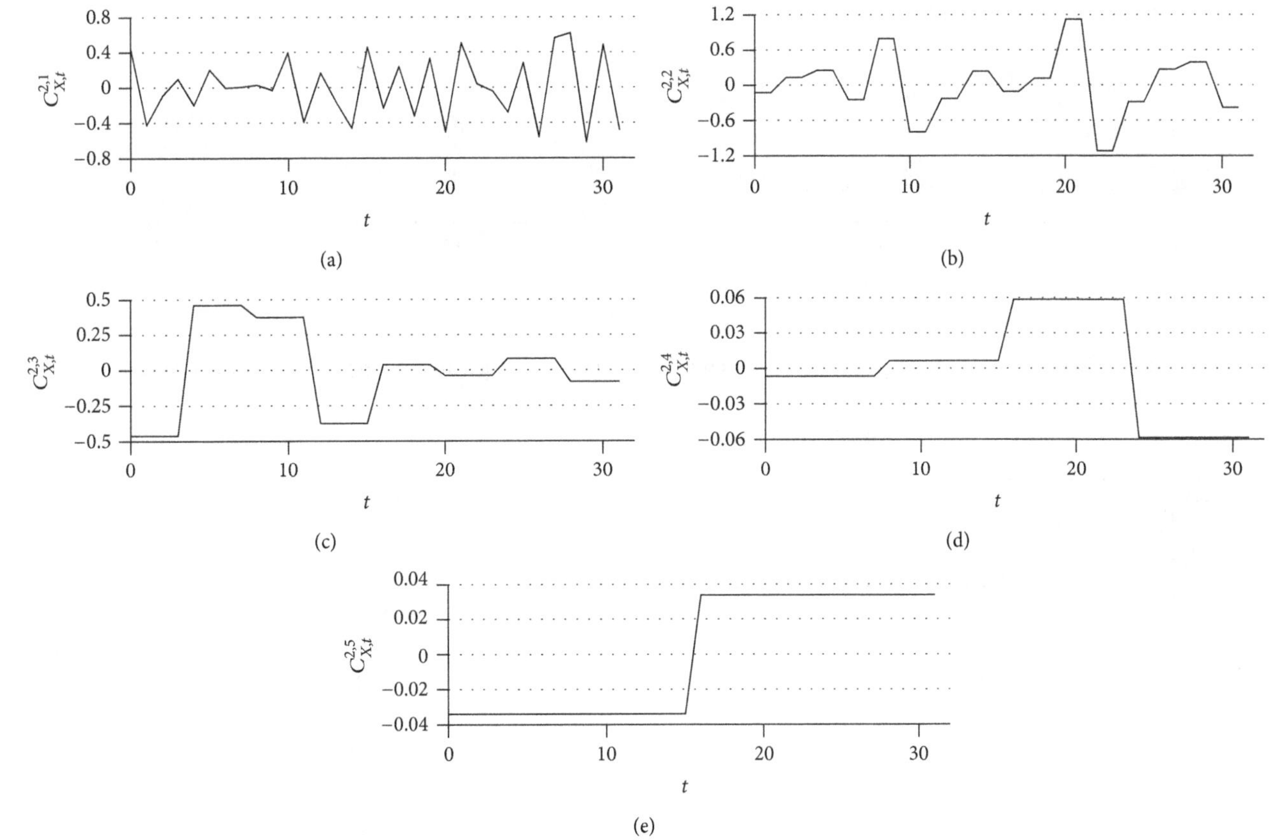

FIGURE 1: Components of an H-SOSS sample of size 32 and $H = 0.9$.

The wavelet-based function is

$$\psi_n(t) = \begin{cases} 1 - \dfrac{1}{n}, & 0 \le t < \dfrac{1}{n}, \\ -\dfrac{1}{n}, & \dfrac{1}{n} \le t < 1, \\ 0, & \text{otherwise}, \end{cases} \qquad (27)$$

and the wavelet functions derived from $\psi_n(t)$ are obtained by three operations: scaling, displacement (similarly to (10)), and a phase shift, that is,

$$\psi_{n,j,k,\theta}(t) = \left\{1 - u\left[n^j(k-1)\right]\right\} \psi_{n,j,k}\left(t - \theta n^{j-1}\right) + u\left[n^j(k-1)\right] \psi_{n,j,k}\left(t + n^j - \theta n^{j-1}\right), \qquad (28)$$

$$\psi_{n,j,k}(t) = n^{-j/2} \psi_n\left(n^{-j} t - k\right),$$

for $j = 1, 2, \ldots, J$, $k = 0, 2, \ldots, n^{J-j} - 1$, and $\theta = 0, 1, 2, \ldots, n-1$. Note that note that $\psi_{n,0,0,0}(t) = \psi_n(t)$.

The function defined by (27) is a generalized form of the Haar wavelet. It is always a rectangle-shaped function, but it is not symmetric about the horizontal axis except for $n = 2$.

Figure 2 shows the basis function for $n = 2$ without phase shift and with a phase shift of 1/2, respectively. Obviously, $\langle X(t), \psi_{2,j,k,0}(t)\rangle = -\langle X(t), \psi_{2,j,k,1}(t)\rangle$, which means that both products give the same information. Redundant information can then be reduced by decimating the sequence of coefficients.

Figure 3 shows the basis function for $n = 3$. Note that the phase shift moves the rectangle of height 2/3 from one-third to another. In this case, there exists also redundant information, as $\psi_{3,0,0,2}(t) = -\psi_{3,0,0,0}(t) - \psi_{3,0,0,1}(t)$ and in the general case $\psi_{n,j,k,\theta}(t) = -\sum_{i \ne \theta} \psi_{n,j,k,i}(t)$, which means that one of the coefficients, for example, the one obtained with the last phase shift, can be discarded. This means that the sequence of coefficients can be downsampled without loss of information. For a sample of length N, it is easy to verify that the number of observations that remains in all components (sequences of coefficients) after the downsampling is $N - 1$, which can be complemented with the sample mean, in the case that this is not zero.

3.2. Wavelet Synthesis of Self-Similar Time Series. A method to synthesize H_1-SOSS from practically any time series, regardless of whether it is or not self-similar or its marginal distribution, is proposed. This method consists of adjusting the sum expressed by (9) with a set of weights, that is,

$$X_{H_1}(t) = \sum_{j=i}^{J} w_j \sum_{k=1}^{2^j} d_X(j,k)\, \psi_{j,k}(t), \qquad (29)$$

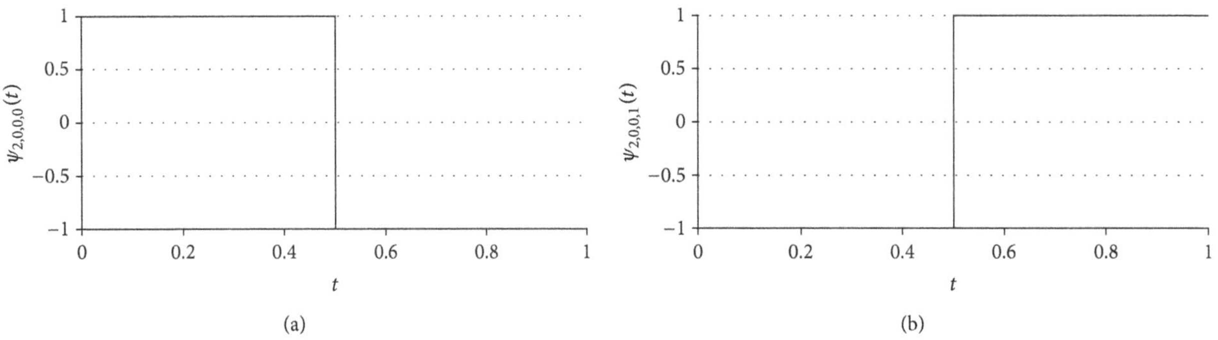

FIGURE 2: Basis functions: (a) $\psi_{2,0,0,0}(t)$ (no phase shift) and (b) $\psi_{2,0,0,1}(t)$ (equal to $\psi_{2,0,0,0}(t)$ with a phase shift of 1/2).

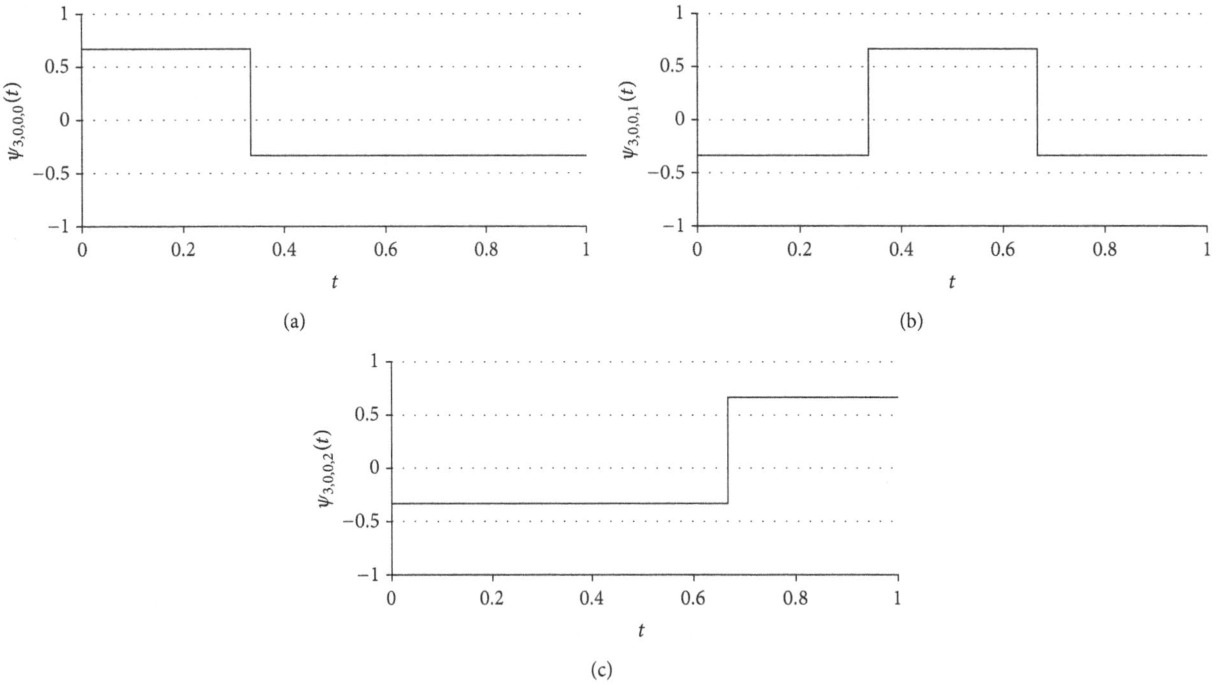

FIGURE 3: Basis functions: (a) $\psi_{3,0,0,0}(t)$ (no phase shift), (b) $\psi_{3,0,0,1}(t)$ ($\psi_{3,0,0,0}(t)$ with a phase shift of 1/3), and (c) $\psi_{3,0,0,2}(t)$ ($\psi_{3,0,0,0}(t)$ with a phase shift of 2/3).

where these weights w_j are defined as

$$w_j = \sqrt{\frac{\widehat{c_f C} \cdot 2^{j(2H_1-1)}}{\widehat{S}_2(j)}}, \tag{30}$$

where $\widehat{S}_2(j)$ and $\widehat{c_f C}$ are the respective estimations of $S_2(j)$ and $c_f C$ (the associated power parameter [4]) from X_t and H_1 is the desired Hurst index of the new synthetic series. It is necessary that $\widehat{S}_2(j) > 0$ for all $j = 1, \ldots, J$.

The weighted sum (29) can also be expressed in terms of the orthogonal components described in Section 3 and defined by (16) as follows:

$$X_t = \sum_{i=1}^{L} w_i C_{X,t}^{n,i}, \tag{31}$$

where the weights w_i are computed as

$$w_i = \sqrt{r^{i-1} \cdot \frac{1-r^L}{1-r} \cdot \frac{\widehat{\sigma}_X^2}{\operatorname{var}\left\{\widehat{C}_{X,t}^{n,i}\right\}}}, \tag{32}$$

where r is defined by (20). Note that the only restriction of (32), similarly to (30), is that the estimated variance $\widehat{C}_{X,t}^{n,i}$ must be nonzero. Even though this synthesis does not depend either on the marginal distribution or the correlation structure of the input signal, it is preferable that this is self-similar (e.g., FGN) and that its Hurst index is close to that of $X_{H_1}(t)$, that is, $H_1 \approx H$.

Pathological behavior can be produced in the output series for some critical conditions, for example, noticeable steps, which may produce a nonstationary signal, result from transforming an SRD, or uncorrelated input signal to an LRD output with H close to 1. Impulses of very large magnitude (outliers) can be also be produced when, for some j, $\widehat{S}_2(j)$ is close to zero.

A similar methodology was developed by Deléchelle et al. [27], to synthesize fractional Gaussian noise by performing a weighted sum of the intrinsic mode functions (IMF) of a white noise process. The advantages of the proposed synthesis compared to that of Deléchelle et al. are that the components defined by (16) are exactly orthogonal, the relation between the weights for the reconstruction sum are mathematically well defined, and the Hurst index of the synthesized time series matches perfectly the wavelet estimator proposed by Abry et al. [4], that is, the estimated H of the synthesized series is unbiased ($E(\widehat{H}) = H$) and has zero variance ($\text{var}(\widehat{H}) = 0$). The disadvantages are that the components are sequences of squared signals (because of the expansion described in Section 3) and not sinusoids, and noticeable steps arise when synthesizing a time series with high Hurst index, for example, close to 1, from an input that is SRD or weakly correlated. A solution for this problem is to apply interpolation (as in EMD) instead of expanding the series in order to produce softer components (sinusoids or polynomial) instead of square type, with the consequence that the Hurst index is no longer exact, but approximated.

4. Estimation of Mean and Variance Self-Similar Time Series

4.1. Sample Mean. The sample mean of a self-similar process is unbiased: its expected value is the process mean, that is, $E(\overline{X}) = \mu_X$, where $\overline{X} = 1/N\sum_{t=1}^{N} X_t$, regardless of the presence of correlation between observations. However, its variance does not depend only on the sample size (N) but also on the degree of self-similarity (H) of the process as follows:

$$\text{var}\left(\overline{X}\right) = \sigma_X^2 N^{2H-2}, \tag{33}$$

which becomes

$$\text{var}\left(\overline{X}\right) = \frac{\sigma_X^2}{N}, \tag{34}$$

(classical estimator) for $H = 0.5$ (uncorrelated observations). Figure 4 shows the *probability distribution function* (PDF) of the sample mean of standardized FGN.

To derive (33), consider that $\overline{X}$, estimated from a sample of size N, behaves exactly the same as the stationary aggregated process $X_k^{(N)}$, defined by (1), and its variance is determined by the definition of second-order self-similarity (7). Expression (33) can be also derived (for $H > 0.5$) from the autocorrelation coefficient $\rho(k) = 0.5[(k+1)^{2H} - 2k^{2H} + (k-1)^{2H}]$ for $k \geq 1$ ($\rho(0) = 1$) and $\text{var}(\overline{X}) = (\sigma_x^2/N^2)\sum_{i,j=1}^{N} \rho(k)$.

Important implications of (33) about the uncertainty of the mean are (1) that it increases with the Hurst index, for example, $\text{var}(\overline{X})$ tends to σ_X^2 as H tends to 1, which makes the sample mean worth a single observation and (2) that it cannot be zero for any case when estimated from a finite-size sample.

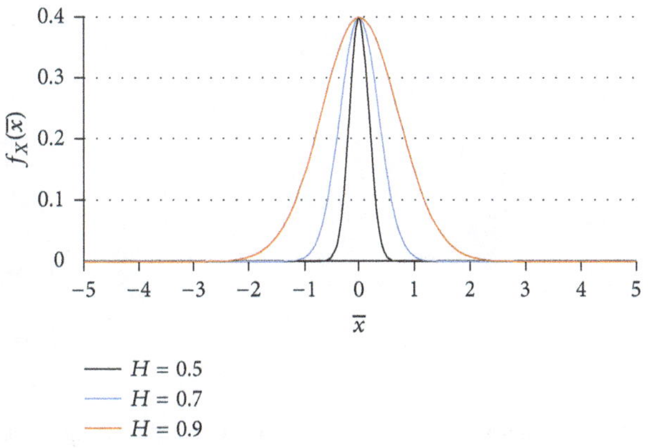

FIGURE 4: Distribution of the sample mean of standardized fractional Gaussian noise processes with $H = \{0.50, 0.70, 0.90\}$ and $N = 32$.

4.2. Sample Variance. For a high number of observations, sample variance is usually calculated as

$$\widehat{\sigma}_X^2 = \frac{1}{N}\sum_{t=1}^{N}\left(X_t - \widehat{\mu}_X\right)^2. \tag{35}$$

It is known that estimator (35) is biased, so for small samples, it is more adequate to use

$$\widehat{\sigma}_X^2 = \frac{1}{N-1}\sum_{t=1}^{N}\left(X_t - \overline{X}\right)^2. \tag{36}$$

Particularly, if the observations X_t are independent and come from a normal distribution, $\widehat{\sigma}_X^2$ is distributed as $\widehat{\sigma}_X^2 \sim (\sigma_X^2/N)\chi_{n-1}^2$, as stated by Cochran's theorem [28, 29]. Formula (36) is the most used estimator of the sample variance [14] but, as Beran indicates in [30], it is needed to know which assumptions this estimator is based on in order to apply it correctly; otherwise, it may be the source of errors that in practice cannot be negligible for all cases.

A self-similar process is uncorrelated only and only if the Hurst index is 0.5. In this particular case, the classical estimator of sample variance, defined by (36), is unbiased.

The expected value of the sample variance defined by (36) is

$$E\left(\widehat{\sigma}_X^2\right) = E\left[\frac{1}{N-1}\sum_{t=1}^{N}\left(X_t - \widehat{\mu}\right)^2\right], \tag{37}$$

which can be expressed as

$$E\left(\widehat{\sigma}_X^2\right) = \frac{N}{N-1}\left[E\left(X_t^2\right) - E\left(\widehat{\mu}^2\right)\right], \tag{38}$$

then,

$$\begin{aligned} E\left(\widehat{\sigma}_X^2\right) &= \frac{N}{N-1}\left\{\text{var}\left(X_t\right) + \left[E\left(X_t\right)\right]^2 \right. \\ &\quad \left. - \left\{\text{var}\left(\widehat{\mu}\right) + \left[E\left(\widehat{\mu}\right)\right]^2\right\}\right\}, \end{aligned} \tag{39}$$

$$E\left(\widehat{\sigma}_X^2\right) = \frac{N}{N-1}\sigma_X^2\left(1 - N^{2H-2}\right).$$

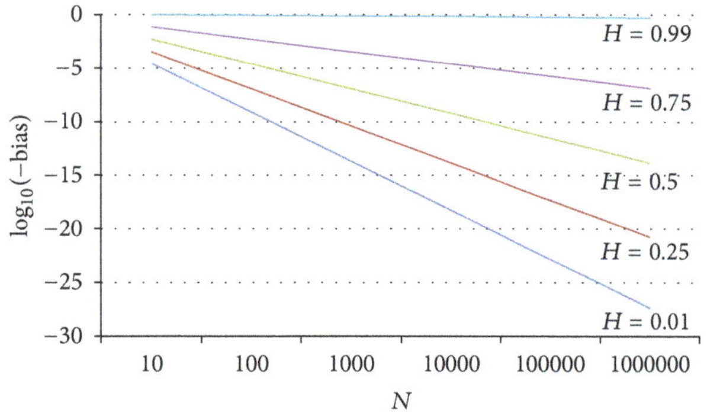

FIGURE 5: Logarithm of the variance bias for different sample sizes.

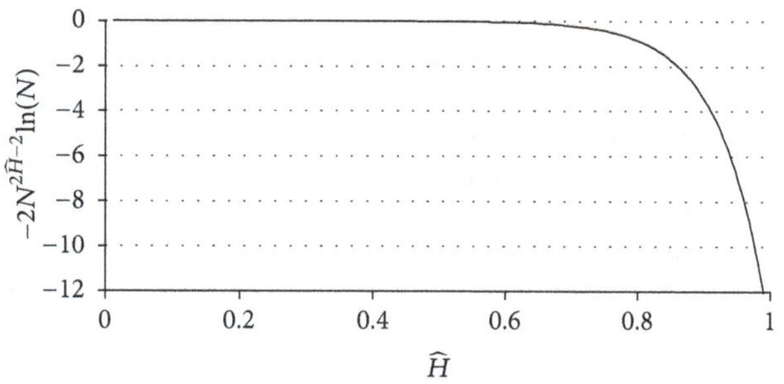

FIGURE 6: Plot $-2N^{2\widehat{H}-2}\ln(N)$ versus $\widehat{H}$.

Expression (39) proves that the classical estimator (36) is biased (i.e., $E(\widehat{\sigma}_X^2) \neq \sigma_X^2$) for $H \neq 0.5$. It is straightforward that the unbiased variance estimator for self-similar processes is then

$$\widehat{\sigma}_X^2 = \frac{1}{N - N^{2H-1}} \sum_{t=1}^{N} \left(X_t - \overline{X}\right)^2, \quad (40)$$

which obviously becomes (36) for $H = 0.5$.

A plot of $\log_{10}(\widehat{\sigma}_X^2 - \sigma_X^2)$ versus N is shown in Figure 5. Note that, for a fixed sample size, as H increases the estimation of the variance by means of the classical estimator, (36) becomes less significant.

Figure 5 exemplifies that as the sample size is greater, the classical variance estimator has less bias; however, as the Hurst index of the process is greater, the bias is also greater (in magnitude). Note that as H approaches to 1, the variance is considerably underestimated, which makes the classical variance estimator useless.

The variance of the estimated variance of a self-similar time series can be approximated by applying the formula proposed by Yunhua in [31] for $k = 0$, that is,

$$\operatorname{var}\left(\widehat{\sigma}_X^2\right) = \frac{\left(2N^{4H-3} + 8H - 7\right)\left(2H^2\right)(2H-1)^2}{N(4H-3)} + \frac{1}{N}. \quad (41)$$

This approximation is close to the variance of $\widehat{\sigma}_X^2$, with the disadvantage that it has a discontinuity in $H = 0.75$. Further work can be developed in order to verify this approximation and to quantify its error.

Let us mention that, although the proposed estimator of the sample variance is unbiased, its performance relies on the estimation of the Hurst index. This dependence is very noticeable as H approaches to 1, as the statistic $1 - N^{2\widehat{H}-2}$ is especially sensitive to the variation of $\widehat{H}$ under that condition. The derivative $d(1-N^{2\widehat{H}-2})/d\widehat{H} = -2N^{2\widehat{H}-2}\ln(N)$ versus H is shown in Figure 6. Note that as the estimated H increases, the estimation of the mean of the sample variance is more variable.

An immediate implication of this is that processes with Hurst index close to 1 must be carefully treated, as slight deviations of the Hurst index estimation derive in a nonnegligible error in the estimation of the process variance.

4.3. Statistics of the Aggregated Process. The aggregated process $X_k^{(m)}$ (defined by (2)) derived from an H-SS process is also H-SS (self-similar with the same Hurst index). It is also true for the case of H-SOSS processes, and this aggregated process is, by definition, identically distributed to the sample mean obtained from a set of $N = m$ observations (m is the aggregation level, as in (2)), that is, $X_k^{(m)} \sim \widehat{\mu}$. Also, the aggregated sample variance obtained with the classical estimator (36) is biased, as it is well known [32].

The variance of the aggregated series of an H-SOSS process must be then estimated with the unbiased formula (40) adapted to the number of observations in the sample, that is,

$$\operatorname{var}\left(\widehat{X}_k^{(m)}\right) = \frac{1}{N_i - N_i^{2H-1}} \sum_{k=1}^{N_i} \left(X_k^{(m)} - \overline{X_k^{(m)}}\right)^2, \quad (42)$$

where N_i is the size of the series after aggregation (i.e., $N_i = N/m$). Note that the estimation of the sample mean from the aggregated sample is also unbiased, that is, $E[\overline{X_k^{(m)}}] = E(X_t)$, and it is more reliable than the mean estimated from a sample of the same size, that is, $\operatorname{var}[\overline{X_k^{(m)}}] = m^{2H-2}\operatorname{var}(\overline{X})$ if the samples are of equal size.

4.4. Statistics of the Orthogonal Components. Let $\{\widehat{X}_t; t = 1, \ldots, N\}$ be a finite-length self-similar time series such that $N = n^J$ ($J < \infty$) and $n \geq 2$ (i.e., N is a power of n); then a set of nonzero J components ($\widehat{C}_{X,t}^{n,j}; j = 1, \ldots, J$) can be obtained as expressed by the analysis (16). As the components are pair-wise orthogonal, the variance of $\widehat{X}_t$ ($\widehat{\sigma}_X^2$) is the sum of a finite number of variances:

$$\widehat{\sigma}_X^2 = \sum_{j=1}^{J} \operatorname{var}\left(\widehat{C}_{X,t}^{n,j}\right). \quad (43)$$

Expressions (43) and (19) imply that the variance of the ith component ($\widehat{C}_{X,t}^{n,i}$) (computed using formula (35)), which has also finite length, is

$$\operatorname{var}\left(\widehat{C}_{X,t}^{n,i}\right) = \frac{1-r}{1-r^J} r^{i-1} \widehat{\sigma}_X^2, \quad (44)$$

and estimating $\operatorname{var}(C_{X,t}^{n,j})$ as in (23), it follows that

$$\operatorname{var}\left(\widehat{C}_{X,t}^{n,i}\right) = \frac{\operatorname{var}\left(C_{X,t}^{n,j}\right)\left(1 - N^{2H-2}\right)}{1 - r^J}. \quad (45)$$

Replacing r^J by N^{2H-2}, it yields

$$\operatorname{var}\left(\widehat{C}^{n,i}_{X,t}\right) = \operatorname{var}\left(C^{n,j}_{X,t}\right). \tag{46}$$

Expression (46) implies that the estimation of the variance of components is unbiased (a desirable property) and, as a consequence, so it is the estimation of the statistic $S_2(j)$. Another implication of (46) is that the estimations of the Hurst index and the power parameter $c_f C$ from the *Logscale Diagram* are unbiased, as have previously been proven by the authors of [11].

4.5. Correlation of the Wavelet Coefficients. Let us describe the autocovariance of the Haar wavelet coefficients, that is, the case for $n = 2$. The structure of this jth component ($C^{2,j}_{X,t}$), as a function of the elements of X_t, is

$$C^{2,j}_{X,\tau} = \frac{x^{(2^{j-1})}_{2\tau-1} - x^{(2^{j-1})}_{2\tau}}{2}. \tag{47}$$

Note that $C^{2,j}_{X,\tau}$ is a downsampled version of $C^{2,j}_{X,t}$, that is, only the first observation of each 2^j of $C^{2,j}_{X,t}$ remains.

Then, two consecutive coefficients are correlated as

$$\begin{aligned}\gamma_{C_j}(k) &= E\left(C^{2,j}_{X,\tau}, C^{2,j}_{X,\tau+k}\right)\\ &= E\left(\frac{x^{(2^{j-1})}_{2\tau-1} - x^{(2^{j-1})}_{2\tau}}{2}\,\frac{x^{(2^{j-1})}_{2(\tau+k)-1} - x^{(2^{j-1})}_{2(\tau+k)}}{2}\right).\end{aligned} \tag{48}$$

Assuming that X_t represents a zero-mean H-SOSS process, and according to the definition (8), $E(x^{(2^{j-1})}_{2\tau-1} x^{(2^{j-1})}_{2(\tau+k)-1}) = E(x^{(2^{j-1})}_{2\tau} x^{(2^{j-1})}_{2(\tau+k)}) = \gamma_X(2k)$, $E(x^{(2^{j-1})}_{2\tau-1} x^{(2^{j-1})}_{2(\tau+k)}) = \gamma_X(2k+1)$ and $E(x^{(2^{j-1})}_{2\tau} x^{(2^{j-1})}_{2(\tau+k)-1}) = \gamma_X(2k-1)$. Then, (48) is calculated as

$$\gamma_{C_j}(k) = \frac{2^{2H-2}\sigma_X^2}{4}\left[2\rho_X(2k) - \rho_X(2k+1) - \rho_X(2k-1)\right], \tag{49}$$

and the correlation coefficient of the jth component is then

$$\rho_{C_j}(k) = \rho_C(k) = \frac{\left[\rho_X(2k+1) - 2\rho_X(2k) + \rho_X(2k-1)\right]}{2^{2H} - 4}. \tag{50}$$

As (50) shows, the correlation structure is the same for all components, that is, $\rho_{C_j}(k)$ is independent of j. Other implications of (50) are that

$$\sum_{k=0}^{\infty}\rho_{C_j}(k) = 1 + \frac{-\rho_X(1) - \rho_X(2k+1) + 2\sum_{i=1}^{2K+1}\left[(-1)^{i-1}\rho_X(i)\right]}{2^{2H} - 4}, \tag{51}$$

which can be approximated as

$$\sum_{k=0}^{\infty}\rho_{C_j}(k) \approx -0.052H^2 - 0.311H + 1.168 < \infty, \tag{52}$$

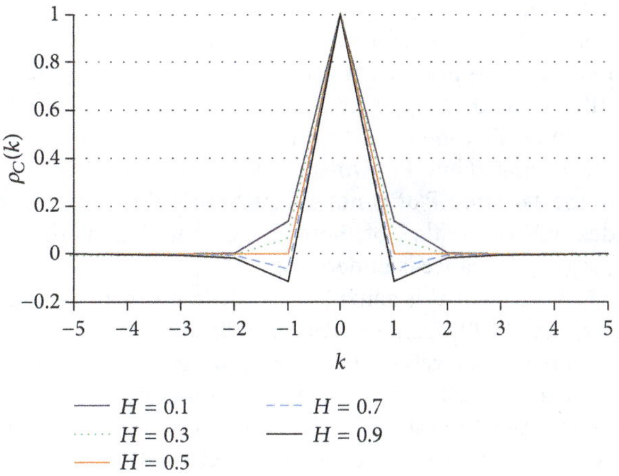

FIGURE 7: Correlation coefficient of the Haar wavelet components.

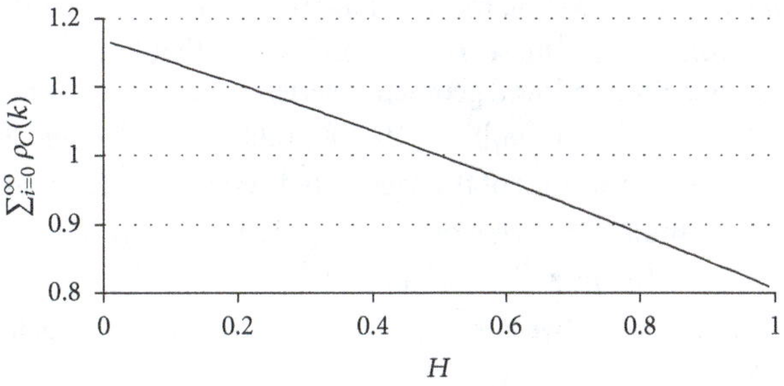

FIGURE 8: Sum of correlation coefficients of the Haar wavelet components.

that is, the coefficients are weakly correlated and the sum of correlations is finite [33]; furthermore, none of the components (sequences of wavelet coefficient) can be a self-similar time series except for $H = 0.5$, for example, components of a white noise process are also white noise processes. A plot of $\rho_{C_j}(k)$ versus k is shown in Figure 7 for $H = \{0.1, 0.3, 0.5, 0.7, 0.9\}$. The sum $\sum_{k=0}^{\infty}\rho_{C_j}(k)$ versus H is shown in Figure 8.

As the maximum magnitude of $\gamma_{C1}(1)$ is $1/8$ (when $H \to 0$), it is said that these coefficients are *quasiuncorrelated*. Note that $\gamma_{C1}(1) = 0$ for $H = 0.5$ and $H \to 1$.

Although the assumption that the estimation of the 1st component variance ($\operatorname{var}(\widehat{C}^{n,1}_{X,t})$) is unbiased is nearly accurate, it may not hold for components of greater order. As the wavelet coefficients $d_X(j,k)$ are almost uncorrelated, the estimation of the component variance (i.e., $\sum_{t=1}^{N/n^i}(\widehat{C}^{n,i}_{X,t})^2$) is approximately $((N - n^i)/N)\operatorname{var}(C^{n,i}_{X,t})$.

5. *Variance Plot*-Based Estimation of the Hurst Index

As described in Section 2.3, the *Variance Plot* is a straight line for self-similar time series but, as many authors have claimed, it underestimates the Hurst index when working with real world data. This is a consequence of the inadequate estimation of the aggregated variance, caused by the application of the classical formula (36) regardless of whether the original process presents any type of correlation. The solution would be then to apply the unbiased formula (42), but it leads to an ill-conditioned problem: the Hurst index needs to be estimated and known at the same time. Then, it is not that the *Variance Plot* is not adequate to estimate the Hurst index, rather the flaw of those implementations is that the aggregated variance is underestimated.

Nevertheless, it is actually possible to estimate the Hurst index analytically from the *Variance Plot*: the key is to choose the aggregation levels so that they form a geometric series, as explained in Section 5.1. Note that a numerical method can also be applied to estimate simultaneously the *Variance Plot* and the Hurst index, but the proposed solution is computationally simpler and more efficient.

5.1. Analytical Solution to the Ill-Conditioned Problem. Let $\{m_i;\ i = 1,\ldots,M\}$ be the set of aggregation levels such that $m_i = am_{i-1} = a^{i-1}m_1$, $a, m_1 \in \mathbb{N}$ and $a > 1$, that is, the levels of aggregation follow a geometric series, for example, $\{m_i\} = \{2,4,8,\ldots,2^M\}$ or $\{m_i\} = \{10,100,1000,\ldots,10^M\}$, and let $\hat{\sigma}^2_{X^{(m_i)}}$ be the variance of the aggregated series $X_k^{(m_i)}$ estimated with formula (35). Obviously, $\hat{\sigma}^2_{X^{(m_i)}}$ is biased, as $\hat{\sigma}^2_{X^{(m_i)}} = \sigma^2_{X^{(m_i)}}[1-(N/m_i)^{2H-2}] = \sigma^2_X m_i^{2H-2}(1-N^{2H-2})$, but then the difference between $\hat{\sigma}^2_{X^{(m_i)}}$ and $\hat{\sigma}^2_{X^{(m_{i+1})}}$ is calculated as follows:

$$\hat{\Delta}_i = \hat{\sigma}^2_{X^{(m_i)}} - \hat{\sigma}^2_{X^{(m_{i+1})}} = \sigma^2_X m_1^{2H-2} a^{(i-1)(2H-2)}\left(1-a^{2H-2}\right),\quad i = 1,\ldots,M-1, \tag{53}$$

and its logarithm

$$\log_a \Delta_i = \log_a\left[\sigma^2_X m_1^{2H-2}\left(1-a^{2H-2}\right)\right] + (i-1)(2H-2). \tag{54}$$

Finally, the slope (s) of the plot $\log_a\Delta_i$ versus i is obtained (e.g., with a weighted least square regression) and H is estimated as $\hat{H} = \hat{s}/2 + 1$. It can be easily proven as $\hat{\sigma}^2_{X^{(m_i)}} = X_t^{(n^{j-1}E)}$ and substituting it in (16) and (19).

The slope is computed by the following weighted formula:

$$\hat{s} = \frac{\sum_{i=1}^{M-1}\left(i\hat{\Delta}_i W_i\right) - \sum_{i=1}^{M-1}\left(iW_i\right)\cdot\sum_{i=1}^{M-1}\left(\hat{\Delta}_i W_i\right)}{\sum_{i=1}^{M-1}\left(i^2 W_i\right) - \left[\sum_{i=1}^{M-1}\left(iW_i\right)\right]^2}, \tag{55}$$

where the weights are such that $\sum_{i=1}^{M-1}(W_i) = 1$ and they are adequate so that $\hat{s}$ has minimal variance, for example, $W_i = W_{i-1}/m_1$.

Note that the *Variance Plot* can be estimated without bias, but the Hurt index is not estimated from it. Furthermore, if the aggregation levels are taken as $m_i = 2^i$, the estimator is exactly the same than the one that uses Haar wavelet. The authors of [34] developed an empirical study of estimation of the Hurst index from series with the presence of trends. They conclude that a method named differenced-variance (a variation of the variance-bases estimator) should not be used for estimating the Hurst index. The proposed solution is also a differenced-variance type method, but it can be used to estimate the Hurst index without bias and with optimal variance. Evidently, when working with real world traces, the *Variance Plot* may differ from the straight line, and an additional bias results from the logarithm as $E[\log(\cdot)] \neq \log[E(\cdot)]$. This bias can be subtracted analogously to (15), that is,

$$\log_a \mathrm{var}\left[X_t^{(m_i)}\right] = \log_a \mathrm{var}\left[X_t^{(m_i)}\right] - g_i, \tag{56}$$

where g_i is the bias defined in(14).

6. Simulation and Measurements

6.1. Estimation of the Sample Mean. In order to verify the equations that describe the mean and variance of the sample mean, a set of zero-mean, unitary variance, and FGN time series of size $N_p = 10^6$ observations are generated using an implementation of the generator proposed by Davies and Harte in [35], each for a different Hurst index for $H = \{0.30, 0.50, 0.70, 0.90\}$. Then, the mean is estimated from blocks of size $N = 100$ and the empirical PDF is obtained from the estimations and compared to the classical (34) and proposed (33) estimators. Figure 9(a) shows that the variance of the estimated mean does not fit the classical model when SRD or LRD is present (Figures 9(a), 9(c), and 9(d)), but only for the uncorrelated case (Figure 9(b)). Only proposed estimator (33) represents adequately this phenomenon for the four cases.

6.2. Estimation of the Sample Variance. The followed procedure to verify the proposed estimator of the sample variance $(\hat{\sigma}^2_X)$ consists of the generation of a set of 100 FGN samples of size 1024 for each value of $H = \{0.05, 0.10, 0.15, \ldots, 0.95\}$ and the estimation of the variance using the classical formula (36) and the proposed estimator (40). The respective mean of both estimations for each set was obtained. For the estimation of the Hurst index the wavelet estimator of Veitch and Abry [11] is used.

Figure 10 shows that the classical formula underestimates the variance noticeably for higher values of H. For $H > 0.95$ the estimated variance is less than half of the process variance. The proposed formula (40) does not underestimate the variance, but for high values of H the estimated variance is significantly different from one realization to another. This variation results from the estimation of the Hurst index, as the statistic $1 - N^{2\hat{H}-2}$ is very sensitive to the variations of $\hat{H}$, which depends, in turn, on the efficiency of the sample generator. This is an indicator that the generator proposed by Davies and Harte may be less accurate as H is closer to 1.

Figure 11 shows the variance of the estimated variance obtained with the proposed formula (40) compared to the approximation proposed by Yunhua in [31], as expected, the variance of the estimation is close to zero (lower than 0.001) for $H < 0.5$, but as the Hurst index increases, it becomes noticeable when $H > 0.75$. This verifies the observation of [31], which says that beyond $H = 0.75$ the precision of the autocorrelation is about one order lower than when $H < 0.75$.

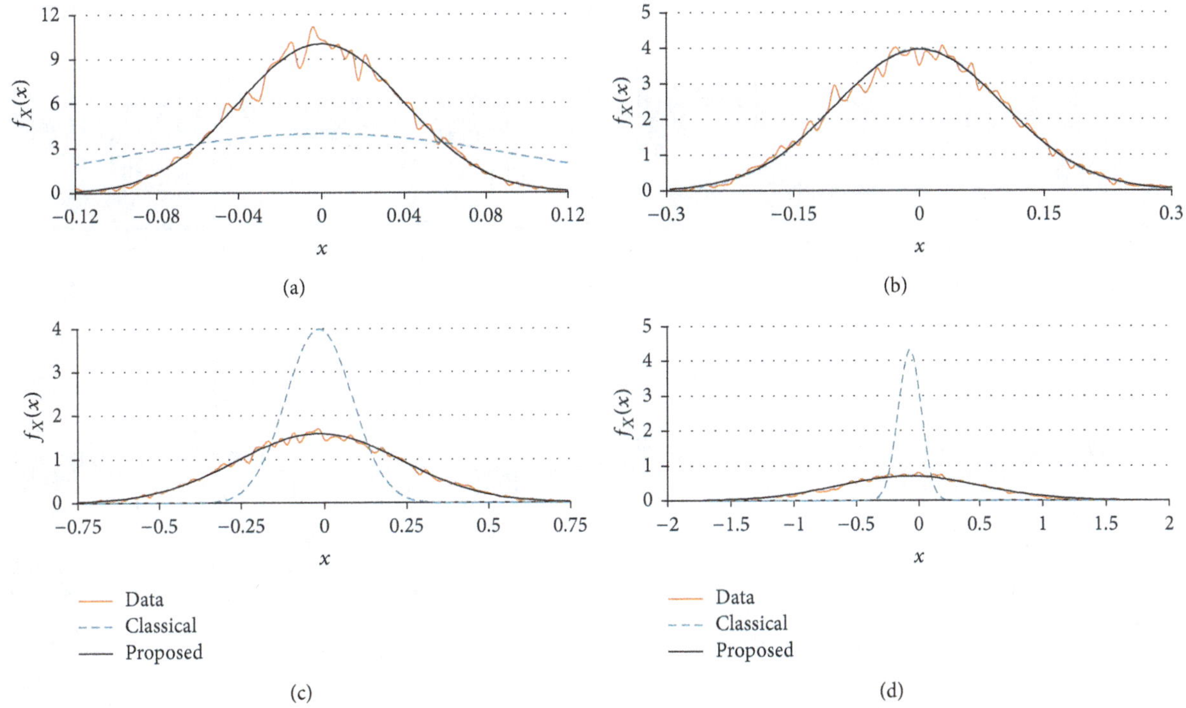

FIGURE 9: Estimation of the sample mean with $n = 100$ for four cases: (a) $H = 0.30$, (b) $H = 0.50$, (c) $H = 0.70$, and (d) $H = 0.90$.

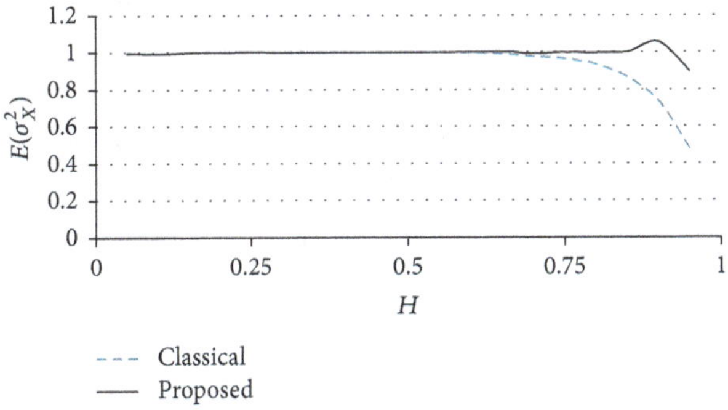

FIGURE 10: Mean of the estimated variance for $n = 1024$ and $H = \{0.05, 0.10, 0.15, \ldots, 0.95\}$.

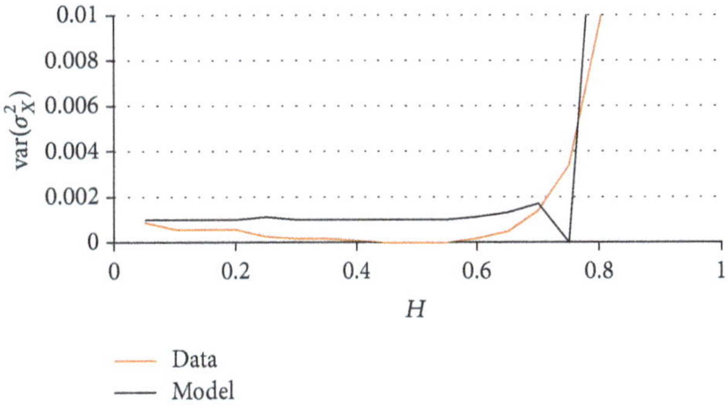

FIGURE 11: Variance of the estimated variance for $n = 1024$ and $H = \{0.05, 0.10, 0.15, \ldots, 0.95\}$.

6.3. Synthesis of H-SOSS Time Series. To exemplify the proposed wavelet-based synthesis (described in Section 3.2), four time series with respective Hurst index 0.3, 0.5, 0.7, and 0.9 were synthesized from an FGN sample of size 1024. The *Logscale Diagram* of the four new time series were obtained and compared to that of the original sample. The plot X_t versus t for each one; the four synthesized series is shown in Figure 12. One can visually check the presence of positive correlation in Figures 12(c) and 12(d).

The *Logscale Diagram* of these artificial series is shown in Figure 13. Note that the original *Logscale Diagram* of the source sample is not a straight line, but it so is for the synthesized series. Also, the estimated Hurst index of this series is $\widehat{H} = 0.56$ and its estimated *Logscale Diagram* is not a straight line, but the estimated Hurst index of the four generated series is exactly the desired, for example, $\widehat{H} = 0.30$ for the series shown in Figure 12(a) and the same for the others, and their respective *Logscale Diagram* is a straight line.

6.4. Voice over IP (VoIP) Measurements. The jitter behavior of Voice over Internet Protocol (VoIP) traffic by means of networks measurements is analyzed. As result of this analysis, detailed characterization and accurate modeling of this Quality of Service (QoS) parameter is provided. Previous studies have revealed that VoIP jitter can be modeled by self-similar processes with short-range dependence (SRD) or long-range dependence (LRD) [36]. The discovery of LRD (a kind of asymptotic fractal scaling) and weak self-similarity in the VoIP jitter data traces was followed by a further work that shows the evidence for multifractal behavior. The discovery of evidence for multifractal behavior is a richer form of scaling

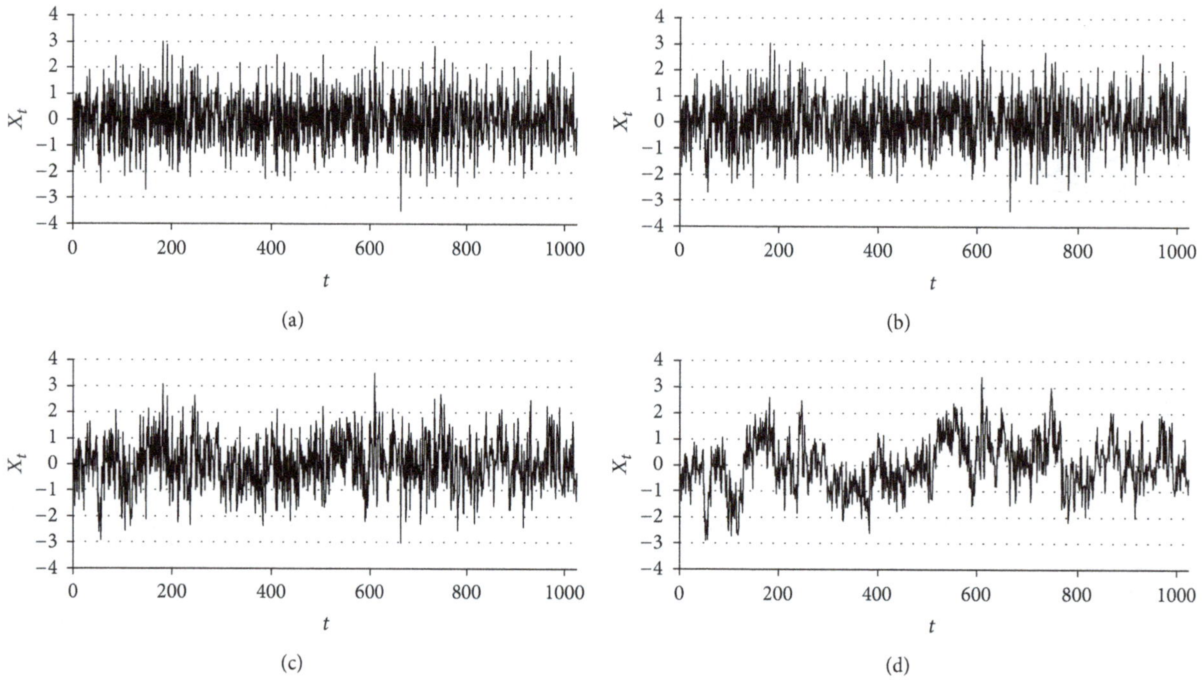

FIGURE 12: Plot versus time of the four synthesized time series. (a) $H = 0.30$, (b) $H = 0.50$, (c) $H = 0.70$, and (d) $H = 0.90$.

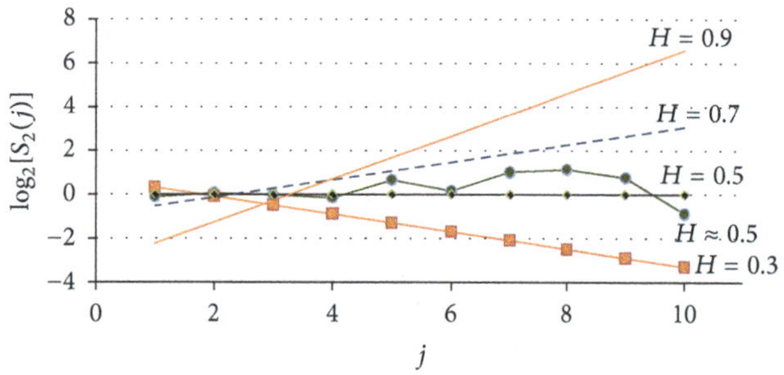

FIGURE 13: *Logscale Diagram* of synthesized time series.

behavior associated with nonuniform local variability, which could lead to a complete and robust model of IP network traffic over all time scales.

Motivated by such concerns, the evidence for multifractal behavior of VoIP jitter data traces is reviewed. In order to accomplish this analysis, the time series of VoIP jitter into a set of time series or components $C^{2,j}_{X,\tau}$ is decomposed as defined by (16). The behavior of these components is used to determine the kind of asymptotic fractal scaling. If the variance of the components of a time series is modeled by a straight line, the time series exhibits monofractal behavior, and a linear regression can be applied in order to estimate the Hurst parameter. On the other hand, if the variance of the components cannot be adequately modeled with a linear model, then the scaling behavior should be described with more than one scaling parameter, that is, the time series exhibits multifractal behavior [23]. In Figures 14 and 15, we show the components behavior of the collected VoIP jitter data traces.

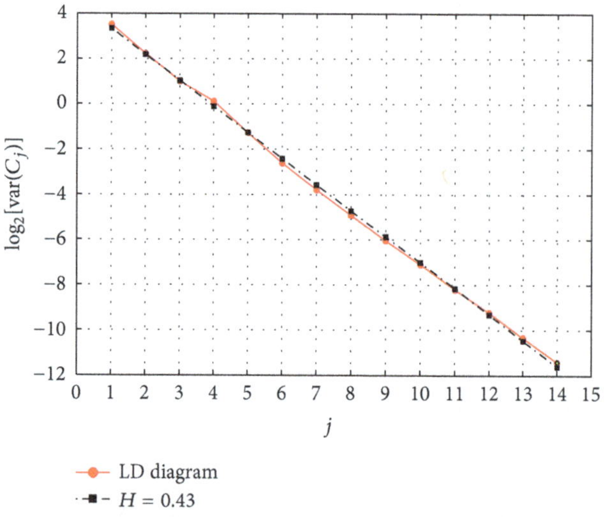

FIGURE 14: Components behavior of VoIP jitter data traces: monofractal behavior.

Figure 14 shows the components behaviors of a VoIP jitter data trace that belongs to the data sets with SRD. It is observed that the variance of the components of this time series is modeled by a straight line; therefore, the time series exhibits monofractal behavior.

Figure 15 shows the components behaviors of a VoIP jitter data trace that belongs to the data sets with LRD. It is observed that the variance of the components of this time series cannot be adequately modeled with a linear model, and the scaling behavior should be described with multiple scaling parameters (biscaling); therefore, this time series exhibits multifractal behavior.

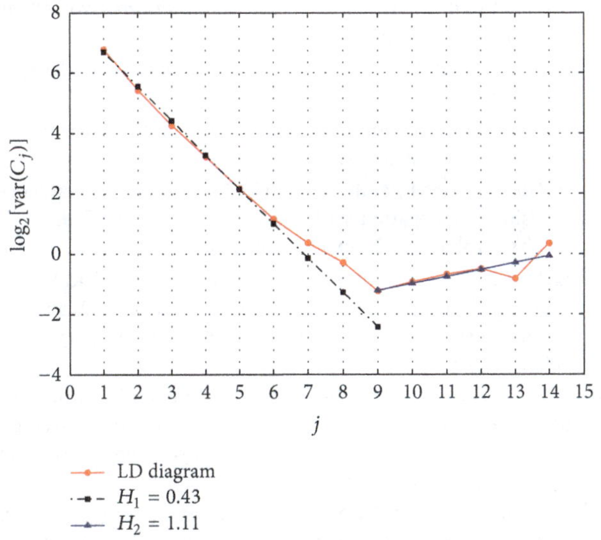

FIGURE 15: Components behavior of VoIP jitter data traces: multifractal behavior.

These results show that VoIP jitter with SRD or LRD exhibit monofractal or multifractal behavior, respectively. This phenomenon explains the behavior of the data traces with SRD and high degree of self-similarity (scale invariance), because the self-similarity is defined for a single scale parameter. On the other hand, the data traces with LRD exhibit weak self-similarity because they have associated nonuniform local variability (multifractal behavior).

The implication of this behavior for VoIP and other interactive multimedia services is that receiver dejitter buffer may not be large enough to mask the jitter with LRD and multifractal characteristics.

7. Conclusion

An orthogonal decomposition, that constitutes a powerful statistic tool to study discrete time series, is presented. The resulting components ($C_{X,t}^{n,j}$) have zero-mean and three desirably properties: (1) they synthesize the source signal without loss, (2) they are pair wise orthogonal, and (3) their variances form a geometric series whose rate is related to the Hurst index.

This decomposition is firstly compared to Haar wavelet based. For a particular case these two coincide, but the proposed one is more general, as other levels of aggregation may occur. In this case, the Haar wavelet is not sufficient and a special class of basis functions are defined, and the wavelet coefficients are obtained by the inner products between the signal under study and a scaled, displaced, and phase shifted versions of the basis function, in contradistinction of other wavelet decomposition, that only apply scaling and displacement. For a fixed scaling and displacement, the last phase shift can be discarded (i.e., the components are downsampled) as it does not provide additional information.

The proposed decomposition can be used to estimate the Hurst index, as the plot $j + \log_n[\mathrm{var}(C_{X,t}^{n,j})]$ versus j is equivalent to the *Logscale Diagram* proposed by Veitch and Abry in [11]. It is a straight line for H-SOSS time series, and the slope is related to the Hurst index so that $s = 2H - 1$.

The components can be also used to synthesize H-SOSS time series by means of the weighted sum defined by (31) and (32), regardless of the distribution of the source signal and whether it is or not self-similar. The synthesis is exact, that is, the Hurst index of the synthesized series is exactly the desired, which is an advantage over the proposed synthesis, that uses IMF from an EMD decomposition, described in [27].

A study of the estimated mean and variance of self-similar time series is also presented. Both statistics, mean and variance, can be estimated without bias, by applying the classic sample mean ($\overline{X} = 1/N\sum_{t=1}^{N} X_t$) and the proposed variance estimator (40). The variance of these estimators depends on the Hurst index of the process. In the case of the sample mean, its variance is an increasing function of H, as expressed by (33), such that $\mathrm{var}(\overline{X}) \in (\sigma_X^2 N^{-2}, \sigma_X^2)$. Note that the sample mean becomes less significant as H approaches to 1. For the variance of $\widehat{\sigma}_X^2$, it can be observed that it has the best performance when $H = 0.5$ (the variance of $\widehat{\sigma}_X^2$ is minimal). For $H < 0.5$, the variance increases as H is lower, but it is still acceptable (e.g., $\mathrm{var}(\widehat{\sigma}_X^2) < 0.001\sigma_X^2$ when $H \to 0$. But when H gets close to 1, the uncertainty of the sample variance increases rapidly, making the estimation of σ_X^2 less significant.

It is demonstrated also (clarifying a popular misunderstanding) that the *Variance Plot* can be used to estimate efficiently the Hurst index. The claims of many researchers about the inefficiency of the *Variance Plot* are the result of an ill-conditioned problem: to estimate the Hurst index the variance of the aggregated series is needed and vice versa, leading to a vicious circle. It is shown how this problem is avoided by estimating the *Variance Plot* using a set of aggregation levels that follow a geometric series and calculating *the differences* between the variances of the corresponding aggregated series; then, a weighted linear regression is applied to estimate the slope ($\widehat{s}$), and the estimated Hurst index is $\widehat{H} = \widehat{s}/2 + 1$. This result gives more significance to the results published by Abry et al. in [26].

Acknowledgment

This work was supported by PROMEP Grant UQROO-PTC-110.

References

[1] W. E. Leland, M. S. Taqqu, W. Willinger, and D. V. Wilson, "On the self-similar nature of Ethernet traffic (extended version)," *IEEE/ACM Transactions on Networking*, vol. 2, no. 1, pp. 1–15, 1994.

[2] W. E. Leland, M. S. Taqqu, W. Willinger, and D. V. Wilson, "On the self-similar nature of Ethernet traffic," in *Proceedings of the ACM SIGCOMM Computer Communication Review*, vol. 23, pp. 183–193, San Francisco, Calif, USA, 1993.

[3] T. Karagiannis, M. L. Molle, and M. Faloutsos, "Long-range dependence: Ten years of internet traffic modeling," *IEEE Internet Computing*, vol. 8, no. 5, pp. 57–64, 2004.

[4] P. Abry, R. Baraniuk, P. Flandrin, R. Riedi, and D. Veitch, "Multiscale nature of network traffic," *IEEE Signal Processing Magazine*, vol. 19, no. 3, pp. 28–46, 2002.

[5] M. S. Taqqu, V. Teverovsky, and W. Willinger, "Is network traffic self-similar or multifractal?" *Fractals*, vol. 5, no. 1, pp. 63–73, 1997.

[6] P. Abry, R. Baraniuk, P. Flandrin, R. Riedi, and D. Veitch, "Multiscale nature of network traffic," *IEEE Signal Processing Magazine*, vol. 19, no. 3, pp. 28–46, 2002.

[7] H. D. J. Jeong, J. S. R. Lee, D. McNickle, and K. Pawlikowski, "Comparison of various estimators in simulated FGN," *Simulation Modelling Practice and Theory*, vol. 15, no. 9, pp. 1173–1191, 2007.

[8] J. Mielniczuk and P. Wojdyłło, "Estimation of Hurst exponent revisited," *Computational Statistics and Data Analysis*, vol. 51, no. 9, pp. 4510–4525, 2007.

[9] P. Shang, Y. Lu, and S. Kamae, "Detecting long-range correlations of traffic time series with multifractal detrended fluctuation analysis," *Chaos, Solitons and Fractals*, vol. 36, no. 1, pp. 82–90, 2008.

[10] G. Horn, A. Kvalbein, J. Blomskøld, and E. Nilsen, "An empirical comparison of generators for self similar simulated traffic," *Performance Evaluation*, vol. 64, no. 2, pp. 162–190, 2007.

[11] D. Veitch and P. Abry, "A wavelet-based joint estimator of the parameters of long-rangé dependence," *IEEE Transactions on Information Theory*, vol. 45, no. 3, pp. 878–897, 1999.

[12] D. Radev and I. Lokshina, "Advanced models and algorithms for self-similar ip network traffic simulation and performance analysis," *Journal of Electrical Engineering*, vol. 61, no. 6, pp. 341–349, 2010.

[13] R. G. Clegg, "A practical guide to measuring the Hurst parameter," in *Proceedings of the 21st UK Performance Engineering Workshop*, School of Computing Science, Technical Report Series, pp. 43–55, 2006.

[14] J. F. Kenney, *Mathematics of Statistics*, Van Nostrand, New York, NY, USA, 1939.

[15] J. F. Kenney and E. S. Keeping, *Mathematics of Statistics*, part 2, Van Nostrand, Princeton, NJ, USA, 2nd edition, 1951.

[16] T. T. Soong, *Fundamentals of Probability and Statistics for Engineers*, John Wiley & Sons, New York, NY, USA, 2004.

[17] B. Tsybakov and N. D. Georganas, "Self-similar processes in communications networks," *IEEE Transactions on Information Theory*, vol. 44, no. 5, pp. 1713–1725, 1998.

[18] I. W. C. Lee and A. O. Fapojuwo, "Stochastic processes for computer network traffic modeling," *Computer Communications*, vol. 29, no. 1, pp. 1–23, 2005.

[19] D. Radev and I. Lokshina, "Self-similar simulation of IP traffic for wireless networks," *International Journal of Mobile Network Design and Innovation*, vol. 2, no. 3-4, pp. 202–208, 2007.

[20] J. M. Ramírez-Corte, V. Alarcon-Aquino, G. Rosas-Cholula, P. Gomez Gil, and J. Escamilla-Ambrosio, "P-300 rhythm detection using anfis algorithm and wavelet feature extraction in eeg signals," in *Proceedings of the World Congress on Engineering and Computer Science*, San Francisco, Calif, USA, October 2010.

[21] M. S. Taqqu, V. Teverovsky, and W. Willinger, "Is network traffic self-similar or multifractal?" *Fractals*, vol. 5, no. 1, pp. 63–73, 1997.

[22] O. Sheluhin, S. Smolskiy, and A. Osin, *Self-Similar Processes in Telecommunications*, Wiley, 2007.

[23] S. Stoev, M. S. Taqqu, C. Park, and J. S. Marron, "On the wavelet spectrum diagnostic for Hurst parameter estimation in the analysis of Internet traffic," *Computer Networks*, vol. 48, no. 3, pp. 423–445, 2005.

[24] M. V. Wickerhauser, *Adapted Wavelet Analysis from Theory to Software*, IEEE Press.

[25] N. Rillo, "Introduction to wavelets theory," http://www.mat.ub.es/~soria/TAD-Wavelets.pdf.

[26] P. Abry, D. Veitch, and P. Flandrin, "Long-range dependence: Revisiting aggregation with wavelets," *Journal of Time Series Analysis*, vol. 19, no. 3, pp. 253–266, 1998.

[27] É. Deléchelle, J. C. Nunes, and J. Lemoine, "Empirical mode decomposition synthesis of fractional processes in 1D- and 2D-space," *Image and Vision Computing*, vol. 23, no. 9, pp. 799–806, 2005.

[28] W. G. Cochran, "The distribution of quadratic forms in a normal system, with applications to the analysis of covariance," *Mathematical Proceedings of the Cambridge Philosophical Society*, vol. 30, no. 2, pp. 178–191, 1934.

[29] Y. Tian and G. P. H. Styan, "Cochran's statistical theorem revisited," *Journal of Statistical Planning and Inference*, vol. 136, no. 8, pp. 2659–2667, 2006.

[30] J. Beran, *Statistics for Long-Memory Processes*, Monographs on Statistics & Applied Probability, Chapman & Hall/CRC, 1994.

[31] R. Yunhua, "Evaluation and estimation of second-order self-similar network traffic," *Computer Communications*, vol. 27, no. 9, pp. 898–904, 2004.

[32] M. Krunz and I. Matta, "Analytical investigation of the bias effect in variance-type estimators for inference of long-range dependence," *Computer Networks*, vol. 40, no. 3, pp. 445–458, 2002.

[33] P. Flandrin, "Wavelet analysis and synthesis of fractional Brownian motion," *IEEE Transactions on Information Theory*, vol. 38, no. 2, pp. 910–917, 1992.

[34] V. Teverovsky and M. Taqqu, "Testing for long-range dependence in the presence of shifting means or a slowly declining trend, using a variance-type estimator," *Journal of Time Series Analysis*, vol. 18, no. 3, pp. 279–304, 1997.

[35] R. B. Davies and D. S. Harte, "Tests for hurst effect," *Biometrika*, vol. 74, no. 1, pp. 95–101, 1987.

[36] H. Toral, *QoS parameters modeling of self-similar VoIP traffic and an improvement to the E model [Ph.D. thesis]*, Electrical Engineering, Telecommunication Section, CINVESTAV, Jalisco, Mexico, 2010.

3

Parameter Identification of Anaerobic Wastewater Treatment Bioprocesses Using Particle Swarm Optimization

Dorin Sendrescu

Department of Automatic Control, University of Craiova, A.I. Cuza 13, 200585 Craiova, Romania

Correspondence should be addressed to Dorin Sendrescu; dorins@automation.ucv.ro

Academic Editor: Carlo Cattani

This paper deals with the offline parameters identification for a class of wastewater treatment bioprocesses using particle swarm optimization (PSO) techniques. Particle swarm optimization is a relatively new heuristic method that has produced promising results for solving complex optimization problems. In this paper one uses some variants of the PSO algorithm for parameter estimation of an anaerobic wastewater treatment process that is a complex biotechnological system. The identification scheme is based on a multimodal numerical optimization problem with high dimension. The performances of the method are analyzed by numerical simulations.

1. Introduction

It is well known that the biotechnology is one of the fields that over the last decades have a high development. Therefore, due to its advantages, the control of industrial bioprocesses has been an important practical problem attracting wide attention. Biotechnology applications can be found especially in agriculture, in food industry, in medicine and pharmaceutical processes, in waste treatment processes, and so forth. A frequent and important challenge in control of such living processes is finding an accurate model of the system. The bioprocesses are highly nonlinear, and their kinetic parameters are usually badly or inadequately known [1]. This problem becomes of great importance in complex systems where critical instability of the process must be avoided. Parameters characterizing the internal behavior of biotechnological systems are usually not directly accessible to measurement. Their measurement is usually approached indirectly as a parameter estimation problem [2]. In this paper a dynamic model describing the internal structure of the system is formulated, and an algorithm based on PSO for parameter estimation is designed.

In recent years, a progress has been made in the area of continuous-time system identification [3]. Even if the most physical systems are naturally continuous, a much more attention has been paid to parameter estimation of discrete-time systems, mainly because they are better suited for numerical implementations. Continuous-time identification makes possible a more direct link to the physical properties and operation of the underlying systems and the direct estimation of physical parameters which have a clear significance. The most common approach for parameter estimation of linear or nonlinear systems is the use of prediction-error identification methods (PEM) [4]. In this category falls the well-known least squares methods or the maximum likelihood methods. In this approach, identification consists in minimization of a scalar-valued function of the model parameter. In general, this function cannot be minimized by analytical methods so the solution has to be found by iterative, numerical techniques. There is an extensive literature on such numerical problems. In classical approach the most used procedures are the quasi-Newton methods and interior point algorithms. The main drawback of these nonlinear parameter optimization techniques is that they are often unreliable; for example, they give no guarantee of converging to a true minimum. The increasing computational power of personal computers and microcontrollers allowed the implementation of several optimization algorithms inspired from natural phenomena. Examples of these algorithms include the Simulated Annealing [5], Genetics Algorithms (GA) [6], or Ant Colony

Optimization [7] algorithms. Particle Swarm Optimization (PSO) [8] is among these nature inspired algorithms. It is inspired by the ability of birds flocking to find food that they have no previous knowledge of its location. Every member of the swarm is affected by its own experience and its neighbors' experiences. Although the idea behind PSO is simple and can be implemented by two lines of programming code, the emergent behavior is complex and hard to completely understand [9, 10].

The most important approaches for the yield and kinetic coefficients estimation of biotechnological systems make use of the state transformations based on the general structure [11]. In this paper we propose an identification method based on particle swarm optimization techniques for these classes of biotechnological systems considering that the unknown parameters can appear in rational relations with measured variables. The paper is organized in the following way. The nonlinear dynamical model of an anaerobic wastewater treatment bioprocess is given in Section 2. Section 3 presents the identification algorithm using the particle swarm optimization techniques. Some numerical simulations are presented in Section 4 and conclusions in Section 5.

2. Nonlinear Dynamical Model of Anaerobic Wastewater Treatment Bioprocesses

A process that takes place in a bioreactor can be described as a set of m biochemical reactions involving n components (with $n \geq m$). The global dynamics can be represented by the following dynamical state-space model [1]:

$$\frac{d\xi}{dt} = K \cdot \varphi(\xi, t) - D\xi + F - Q, \tag{1}$$

where $\xi \in \Re^{n\times 1}$.

This model describes the behavior of an entire class of biotechnological processes and is referred to as the general dynamical state-space model of this class of bioprocesses [9]. In (1), the term $K \cdot \varphi(\xi, t)$ is the rate of consumption and/or production of the components in the reactor; that is, the reaction kinetics and the term $-D\xi + F - Q$ represents the exchange with the environment. The strongly nonlinear character of the model (1) is given by the reaction kinetics. In many situations, the yield coefficients, the structure, and the parameters of the reaction rates are partially known or unknown. Many of the evolved control methods for these kinds of systems—like model predictive control and robust or adaptive control—are based on good initial estimates of the yield and kinetic parameters.

Anaerobic digestion is a multistage biological wastewater treatment process whereby bacteria, in the absence of oxygen, decompose organic matter to carbon dioxide CO_2, methane CH_4, and water [12]. Four metabolic paths can be identified in this process: two for acidogenesis and two for methanation (see Figure 1). In the first acidogenic path, glucose (or another complex substrate) from the wastewater is decomposed into volatile fatty acids (acetates, propionic acid), hydrogen H_2, and inorganic carbon by acidogenic bacteria.

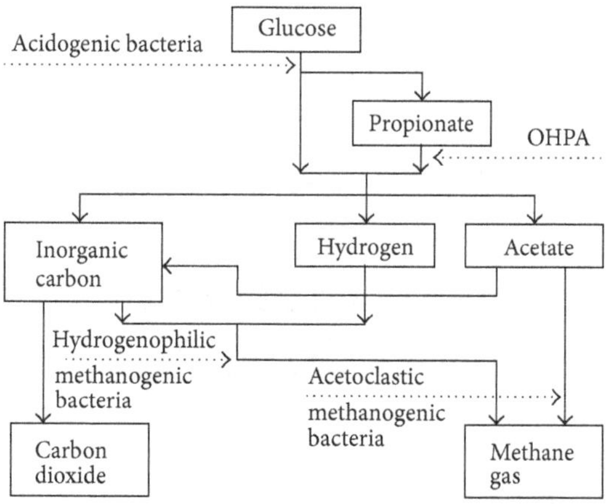

FIGURE 1: A schematic view of a typical anaerobic digestion process.

In the second acidogenic path, Obligate Hydrogen Producing Acetogens (OHPA) decompose propionate into acetate, H_2, and inorganic carbon. In the first methanation path, acetate is transformed into CH_4 and CO_2 by acetoclastic methanogenic bacteria, while, in the second methanation path, H_2 combines inorganic carbon to produce CH_4 under the action of hydrogenophilic methanogenic bacteria. The reaction scheme of this complex bioprocess involves 4 reactions and 10 components.

Since the anaerobic digestion is a complex bioprocess, this dynamical model being described by ten differential equations, for control purpose an appropriately reduced order model can be used. Using the singular perturbation method, the following reduced order dynamical model can be obtained:

$$\dot{X}_1 = \varphi_1 - DX_1, \tag{2}$$

$$\dot{S}_1 = -k_1\varphi_1 - DS_1 + DS_{\text{in}}, \tag{3}$$

$$\dot{X}_2 = \varphi_2 - DX_2, \tag{4}$$

$$\dot{S}_2 = k_3\varphi_1 - k_2\varphi_2 - DS_2, \tag{5}$$

$$\dot{S}_5 = k_4\varphi_1 + k_5\varphi_2 - DS_5 - Q_{CO_2}, \tag{6}$$

$$\dot{P} = k_6\varphi_2 + k_7\varphi_1 - DP - Q_P, \tag{7}$$

where X_1, X_2 are acidogenic and acetoclastic methanogenic bacteria, respectively, and S_1, S_2, S_5 are glucose, acetate and inorganic carbon, respectively, and P is methane; φ_1, φ_2 are the rates of first acidogenic reaction and methanation reaction respectively, $Q_P = c_pP$ with $c_p > 0$ and Q_{CO_2} represent gaseous outflow rates of CH_4 and CO_2, respectively, S_{in} is the influent substrate concentration, D the dilution rate, and k_i $(i = 0, \ldots, 7)$ are yield coefficients. Each reaction rate is a growth rate and may be written as $\varphi_i = \mu_i X_i$, $i = 1, 2$, where μ_i, $i = 1, 2$, is the specific growth rate of reaction i.

Defining the state vector as $\xi = [X_1\ S_1\ X_2\ S_2\ S_5\ P]^T$, the model (2)–(7) can be written in matrix form as

$$\begin{bmatrix} \dot{X}_1 \\ \dot{S}_1 \\ \dot{X}_2 \\ \dot{S}_2 \\ \dot{S}_5 \\ \dot{P} \end{bmatrix} = \begin{bmatrix} 1 & 0 \\ -k_1 & 0 \\ 0 & 1 \\ k_2 & -k_3 \\ k_4 & k_5 \\ k_6 & k_7 \end{bmatrix} \begin{bmatrix} \varphi_1 \\ \varphi_2 \end{bmatrix} - D \begin{bmatrix} X_1 \\ S_1 \\ X_2 \\ S_2 \\ S_5 \\ P \end{bmatrix} + \begin{bmatrix} 0 \\ DS_{\text{in}} \\ 0 \\ 0 \\ 0 \\ 0 \end{bmatrix} - \begin{bmatrix} 0 \\ 0 \\ 0 \\ 0 \\ Q_{CO_2} \\ Q_P \end{bmatrix}, \quad (8)$$

or in compact form as

$$\dot{\xi} = K\varphi(\xi) - D\xi + F - Q, \quad (9)$$

where $F = [0\ DS_{\text{in}}\ 0\ 0\ 0\ 0]^T$ is the vector of inflow rates, $Q = [0\ 0\ 0\ 0\ Q_{CO_2}\ Q_P]^T$ is the vector of gaseous outflow rates, $\varphi = [\varphi_1\ \varphi_2]^T$ is the vector of reaction rates, which can be written as $\varphi(\xi) = G(\xi)\alpha(\xi)$, with $G(\xi)$ being a diagonal matrix whose entries are products of the component concentrations involved in each reaction and $\alpha = [\alpha_1\ \alpha_2]^T$ the vector of specific reaction rates, and K is the yield coefficient's matrix. The matrices K and G have the following structure:

$$K = \begin{bmatrix} 1 & -k_1 & 0 & k_2 & k_4 & k_6 \\ 0 & 0 & 1 & -k_3 & k_5 & k_7 \end{bmatrix}^T,$$
$$G = \begin{bmatrix} X_1 S_1 & 0 \\ 0 & X_2 S_2 \end{bmatrix}. \quad (10)$$

The most difficult task for the construction of the dynamical model is the modeling of the reaction kinetics [13]. The form of kinetics is complex, nonlinear, and in many cases unknown. In our study one considers that reaction rates are given by the Monod law

$$\varphi_1(\xi) = \mu_1^* \frac{S_1 \cdot X_1}{K_{M_1} + S_1}, \quad (11)$$

and the Haldane kinetic model

$$\varphi_2(\xi) = \mu_2^* \frac{S_2 \cdot X_2}{K_{M_2} + S_2 + S_2^2/K_i}, \quad (12)$$

where K_{M_1}, K_{M_2} are Michaelis-Menten constants, μ_1^*, μ_2^* represent specific growth rate coefficients, and K_i is the inhibition constant.

For simplicity, we will denote the unknown plant parameters by the vector

$$\theta = [\theta_1\ \theta_2 \cdots \theta_{12}]^T, \quad (13)$$

where

$$\begin{gathered} \theta_1 = k_1; \quad \theta_2 = k_2; \quad \theta_3 = k_3; \\ \theta_4 = k_4; \quad \theta_5 = k_5 \quad \theta_6 = k_6 \quad \theta_7 = k_7 \\ \theta_8 = \mu_1^*; \quad \theta_9 = \mu_2^*; \quad \theta_{10} = K_{M_1}; \\ \theta_{11} = K_{M_2}; \quad \theta_{12} = K_i. \end{gathered} \quad (14)$$

3. Parameter Estimation Using PSO

At the beginning of parameter estimation, the input and output data are known and the real system parameters are assumed as unknown. The identification problem is formulated in terms of an optimization problem in which the error between an actual physical measured response of the system and the simulated response of a parameterized model is minimized. The estimation of the system parameters is achieved as a result of minimizing the error function by the PSO algorithm.

3.1. Problem Statement. Consider the following n-dimensional nonlinear system:

$$\frac{d\xi(t)}{dt} = f(\xi, t; \theta), \quad (15)$$

where $\xi \in R^n$ is the state vector, $\theta \in R^m$ is the unknown parameters vector, and f is a given nonlinear vector function.

To estimate the unknown parameters in (15), a parameter identification system is defined as follows:

$$\frac{d\hat{\xi}(t)}{dt} = f(\hat{\xi}, t; \hat{\theta}), \quad (16)$$

where $\hat{\xi} \in R^n$ is the estimated state vector and $\hat{\theta} \in R^m$ is the estimated parameters vector.

The objective function defined as the mean squared errors between real and estimated responses for a number N of given samples is considered as fitness of estimated model parameters:

$$V = \frac{1}{N+D} \sum_{j=1}^{D} \sum_{k=1}^{N} \left(\xi_j^k - \hat{\xi}_j^k\right)^2, \quad (17)$$

where D is the number of measurable states, N is the data length used for parameter identification, whereas ξ_j^k and $\hat{\xi}_j^k$ are the real and estimated values of state j at time k, respectively.

This objective function is a difficult function to minimize because there are many local minima, and the global minimum has a very narrow domain of attraction. Our goal is to determine the system parameters, using particle swarm optimization algorithms in such a way that the value of V is minimized, approaching zero as much as possible.

3.2. Overview of Basic PSO Algorithms. During the last decade, PSO algorithms have gained much attention and wide applications in different fields due to their effectiveness in performing difficult optimization issues, as well as simplicity of implementation and ability to fast converge to a reasonably good solution [14–16]. PSO is a population-based heuristic global optimization technique, first introduced by Kennedy and Eberhart [8] and referred to as a swarm-intelligence technique. It is motivated from the simulation of social behavior of animals such as bird flocking, fish schooling, and swarm. In this algorithm, the population is called a swarm, and the trajectory of each particle in the search space is controlled through the medium of a term called "velocity," according to its own flying experience and swarm experience in the search space. Mathematical description of basic PSO and some important variants is presented in the following.

Candidate solutions of a population called particles coexist and evolve simultaneously based on knowledge sharing with neighboring particles. Each particle represents a potential solution to the optimization problem, and it has a fitness value decided by optimal function. Supposing that search space is D-dimensional, each individual is treated as a particle in the D-dimensional search space. The position and rate of position change for ith particle can be represented by D-dimensional vector, $x_i = (x_{i1}, x_{i2}, \ldots, x_{iD})$ and $v_i = (v_{i1}, v_{i2}, \ldots, v_{iD})$, respectively. The best position previously visited by the ith particle is recorded and represented as $p_i = (p_{i1}, p_{i2}, \ldots, p_{iD})$, called *pbest*. The swarm best position previously visited by all the particles in the population is represented as $p_g = (p_{g1}, p_{g2}, \ldots, p_{gD})$, called *gbest*. Then particles search their best position, which are guided by swarm information p_g and their own information p_i. Each particle modifies its velocity to find a better solution (position) by applying its own flying experience (i.e., memory of the best position found in earlier flights) and the experience of neighboring particles (i.e., the best solution found by the population). Each particle position is evaluated by using fitness function and updates its position and velocity according to the following equations:

$$v_i^{k+1} = \omega \cdot v_i^k + c_1 r_1 \left(pbest_i^k - x_i^t\right) + c_2 r_2 \left(gbest_i^k - x_i^k\right),$$
$$x_i^{k+1} = x_i^k + v_i^{k+1}, \quad (18)$$

where k is iteration number, ω is inertia weight, c_1 and c_2 are two acceleration coefficients regulating the relative velocity toward local and global best positions, and r_1 and r_2 are two random numbers from interval $[0, 1]$. Many effects have been made over the last decade to determinate the inertia weight. Various studies has shown that under certain conditions convergence is guaranteed to a stable equilibrium point. These conditions include $\omega > (c_1 + c_2)/2 - 1$ and $0 < \omega < 1$. The technique originally proposed was to bound velocities so that each component of v_i is kept within the range $[-V_{max}, +V_{max}]$.

Unfortunately, this simple form of PSO suffers from the premature convergence problem, which is particularly true in complex problems since the interacted information among particles in PSO is too simple to encourage a global search. Many efforts have been made to avoid the premature convergence. One solution is the use of a constriction factor to insure convergence of the PSO, introduced in [17]. Thus, the expression for velocity has been modified as

$$v_i^{k+1} = h \cdot \left[v_i^k + c_1 r_1 \left(pbest_i^k - x_i^t\right) + c_2 r_2 \left(gbest_i^k - x_i^k\right)\right],$$
$$x_i^{k+1} = x_i^k + v_i^{k+1}, \quad (19)$$

where h represents the constriction factor and is defined as

$$h = \frac{2}{\left|2 - \alpha - \sqrt{\alpha^2 - 4\alpha}\right|}, \quad (20)$$

$$\alpha = c_1 + c_2 > 4. \quad (21)$$

In this variant of the PSO algorithm, h controls the magnitude of the particle velocity and can be seen as a dampening factor. It provides the algorithm with two important features. First, it usually leads to faster convergence than standard PSO. Second, the swarm maintains the ability to perform wide movements in the search space, even if convergence is already advanced, but a new optimum is found. Therefore, the constriction PSO has the potential to avoid being trapped in local optima while possessing a fast convergence. It was shown to have superior performance compared to a standard PSO.

It is shown that a larger inertia weight tends to facilitate the global exploration and a smaller inertia weight achieves the local exploration to fine-tune the current search area [18]. The best performance could be obtained by initially setting ω to some relatively high value (e.g., 0.9), which corresponds to a system where particles perform extensive exploration, and gradually reducing ω to a much lower value (e.g., 0.4), where the system would be more dissipative and exploitative and would be better at homing into local optima. In [19], a linearly decreased inertia weight ω over time is proposed, where ω is given by the following equation:

$$\omega = \left(\omega_i - \omega_f\right) \cdot \frac{k_{max} - k}{k_{max}} + \omega_f, \quad (22)$$

where ω_i, ω_f are starting and final values of inertia weight, respectively; k_{max} is the maximum number of the iteration, and k is the current iteration number. It is generally taken that starting value $\omega_i = 0.9$ and final value $\omega_f = 0.4$ [20].

On the other hand, in [21] was introduced PSO with time-varying acceleration coefficients. The improvement has the same motivation and the similar techniques as the adaptation of inertia weight. In this case, the cognitive coefficient c_1 is decreased linearly and the social coefficient c_2 is increased linearly over time as follows:

$$c_1 = \left(c_{1f} - c_{1i}\right) \cdot \frac{k_{max} - k}{k_{max}} + c_{1i},$$
$$c_2 = \left(c_{2f} - c_{2i}\right) \cdot \frac{k_{max} - k}{k_{max}} + c_{2i}, \quad (23)$$

where c_{1i} and c_{2i} are the initial values of the acceleration coefficients c_1 and c_2; c_{1f} and c_{2f} are the final values of the acceleration coefficients c_1 and c_2, respectively. Usually, $c_{1i} = 2.5$, $c_{2i} = 0.5$, $c_{1f} = 0.5$, and $c_{2f} = 2.5$.

3.3. Identification Algorithm. Considering all the states of the nonlinear system (8) at the sampling moments $k * Ts$ (Ts = sampling period) known, the identification algorithm has the following steps.

Step 1. Initialize a population of particles with random positions and velocities on D dimensions in search space.

Step 2. For each particle, evaluate the desired optimization fitness function (17) in D variables.

Step 3. Compare particle's fitness evaluation with its *pbest*. If current value is better than *pbest*, then set *pbest* equal to the current value x_i in D-dimensional space.

Step 4. Identify the particle in swarm with the best success so far, and assign its index to the variable *gbest*.

Step 5. Change the velocity and position of the particle according to (19).

Step 6. *if* a criterion is met (usually a sufficiently good fitness or a maximum number of iterations)
then Stop;
else
go to Step 2.

3.4. "Classical" Identification Procedure. The "classical" approach for identification of yield coefficients is a two-step procedure under the assumption that full state measurements are available [1]. This method is based on a state transformation that allows reformulating the dynamical model into separate submodels. The first submodel depends only on the reaction structure and is independent of the kinetics. It can be linearly reparametrized and used for the identification of the yield coefficients by means of linear regressions, provided suitable identifiability conditions are satisfied. We present briefly this method for estimating the yield coefficients, and we use it for comparison in the next section.

The general dynamical model given in (2) represents a particular class of nonlinear state-space models. The nonlinearity lies in the reaction rates $\varphi_i(\xi)$ that are (nonlinear) functions of the state variables. These functions enter the model in the form $K\varphi_i(\xi)$ (where K is a constant matrix), which is a set of linear combinations of the same nonlinear functions $\varphi_i(\xi)$. This particular feature can be exploited to separate the nonlinear part from the linear part of the model by an adequate linear state transformation. More precisely, one chooses a nonsingular partition

$$\begin{pmatrix} K_a \\ K_b \end{pmatrix} = TK, \tag{24}$$

with $K_a \in R^{p\times m}$ of full row rank matrix (i.e., p = rank(K)), $K_b \in R^{(n-p)\times m}$, and T a permutation matrix. The induced partitions of the vectors ξ and u are

$$\begin{pmatrix} \xi_a \\ \xi_b \end{pmatrix} = T\xi, \qquad \begin{pmatrix} u_a \\ u_b \end{pmatrix} = Tu. \tag{25}$$

Model (2) is then partitioned into two submodels

$$\begin{aligned} \frac{d\xi_a}{dt} &= K \cdot \varphi(\xi_a, t) - D\xi_a + u_a, \\ \frac{d\xi_b}{dt} &= K \cdot \varphi(\xi_b, t) - D\xi_b + u_b. \end{aligned} \tag{26}$$

Then with the state transformation

$$\begin{aligned} \xi_a &= \xi_a, \\ z &= C\xi_a + \xi_b, \end{aligned} \tag{27}$$

one transforms the initial model into

$$\begin{aligned} \dot{\xi}_a &= K_a\varphi(\xi_a, z - C\xi_a) - D\xi_a + u_a, \\ \dot{z} &= -Dz + Cu_a + u_b, \end{aligned} \tag{28}$$

where the $(n-p) \times p$ matrix C is the unique solution of

$$CK_a + K_b = 0. \tag{29}$$

That is,

$$C = -K_b K_a^*, \tag{30}$$

where K_a^* is a generalized inverse or pseudoinverse of K_a. The subsystem (29) can be augmented with an equation derived from (8) as follows:

$$\begin{aligned} \dot{z} &= -Dz + Cu_a + u_b, \\ \xi_b &= z - C\xi_a. \end{aligned} \tag{31}$$

It can be considered as a linear time-varying (if d varies in the course of time) model with state z, input (ξ_a, u_a, and u_b), and output ξ_b. It is nonlinearly parametrized by the yield coefficients but linearly reparametrized by the nonzero entries of C. When data of the signals ξ_a, u_a, u_b, and ξ_b are available, the auxiliary model (31) can be used to identify the yield coefficients independently of the knowledge of the reaction rates. The model (31) can be used to perform the identification of the nonzero entries of C by a linear regression technique, with the yield coefficients k_i recovered afterwards from (29).

For the estimation of kinetic parameters the main approach is the use of a parameter observers (high gain observers, regressive parameter estimator, sliding mode observers, etc.). All these techniques have numerous tuning parameters and are difficult to implement.

4. Simulation Results and Discussion

The efficacy of our approach is shown by numerical simulations on an interval of 30 hours. The model given by relation (9) was integrated using a fourth-order Runge-Kutta routine with a sampling period of 1 minute and with initial conditions: IC = $[2.5 \;\; 6 \;\; 0.2 \;\; 10 \;\; 0.5 \;\; 0]$.

The influence of sampling period and type of the optimization algorithm and of noisy measurements are analyzed. To compare statistical performances of the different

TABLE 1: Influence of the sampling period.

Estimated parameter	"Real" value	Sampling period		
		1 min	10 min	30 min
θ_1	5.5	5.5009	5.4811	5.4798
θ_2	1	0.9982	1.0111	1.0637
θ_3	16	15.9110	15.9328	14.5629
θ_4	10	10.0016	9.9707	9.9571
θ_5	1.5	1.4938	1.4626	1.3078
θ_6	0.2	0.2003	0.2011	0.1873
θ_7	3	2.9823	2.9581	2.6947
θ_8	0.2	0.2002	0.1974	0.1949
θ_9	0.6	0.5994	0.6010	0.6188
θ_{10}	0.75	0.7539	0.6754	0.6483
θ_{11}	4	3.9731	3.9007	3.0265
θ_{12}	20	20.3973	19.4738	20.3947
NMSE	**0**	**3.0717*e* − 05**	**5.6623*e* − 04**	**1.2143*e* − 01**

TABLE 2: Influence of the algorithm type.

Estimated parameter	"Real" value	Type of optimization algorithm		
		PSO-1	PSO-2	PSO-3
θ_1	5.5	5.5613	5.5849	5.4338
θ_2	1	1.0405	1.0083	0.8406
θ_3	16	15.2050	14.9971	15.6475
θ_4	10	10.0848	10.1252	9.9468
θ_5	1.5	1.5721	1.6422	1.1698
θ_6	0.2	0.2022	0.2040	0.3005
θ_7	3	2.8006	2.7916	2.7569
θ_8	0.2	0.1990	0.1975	0.2286
θ_9	0.6	0.5993	0.6311	0.6594
θ_{10}	0.75	0.9991	1.1153	0.4728
θ_{11}	4	3.5221	3.3470	4.0707
θ_{12}	20	20.9257	20.5048	20.0971
NMSE	**0**	**3.0717*e* − 04**	**0.0283**	**0.3845**

approaches the empirical normalized mean square error (NMSE) was used, that is defined as

$$\text{NMSE} = \frac{1}{N}\sum_{j=1}^{N}\text{NMSE}\left(\widehat{\theta}_j\right), \quad (32)$$

with $\text{NMSE}(\widehat{\theta}_j) = ((\widehat{\theta}_j - \theta_j^*)/\theta_j^*)^2$, where N is the number of estimated parameters, $\widehat{\theta}_j$ is the jth element of the estimated parameter vector while the "$*$" superscript denotes the true value of the parameter.

In order to study the sensitivity of the estimation method to the sampling period and to the type of PSO algorithm, and to the noise, the following parameters were used.

Sampling period: $T_s \in \{1 \text{ min}, 10 \text{ min}, 30 \text{ min}\}$.

Types of the optimization algorithm are as follows:

PSO_1: algorithm based on relation (19) with h defined by relation (20), and $c_1 = c_2 = 2.1$;

TABLE 3: Influence of the noise level.

Estimated parameter	"Real" value	Noise level		
		50 dB	40 dB	30 dB
θ_1	5.5	5.4001	5.3697	5.3362
θ_2	1	1.0066	0.6628	0.4180
θ_3	16	15.5721	16.6508	16.3867
θ_4	10	9.8502	9.8463	9.7771
θ_5	1.5	1.1817	1.1428	0.9150
θ_6	0.2	0.1867	0.2769	0.2711
θ_7	3	2.9626	2.9178	3.1112
θ_8	0.2	0.2042	0.2350	0.3020
θ_9	0.6	0.6030	0.6254	0.5744
θ_{10}	0.75	0.3306	0.2379	0.0058
θ_{11}	4	3.7860	4.7660	4.5629
θ_{12}	20	20.8802	20.9267	19.7113
NMSE	**0**	**3.6827*e* − 3**	**0.0827**	**0.1625**

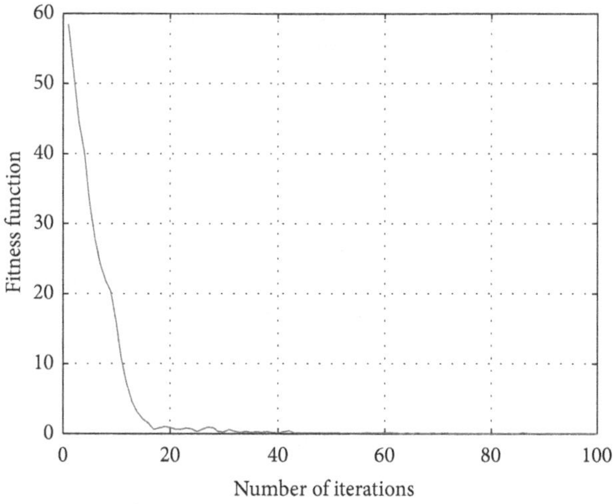

FIGURE 2: Convergence rate for PSO_1 algorithm.

PSO_2: algorithm based on relation (18) with ω, c_1, and c_2 defined by relations (22) and (23) and $\omega_i = 0.9$, $\omega_f = 0.4$, $c_{1i} = 2.5$; $c_{2i} = 0.5$; $c_{1f} = 0.5$, $c_{2f} = 2.5$;

PSO_3: algorithm based on relation (18) with ω, c_1, andc_2 constants: $\omega = 0.75, c_1 = 1.8, c_2 = 2.2$, and $V_{\max} = 10$.

Noise: {zero-mean white: SNR = 50 dB, 40 dB, 30 dB}.

The results of these simulations are presented in Tables 1, 2, and 3. The simulations were performed with a number of particles between 50 and 120. All the presented results are obtained for a number of 80 particles. For a greater number of particles the accuracy of the estimates was not better. In Figure 2 is presented the convergence rate for PSO_1 algorithm with a sampling period of 1 min.

The results presented in Tables 1–3 suggest that our proposed parameter estimation technique yields consistent results. The method is very simple, easily completed and needs fewer parameters, which made it fully developed. However, the research on the PSO is still at the beginning, and a lot of problems are to be resolved. Research on the topology of the new pattern particle swarm which has a better function can be carried out. The neighbouring topology of the different particle swarms is based on the imitation of the different societies. It is meaningful to the use and spread of the algorithm to select the proper topology to enable PSO have the best property and do the research on the suitable ranges of different topologies. Blending PSO with the other intelligent optimization algorithms means combining the advantages of the PSO with the advantages of the other intelligent optimization algorithms to create the compound algorithm that has practical value. For example, the particle swarm optimization algorithm can be improved by the Simulated Annealing approach. Rapid swarm convergence is one of the main advantages of PSO, but this can also be problematic since, if an early solution is suboptimal, the swarm can easily stagnate around it without any pressure to continue exploration. Overall the results indicate that PSO algorithms can be used in the optimization of parameters during model identification.

5. Conclusions

The paper presents a particle swarm optimization based identification procedure for offline estimation of yield and kinetic coefficients in an anaerobic wastewater treatment bioprocess. The identification scheme is formulated in terms of an optimization problem where the error between an actual physical measured response of the system and the simulated response of a parameterized model is minimized. This function is multimodal, and classical iterative methods fail to find the global optimum. The estimation of the system parameters is achieved as a result of minimizing the error function by the PSO algorithm.

The simulations evaluated the effects of sampling period and some basic variants of PSO algorithm and of the noisy measurements. The proposed strategy can still converge to accurate results even in the presence of measurement noise, as illustrated by the numerical study. The PSO algorithm has a simpler procedure and higher computational efficiency than other optimization techniques.

Nomenclature

c_1, c_2: Acceleration coefficients
D: Dilution rate (h^{-1})
F: Vector of feeding rates
F_{in}: Input feed rate (Lh^{-1})
K: Matrix of yield coefficients
$k_i, i = \overline{1,7}$: Yield coefficients
Q: Vector of rates of removal of the components in gaseous form
S_1: Glucose concentration (g/L)
S_2: Acetate concentration (g/L)
S_5: Inorganic carbon concentration (g/L)
X_1: Acidogenic concentration (g/L)
X_2: Acetoclastic methanogenic concentration (g/L)
P: Methane concentration (g/L)
S_{in}: Glucose concentration on the feed (g/L)
Φ: Vector of reaction rates (reaction kinetics)
μ_1^*, μ_2^*: Maximal specific growth rates (h^{-1})
ξ: State vector
$\mathfrak{R}$: The set of real numbers
Ω: Inertia weight
θ: The vector of unknown parameters
$\theta_i, i = \overline{1,9}$: The unknown parameters.

Acknowledgments

This work was supported by CNCSIS-UEFISCSU, Project no. PN II-RU TE 106, 9/04.08.2010.

References

[1] G. Bastin and D. Dochain, *On-Line Estimation and Adaptive Control of Bioreactors*, Elsevier, New York, NY, USA, 1990.

[2] L. Chen, *Modelling, identifiability and control of complex biotechnological systems [Ph.D. thesis]*, Université Catholique de Louvain, Louvain-la-Neuve, Belgium, 1992.

[3] L. M. Li and S. A. Billings, "Continuous time non-linear system identification in the frequency domain," *International Journal of Control*, vol. 74, no. 11, pp. 1052–1061, 2001.

[4] L. Ljung, *System Identification—Theory for the User*, Prentice-Hall, Upper Saddle River, NJ, USA, 2nd edition, 1999.

[5] S. Kirkpatrick, C. D. Gelatt, Jr., and M. P. Vecchi, "Optimization by simulated annealing," *Science*, vol. 220, no. 4598, pp. 671–680, 1983.

[6] J. M. Holland, *Adaptation in Natural and Artificial Systems*, University of Michigan Press, Ann Arbor, Mich, USA, 1975.

[7] A. Colorni, M. Dorigo, and V. Maniezzo, "Distributed optimization by ant colonies," in *Proceedings of the European Conference on Artificial Life (ECAL '91)*, Elsevier, 1991.

[8] J. Kennedy and R. C. Eberhart, "Particle swarm optimization," in *Proceedings of the IEEE International Conference on Neural Networks*, pp. 1942–1948, 1995.

[9] J. Kennedy, "Particle swarms: optimization based on sociocognition," in *Recent Developments in Biologically Inspired Computing*, The Idea Group, 2004.

[10] H. Li, H. Zhong, Z. Yan, and X. Zhang, "Particle swarm optimization algorithm coupled with finite element limit equilibrium method for geotechnical practices," *Mathematical Problems in Engineering*, vol. 2012, Article ID 498690, 14 pages, 2012.

[11] O. Bernard and G. Bastin, "Identification of reaction networks for bioprocesses: determination of a partially unknown pseudo-stoichiometric matrix," *Bioprocess and Biosystems Engineering*, vol. 27, no. 5, pp. 293–301, 2004.

[12] D. Selişteanu, E. Petre, and V. B. Răsvan, "Sliding mode and adaptive sliding-mode control of a class of nonlinear bioprocesses," *International Journal of Adaptive Control and Signal Processing*, vol. 21, no. 8-9, pp. 795–822, 2007.

[13] S. F. Azevedo, P. Ascencio, and D. Sbarbaro, "An adaptive fuzzy hybrid state observer for bioprocesses," *IEEE Transactions on Fuzzy Systems*, vol. 12, pp. 641–651, 2004.

[14] I. Hassanzadeh and S. Mobayen, "Controller design for rotary inverted pendulum system using evolutionary algorithms," *Mathematical Problems in Engineering*, vol. 2011, Article ID 572424, 17 pages, 2011.

[15] T.-S. Zhan and C.-C. Kao, "Modified PSO method for robust control of 3RPS parallel manipulators," *Mathematical Problems in Engineering*, vol. 2010, Article ID 302430, 25 pages, 2010.

[16] P. Umapathy, C. Venkataseshaiah, and M. S. Arumugam, "Particle swarm optimization with various inertia weight variants for optimal power flow solution," *Discrete Dynamics in Nature and Society*, vol. 2010, Article ID 462145, 15 pages, 2010.

[17] M. Clerc and J. Kennedy, "The particle swarm: explosion, stability, and convergence in a multi-dimensional complex space," *IEEE Transactions on Evolutionary Computation*, vol. 6, no. 1, pp. 58–73, 2002.

[18] X. Qian, M. Cao, Z. Su, and J. Chen, "A hybrid particle swarm optimization (PSO)-simplex algorithm for damage identification of delaminated beams," *Mathematical Problems in Engineering*, vol. 2012, Article ID 607418, 11 pages, 2012.

[19] R. C. Eberhart and Y. Shi, "Comparing inertia weights and constriction factors in Particle Swarm Optimization," in *Proceedings of the Congress on Evolutionary Computating*, pp. 84–88, July 2000.

[20] Y. Shi and R. C. Eberhart, "Empirical study of particle swarm optimization," in *Proceedings of the IEEE Congress on Evolutionary Computation*, pp. 1945–1950, IEEE Press, 1999.

[21] A. Ratnaweera, S. K. Halgamuge, and H. C. Watson, "Self-organizing hierarchical particle swarm optimizer with time-varying acceleration coefficients," *IEEE Transactions on Evolutionary Computation*, vol. 8, no. 3, pp. 240–255, 2004.

4

Travel Time Model for Right-Turning Vehicles of Secondary Street at Unsignalized Intersections

Feng Yu-Qin,[1] Leng Jun-Qiang,[2,3] Wang Peng,[4] He Yi,[3] and Zhang Gui-e[3]

[1] *School of Automobile and Traffic Engineering, Heilongjiang Institute of Technology, Harbin 150050, China*
[2] *School of Management, Harbin Institute of Technology, Harbin 150001, China*
[3] *School of Automobile Engineering, Harbin Institute of Technology, Weihai 264209, China*
[4] *Representative Office of PLA in Harbin Railway Administration, Harbin 150001, China*

Correspondence should be addressed to Leng Jun-Qiang; lengjunq@tom.com

Academic Editor: Wuquan Li

The travel time of right-turning vehicles on secondary street at unsignalized intersection is discussed in this paper. Under the assumption that the major-street through vehicles' headway follows Erlang distribution and secondary-street right-turning vehicles' headway follows Poisson distribution. The right-turning vehicles travel time model is established on the basis of gap theory and M/G/1 queue theory. Comparison is done with the common model based on the assumption that the major-street vehicles' headway follows Poisson distribution. An intersection is selected to verify each model. The results show that the model established in this paper has stronger applicability, and its most relative error is less than 15%. In addition, the sensitivity analysis has been done. The results show that right-turning flow rate and major-street flow rate have a significant impact on the travel time. Hence, the methodology for travel time of right-turning vehicles at unsignalized intersection proposed in this paper is effective and applicable.

1. Introduction

As a bottleneck of urban road network, intersections are the emphasis in traffic management and control [1–3]. The travel time of vehicles at intersections is the basis to evaluate road traffic efficiency and the Intelligent Traffic System (ITS) applications. It is also one of the breakthrough points to calculate the vehicles delay at intersections [4, 5]. Based on the analysis of the conflict disciplines of the four-phase signalized intersection between right-turning vehicles and straight-going bicycles, Liang et al. established a theoretical model and a binary regression model of right-turning vehicles travel time. The former was based on gap theory and queue theory, and the latter was established on the basis of field observation data. In the binary regression model, it takes right-turning vehicles flow rate and bicycle through flow rate as independent variables. Meanwhile, their application conditions were studied, respectively [6]. Smith and Walsh established motor vehicles and nonmotor vehicles delay models under different mixed traffic conditions and traffic control methods [7]. Liu et al. took the basic link travel time and intersection delay as the basic unit of vehicle travel time in urban road network [8]. Ban et al. believed that vehicle travel time at signalized intersections contained discontinuities and nonsmoothness, and intersection delay pattern can be used to estimate real-time queue length at intersection [9]. Ahmed studied the travel time of indirect right-turning vehicles at intersection with a GPS device [10]. Lu et al. established the model of travel time at intersection with microscopic simulation data. The model considered the signal control effect at a variety of traffic volume combinations [11].

As mentioned above, there are more studies about vehicles delay at intersections, while few researches are about travel time of different turning movements at intersections [12–15]. Typically, Liang et al. established a right-turning motor vehicles' travel time model with the consideration of conflicts between motor vehicles and nonmotor vehicles at signalized intersections, and it provides a good reference for the following researches. The model was built on strict assumptions. For instance, nonmotor flow and right-turning motor flow were both subject to Poisson distributions. Nonmotor flow only contained bicycles, the bicycle flow was only in one line, and the service time followed a negative exponential distribution [6]. Poisson distribution is applicable to the random vehicle arriving, generally with flow rate less than 500 veh/h per lane. As urban road traffic is more and more congested while Poisson distribution is just applicable to free flow, it is hard to meet the actual traffic circumstances. As a consequence, Poisson distribution will be invalid when the flow rate is more than 500 veh/h per lane. Thus, it limits the application of the model to a large extent. Conversely, Erlang distribution, as a general distribution, has a high applicability to all traffic conditions.

Accordingly, this paper aims to establish the model of the travel time of right-turning vehicles on secondary-street at unsignalized intersections in congested urban traffic condition. This provides theoretical support for improving the traffic management, intersection control, designation, evaluation, and ITS application.

This paper is organized as follows. The first part gives a general introduction. The second part presents the model establishment methodology. Then, Section 3 takes an intersection as a numerical example to validate the models. In Section 4, the model sensitivity is analyzed. The final section then concludes the paper and gives suggestions for further study.

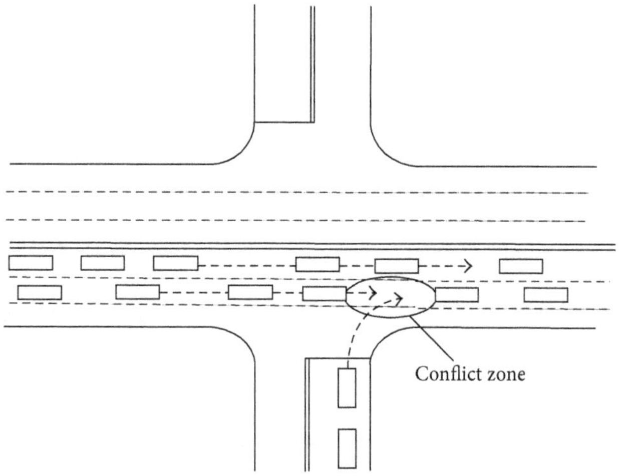

Figure 1: Service desk at unsignalized intersection.

2. Model Establishment

2.1. Research Conditions Settings. Convenient for the study, the following assumptions were made in this paper.

(1) The following distribution models are usually used to describe traffic flow headway: negative exponential distribution, Weibull distribution, and Erlang distribution. The negative exponential distribution can be applied to the situation that vehicles arrive at random, generally in which the flow rate per hour per lane is less than 500 veh/h. Weibull distribution has a broader application range. Erlang distribution is also a general probability distribution of headway. Erlang distribution could be obtained by calculating the parameter l. Considering the applicability and conveniences of calibration, Erlang distribution is usually used to describe headway of major-street traffic flow. The negative exponential distribution is more suitable when the right-turning flow on the secondary street is not large.

(2) The vehicles on secondary street follow the right-in and right-out principle, and pedestrian is prohibited to cross, which is the management and control strategy of the unsignalized intersection.

(3) Drivers abide by the principle of the major-street priority strictly, and there is no grab-line phenomenon.

(4) In this research, the gap theory is adopted. The gap theory can be described as follows: at an unsignalized intersection, vehicles on the major street have the priority to cross, while the secondary-street vehicles have to wait for gaps long enough to cross. Therefore, the maximum flow rate of the secondary-street may be concluded by calculating the number of gaps provided by the major-street flow. The gap theory is used under the situation that secondary-street vehicles pass through the major-street flow perpendicularly at intersections. However, the actual right-turning movements at intersections do not cross the counter flow in perpendicularity. In this case, it is assumed that the right-turning movement crossed them perpendicularly.

(5) Queue theory: a system subjects to M/G/1, that is, vehicles arrival follows Poisson distribution, service time follows random distribution, and a single reception desk obeys the rule that the first comer should be served first under the condition that right-turning vehicles are waiting for services in the queue system. The service time is not definite, and there is only one right-turning lane.

2.2. Model Establishment. Based on the above research conditions settings, the conflict zone between right-turning vehicles on the secondary-street and opposing through vehicles on the major street can be seen as a single service desk as shown in Figure 1. The secondary-street right-turning vehicles receive service, and headway of opposing through flow provides service. As mentioned in the above assumptions, arrival of right-turning flow follows Poisson distribution, the service time follows Poisson distribution, and a single reception desk system obeys the rule of first comer to be served first, namely, the M/M/1 system.

In this research, the critical accepted headway of right-turning vehicles is τ_c, headway of the major-street through vehicles is h, and the following headway of right-turning vehicles is h_f. Considering that there is a vehicle queue on secondary-street, then we will have different conditions: when h is less than τ_c, no right-turning vehicles can merge into major street; when h is between τ_c and $\tau_c + h_f$, only one vehicle can cross; when h is between $\tau_c + h_f$ and $\tau_c + 2h_f$, two vehicles may merge into major street; when $\tau_c + (n-1)h_f \leq$

TABLE 1: Probability of headway.

Headway	Vehicles number/i	Probability/P_i
$\tau_c \le h < \tau_c + h_f$	1	$\sum_{i=0}^{l-1}(\lambda l\tau_c)^i \frac{e^{-\lambda\tau_c}}{i!} - \sum_{i=0}^{l-1}(\lambda l(\tau_c + h_f))^i \frac{e^{-\lambda(\tau_c+h_f)}}{i!}$
$\tau_c + h_f \le h < \tau_c + 2h_f$	2	$\sum_{i=0}^{l-1}(\lambda l(\tau_c + h_f))^i \frac{e^{-\lambda(\tau_c+h_f)}}{i!} - \sum_{i=0}^{l-1}(\lambda l(\tau_c + 2h_f))^i \frac{e^{-\lambda(\tau_c+2h_f)}}{i!}$
$\tau_c + 2h_f \le h < \tau_c + 3h_f$	3	$\sum_{i=0}^{l-1}(\lambda l(\tau_c + 2h_f))^i \frac{e^{-\lambda(\tau_c+2h_f)}}{i!} - \sum_{i=0}^{l-1}(\lambda l(\tau_c + 3h_f))^i \frac{e^{-\lambda(\tau_c+3h_f)}}{i!}$
$\vdots$	$\vdots$	$\vdots$
$\tau_c + (n-1)h_f \le h < \tau_c + nh_f$	n	$\sum_{i=0}^{l-1}(\lambda l(\tau_c + (n-1)h_f))^i \frac{e^{-\lambda(\tau_c+(n-1)h_f)}}{i!} - \sum_{i=0}^{l-1}(\lambda l(\tau_c + nh_f))^i \frac{e^{-\lambda(\tau_c+nh_f)}}{i!}$

$h < \tau_c + nh_f$, n vehicles may merge into major-street. The headway of major-street flow follows Erlang distribution, and it can be written as

$$P(h \ge t) = \sum_{i=0}^{l-1} (\lambda l t)^i \frac{e^{-\lambda t}}{i!}, \quad (1)$$

where λ is the arriving rate of major-street vehicles, veh/s; l is the distribution parameter, if $l = 1$, the formula above could be simplified as negative exponential distribution; while $l = \infty$, headway will follow uniform distribution. In practice, l could be determined by m and S^2 with rounding-off method. The formula may be expressed as

$$l = \operatorname{int}\left(\frac{m^2}{S^2}\right), \quad (2)$$

where m is the observed average headway of major-street flow, s; S^2 is the variance of observed headway of major-street flow

$$P(h \ge t) = \sum_{i=0}^{l-1} (\lambda l t)^i \frac{e^{-\lambda t}}{i!}. \quad (3)$$

Consequently, the probability of different headway on major street can be expressed as

$$\begin{aligned} P_i &= P\left(\tau_c + (i-1)h_f \le h \le \tau_c + ih_f\right) \\ &= P\left(h \ge \tau_c + (i-1)h_f\right) - P\left(h \ge \tau_c + ih_f\right), \end{aligned} \quad (4)$$

where P_i is the probability of headway on the major street that allows i vehicles on the secondary street to merge into major-street flow. The probability of different headway is shown in Table 1.

The number of right-turning vehicles may merge into major-street flow per unit time and can be expressed as

$$N = \sum_{i=1}^{n} i \cdot \lambda \cdot P_i$$

$$N = \sum_{m=0}^{n-1}\sum_{i=0}^{l-1} \lambda\left(\lambda l\left(\tau_c + mh_f\right)\right)^i \frac{e^{-\lambda(\tau_c+mh_f)}}{i!} - n\sum \lambda\left(\lambda l\left(\tau_c + nh_f\right)\right)^i \frac{e^{-\lambda(\tau_c+nh_f)}}{i!}, \quad (5)$$

where N is the number of right-turning vehicles and may merge into major-street flow per unit time, veh/s; i is the number of right-turning vehicles and may merge into major-street flow in one headway; P_i is the probability of the headway and may provide i right-turning vehicles to merge into major-street flow; the other parameters have the same meanings as previously mentioned.

From the M/G/1 queue system, the average travel time of each vehicle in the queue is

$$T = \frac{\rho^2 + \lambda_1^2\delta^2}{2\lambda_1(1-\rho)} + \frac{1}{\mu}, \quad (6)$$

where λ_1 is the average arrival rate of right-turning flow, veh/s; μ is the service rate of right-turning flow, $\mu = N$, veh/s; δ^2 is the variance of vehicles service time; ρ is the service intensity or traffic intensity, $\rho = \lambda_1/\mu$, which reflects the traffic conditions. If $\rho < 1$, it means that the flow is stable, and each traffic condition will be repeated with a certain probability. If $\rho \ge 1$, the flow is unstable, and queue will become longer and longer.

When $\delta^2 = 0$, service time follows uniform distribution, which may be expressed as

$$T = \frac{\rho^2}{2\lambda_1(1-\rho)} + \frac{1}{\mu}. \quad (7)$$

When $\delta^2 = 1/\mu^2$, service time is subject to negative exponential distribution, which can be written as

$$T = \frac{\rho^2}{\lambda_1(1-\rho)} + \frac{1}{\mu}. \quad (8)$$

2.3. Existing Model. Most of the present researches assume that traffic flow arrival is subject to Poisson distribution, and the queue system is an M/M/1 one. The corresponding models are as follows.

The service rate of right-turning flow may be expressed as

$$\mu = \frac{\lambda e^{-\lambda\tau_c}}{1 - e^{-\lambda h_f}}. \quad (9)$$

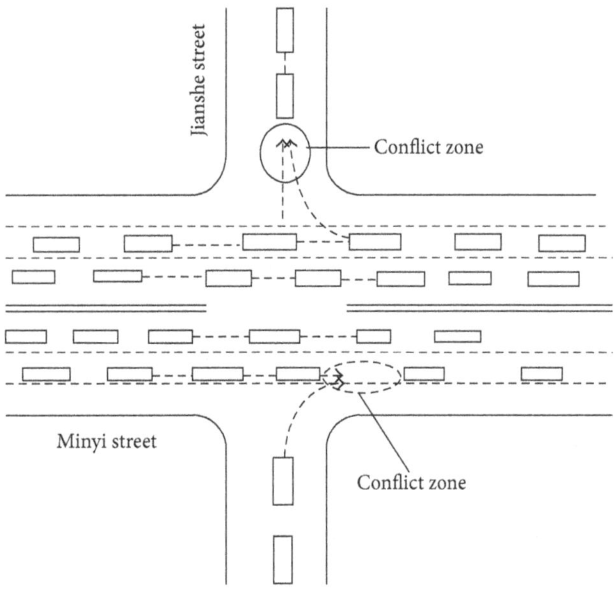

FIGURE 2: The sketch of study case.

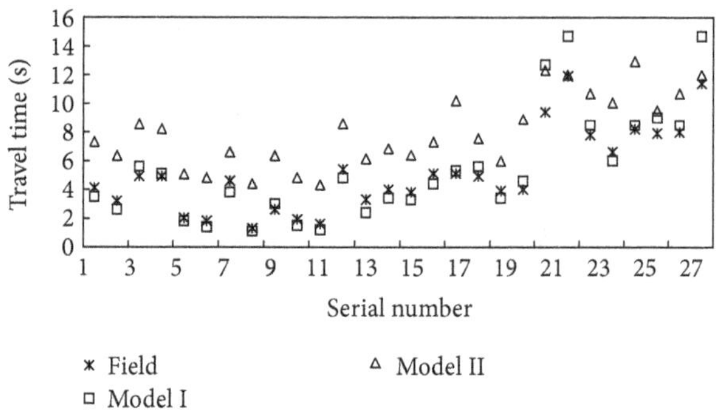

FIGURE 3: Model validation curve.

The travel time of right-turning vehicles merging into the main-street flow can be written as

$$T = \frac{1}{\mu - \lambda_1} = \frac{1}{\left(\lambda e^{-\lambda\tau_c} / \left(1 - e^{-\lambda h_f}\right)\right) - \lambda_1}. \quad (10)$$

The parameters have the same meanings as previously mentioned.

3. Models Validation

There are some differences between the application premise of gap theory and the research conditions in this paper. For example, the merging of right-turning vehicles into the major-street flow is not perpendicular, and the drivers do not strictly abide by the rule of major-street priority, especially when the right-turning flow rate is a bit large. Therefore, there still exists the phenomenon of grabbing line. It can be seen from the above that there exist some objective conditions that do not correspond with the practical situations. Thus, it is necessary to validate the models.

A real intersection of Harbin in China was selected to verify the accuracy of the above model based on Erlang distribution (Model-I) and the one based on Poisson distribution (Model-II). This unsignalized intersection lies in Nangang district of Harbin, and its geometric condition is shown in Figure 2. We conducted field observation in both rush and nonrush hours for one week. The following parameters were observed: flow rate and following headway of the major-street through vehicles, arrival rate and following headway of the secondary-street right-turning vehicles, and the time consumed to merge into the major-street flow and the critical headway accepted by right-turning vehicles. According to the field observation, critical headway τ_c evaluated by maximum likelihood estimation is 3.9 s, and h_f is 2.1 s. The above observation data and the results are applied to (8) and (10), and the results are shown in Table 2. The comparison curves of Model-I, Model-II, and the field observation data are shown in Figure 3.

It can be seen from the results that Model-I has better performance than Model-II, and most times the relative errors of Model-I are less than 15%, with the maximum one 35.11%. The higher accuracy is owing to the reasonable selected headway distribution. The main distinctions between Model-I and Model-II are the distribution models of the major-street traffic flow. Erlang distribution and Poisson distribution are used in Model-I and Model-II, respectively. It is well known that Poisson distributions are applicable to the random vehicle arriving, generally with flow rate less than 500 veh/h per lane. As a matter of fact, urban road traffic is more and more congested, while Poisson distribution is just applicable to free flow, and it is hard to meet the actual traffic circumstances. As a result, the Poisson distribution will be invalid when the flow rate is more than 500 veh/h per lane. Thus, it limits the application of Model-II. Conversely, Erlang distribution, as a general distribution, has a high applicability to all traffic conditions. Consequently, Model-I has better performance than Model-II.

The relative error of Model-I is relevant to the flow rate, indicating that the relative error shows an obvious rising trend with the increase of the flow rate. For instance, the most relative error is more than 15%, and the actual travel time is shorter than the theoretical value, while the flow rate of the major-street speeds up to 0.4 veh/s, and the flow rate of the right-turning vehicles reaches 0.04 veh/s. The reasons are as follows: with the increase of the right-turning vehicles flow rate, the number of headway long enough for right-turning vehicles to merge into major street decreases. When the flow rate of right-turning vehicles is increasing at the moment, there would be a formation of queues. Therefore, the right-turning vehicles may go after the front car to grab into the major street, and then the right-turning travel time will reduce, and the relative error will increase.

TABLE 2: Model validation data.

Serial number	τ_c (s)	h_f (s)	l	λ_1 (veh/s)	λ (veh/s)	Field (s)	Model-I		Model-II	
							Theoretical (s)	Relative error (%)	Theoretical (s)	Relative error (%)
1	3.9	2.1	2	0.03	0.36	4.1	3.5	14.63	7.3	78.44
2	3.9	2.1	2	0.03	0.32	3.2	2.6	18.75	6.3	98.11
3	3.9	2.1	3	0.04	0.38	4.9	5.6	14.29	8.5	74.39
4	3.9	2.1	3	0.04	0.37	4.9	5.1	4.08	8.2	67.57
5	3.9	2.1	2	0.02	0.27	2	1.8	10.00	5.1	152.95
6	3.9	2.1	2	0.03	0.24	1.8	1.4	22.22	4.8	167.40
7	3.9	2.1	3	0.02	0.35	4.6	3.8	17.39	6.6	43.28
8	3.9	2.1	2	0.04	0.2	1.3	1.1	15.38	4.4	238.38
9	3.9	2.1	3	0.03	0.32	2.6	3.0	15.38	6.3	143.82
10	3.9	2.1	3	0.03	0.24	1.9	1.5	21.05	4.8	153.33
11	3.9	2.1	3	0.02	0.22	1.6	1.2	25.00	4.3	169.25
12	3.9	2.1	3	0.05	0.36	5.4	4.8	11.11	8.6	58.70
13	3.9	2.1	2	0.03	0.31	3.3	2.4	27.27	6.1	85.46
14	3.9	2.1	2	0.02	0.36	4	3.4	15.00	6.8	70.43
15	3.9	2.1	2	0.01	0.36	3.8	3.3	13.16	6.4	67.95
16	3.9	2.1	3	0.03	0.36	5.1	4.4	13.73	7.3	43.45
17	3.9	2.1	2	0.05	0.40	5.1	5.3	3.92	10.2	100.27
18	3.9	2.1	3	0.02	0.39	4.9	5.6	14.29	7.6	54.14
19	3.9	2.1	3	0.01	0.34	3.9	3.4	12.82	6.0	53.62
20	3.9	2.1	2	0.04	0.39	4	4.6	15.00	8.9	122.42
21	3.9	2.1	3	0.05	0.44	9.4	12.7	35.11	12.3	31.06
22	3.9	2.1	3	0.04	0.46	11.9	14.7	23.53	12.0	0.64
23	3.9	2.1	3	0.05	0.41	7.8	8.5	8.97	10.7	37.06
24	3.9	2.1	2	0.04	0.42	6.6	6.0	9.09	10.1	52.59
25	3.9	2.1	2	0.05	0.45	8.2	8.5	3.66	12.9	57.81
26	3.9	2.1	3	0.03	0.43	7.9	9.0	13.92	9.5	20.35
27	3.9	2.1	3	0.05	0.41	8	8.5	6.25	10.7	33.63
28	3.9	2.1	3	0.04	0.46	11.4	14.7	28.95	12.0	5.05

4. Sensitivity Analyses

It is proved in the above section that the model established in this article has an overall better performance. The relationship between parameters and travel time was explored in this section. Specifically speaking, these variables include major-street through flow rate and secondary-street right-turning flow rate.

The influence of the variables is shown in Figure 4. To study the influence of major-street through flow rate and secondary-street right-turning flow rate, the travel times at different flow rates are shown in Figure 4. The trends in this figure indicate the following, (1) Right-turning vehicles travel times increase with the rising of the right-turning flow rate and the conflicting through flow rate. (2) When the conflicting through flow rate is low, the right-turning flow rate has a slight effect on right-turning vehicles travel times. With the increase of through flow rate, the right-turning flow rate would have a more obvious effect on travel time. Simultaneously, the threshold is about 0.30 veh/s. These findings indicate two facts: the larger the number of the opposing through vehicles is, the less acceptable the headway

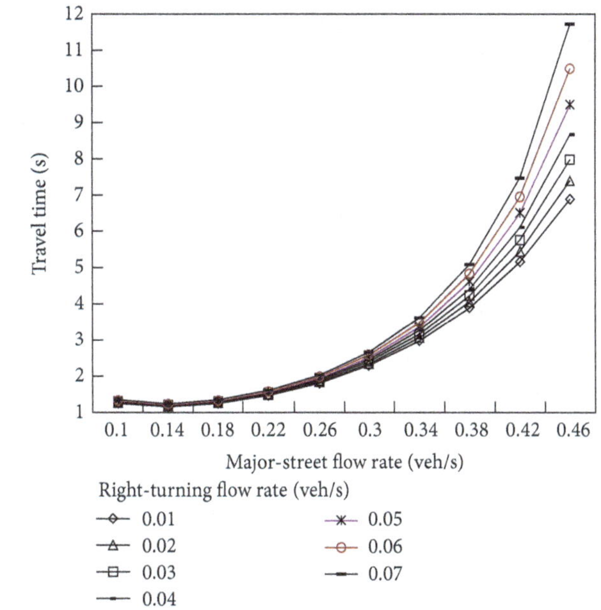

FIGURE 4: Influence of flow rate on right-turning vehicles travel time.

would be; the larger the right-turning flow rate is, the longer the queue and the longer the travel time would be.

5. Conclusions

(1) The travel time model based on Erlang distribution for major-street flow and M/G/1 queue system has a better applicability. The relative error has no obvious linear relationship with flow rate, and the error would still be stable, until the flow rate of the major street speeds up to 0.4 veh/s and the right-turning vehicles flow rate reaches 0.04 veh/s. Meanwhile, when the right-turning flow rate and major-street flow rate increase, the error also increases. This result shows the applicability of the model.

(2) The factors that have significant influence on the travel time are major-street flow rate and right-turning vehicles arrival rate. The travel time increases with the growth of them. In fact, there are other factors that have influence on the travel time, such as the intersection control pattern. Whether it is signalized or unsignalized, or it is interrupted by the pedestrian and the vehicles compositions or not, the intersection control pattern will always affect the travel time. Up to now, there is no such research on intersection control pattern, which we should spare no efforts to do within the coming future.

Acknowledgments

This research was supported by China Postdoctoral Science Foundation (2011M500676), National Education Ministry Humanities and Social Sciences Foundation (12YJCZH097), Heilongjiang Institute of Technology Doctoral Science Foundation (2011BJ05) and Technology Research and Development Program of Shandong (2012G0020129). The authors are also grateful to the anonymous referees for their helpful comments and constructive suggestions on an earlier version of the paper.

References

[1] S. M. Madanat, M. J. Cassidy, and M. H. Wang, "Probabilistic delay model at stop-controlled intersection," *Journal of Transportation Engineering*, vol. 120, no. 1, pp. 21–36, 1994.

[2] M. A. Qureshi and L. D. Han, "Delay model for right-turn lanes at signalized intersections with uniform arrivals and right turns on red," *Transportation Research Record*, vol. 1776, pp. 143–150, 2001.

[3] G. Abu-Lebdeh, R. F. Benekohal, and B. Al-Omari, "Models for right-turn-on-red and their effects on intersection delay," *Transportation Research Record*, vol. 1572, pp. 131–139, 1997.

[4] B. Hoeschen, D. Bullock, and M. Schlappi, "Estimating intersection control delay using large data sets of travel time from a global positioning system," *Transportation Research Record*, vol. 1917, pp. 18–27, 2005.

[5] T. Hagiwara, H. Hamaoka, T. Yaegashi, K. Miki, I. Ohshima, and M. Naito, "Estimation of time lag between right-turning vehicles and pedestrians approaching from the right side," *Transportation Research Record*, vol. 2069, pp. 65–76, 2008.

[6] C. Y. Liang, C. G. Wang, Z. Shen, and D. H. Wang, "Calculation method of travel time of right-turn vehicle at motor- and nonmotor-vehicle mixed traffic intersection," *Journal of Jilin University*, vol. 37, no. 5, pp. 1053–1057, 2007.

[7] R. L. Smith and T. Walsh, "Safety impacts of bicycle lanes," *Transportation Research Record*, vol. 1168, pp. 49–56, 1988.

[8] H. Liu, H. J. Van Zuylen, H. Van Lint, Y. Chen, and K. Zhang, "Prediction of urban travel times with intersection delays," in *Proceedings of the 8th International IEEE Conference on Intelligent Transportation Systems (ITSC '05)*, pp. 1062–1067, September 2005.

[9] X. G. Ban, P. Hao, and Z. B. Sun, "Real time queue length estimation for signalized intersections using travel times from mobile sensors," *Transportation Research C*, vol. 19, pp. 1133–1156, 2011.

[10] K. Ahmed, "Evaluation of low cost technique "indirect right turn" to reduce congestion at urbanized signalized intersection in developing countries," in *Proceedings of the 6th International Symposium on Highway Capacity and Quality of Service (ISHC '11)*, vol. 16, pp. 568–577, July 2011.

[11] C. Lu, F. Zhao, and M. Hadi, "A travel time estimation method for planning models considering signalized intersections," in *Proceedings of the 10th International Conference of Chinese Transportation Professionals—Integrated Transportation Systems: Green, Intelligent, Reliable (ICCTP '10)*, pp. 1993–2000, August 2010.

[12] J. A. Bonneson, "Delay to major-street through vehicles due to right-turn activity," *Transportation Research A*, vol. 32, no. 2, pp. 139–148, 1998.

[13] Y. Wang and N. L. Nihan, "Estimating the risk of collisions between bicycles and motor vehicles at signalized intersections," *Accident Analysis and Prevention*, vol. 36, no. 3, pp. 313–321, 2004.

[14] C. Quiroga, M. Perez, and S. Venglar, "Tool for measuring travel time and delay on arterial corridors," in *Proceedings of the 7th International Conference on: Applications of Advanced Technology in Transportation*, pp. 600–607, August 2002.

[15] J. Q. Leng, Y. Q. Feng, J. Zhai, L. Bao, and Y. He, "Travel time model of left-turning vehicles at signalized intersection," *Mathematical Problem in Engineering*, vol. 2010, Article ID 473847, 10 pages, 2012.

5

Numerical Study on Initial Field of Pollution in the Bohai Sea with an Adjoint Method

Chunhui Wang,[1,2] Xiaoyan Li,[1,3] and Xianqing Lv[1]

[1] *Laboratory of Physical Oceanography, Ocean University of China, Qingdao 266100, China*
[2] *Key Laboratory of Marine Spill Oil Identification and Damage Assessment Technology, The Organization of North China Sea Monitoring Center, Qingdao 266033, China*
[3] *Institute of Oceanology, Chinese Academy of Science, Qingdao 266071, China*

Correspondence should be addressed to Xianqing Lv; xqinglv@ouc.edu.cn

Academic Editor: Bing Chen

Based on the simulation of a marine ecosystem dynamical model in the Bohai Sea, routine monitoring data are assimilated to study the initial field of pollution by using the adjoint method. In order to reduce variables that need to be optimized and make the simulation results more reasonable, an independent grid is selected every four grids both in longitude and latitude, and only the pollutant concentrations of these independent grids needed to be optimized while the other grids were calculated by interpolation method. Based on this method, the stability and reliability of this model were proved by a set of twin experiments. Therefore, this model could be applied in real experiment to simulate the initial field of the total nitrogen (totalN) in May, 2009. Moreover, the distribution of totalN in any time step could be calculated by this model, and the monthly mean distribution in May in the Bohai Sea could be obtained.

1. Introduction

The Bohai Sea is the only inland sea in China with a maritime area being 7.7 × 10^4 km^2. Its mean depth is 18 m, and the deepest point is located in the west of the Lao Tie Shan channel. Because the runoff of the Yellow River, the Haihe River and the Liao River are all discharged into the Bohai Sea; the organic pollutants are tremendous. Unfortunately, water exchange of the Bohai Sea is very weak and its physical self-cleaning capacity is poor due to its special geographical position. The Bohai Sea is surrounded by land on three sides, and it is only connected with the Yellow Sea through the Bohai strait on the east side. Therefore, it is hard to recover if the Bohai Sea is polluted. A mount of industrial effluent, living sewage, and aquaculture waste water are released into the Bohai Sea with rapid development of the economy along the Bohai Sea, which can cause accumulation of nutrient substances, such as nitrogen and phosphorus. The accumulation of nutrient substance brings a series of ecoenvironmental degradation, including red tide, decrease of seawater oxygen content, decline of biodiversity, and decrease in fish catch. In order to protect and recover the ecoenvironment of the Bohai Sea and to coordinate and improve coastal economy, an accurate simulation of the time-varying pollutant distribution is needed. Only in this way can we achieve rational utilization of marine resources and sustainable development of economy.

Recently, more and more numerical models (e.g., Chen et al. [1], Duan and Nanda [2], Lee and Seo [3], Liu et al. [4], Gupta et al. [5], Periáñez [6, 7], Rajar et al. [8], Rajar and Cetina [9]) for simulating pollutant dispersion have been actually developed since they can be used for decision making after releases of contaminants into the marine environment (Periáñez [10]). Gupta et al. [5] applied a two-dimensional numerical model considering organized wastewater discharges to determine the wastewater assimilative capacity of Thane creek. They found that on the basis of monitoring and simulation, the water quality had been deteriorated significantly due to limited flushing capacity. The volumetric load in the creek needed to be restricted because projected wastewater flows and loads for 2015 were above the assimilative capacity of the creek. Huang et al. [11] investigated the distribution characteristics of heavier or lighter pollutants released at different cross-sectional positions of a wide river with a well-tested three-dimensional numerical model, and their findings assisted in cost-effective countermeasures to be taken for accidental or planned pollutant releases into a wide river. All the three major Siberian rivers, including Ob, Yenisei, and Lena, flow northward into Arctic, and they are supposed to be important sources for various contaminants, so Harms et al. [12] applied a three-dimensional coupled ice-ocean-models of different horizontal resolution to simulate the dispersion of water from these rivers. The model study confirmed that contaminant transport through sediment-laden sea ice offers a short and effective pathway for pollutant transport from Siberian River to the Barents and Nordic Seas. Different methodologies for coupling hydrodynamic submodels with mass transport submodels into integrated water quality models are described in a companion paper (Rajar et al. [8]). The conclusion is that the choice of the methodology depends on the space and time scales, on the prevalent forcing factors and on the nature of the contaminant.

The initial condition has dramatic influence on the simulated results when we use model both in meteorology and oceanography. But in most cases, we do not know the initial field in advance. Therefore, it is important to simulate the initial field accurately. For example, the pollutant distribution characteristics are very different if either the pollutant density or the release location is changed when pollutants are released into a river (Huang et al. [11]). A suite of experimental results of Pe ng and Xie [13] show that although forecast errors due to deficiencies in model physics or numerics cannot always be effectively corrected through improving initial conditions alone, the four-dimensional variational data assimilation algorithm based on Princeton Ocean Model (POM) leads to effective convergence between the forecasts and the "observations" by finding an "optimal" initial condition for the storm surge forecasting. Allen et al. [14] found the combination of source location, source strength, and surface wind direction that best matched the dispersion model output to the receptor data by using genetic algorithm (GA) and demonstrated that the GA was capable of computing the correct solution as long as the magnitude of the noise did not exceed that of the receptor data.

In order to obtain the initial condition or the average distribution of pollutant concentration in a certain period, the traditional way is to use all the observations within this period by interpolation method, such as Cressman and Kriging. However, this method only takes the spatial information of the observations into consideration but ignores the time information, which may lead to a big difference between the simulation results and the actual situation. Adjoint method minimizes a predetermined cost function which defines differences between model-derived quantities and measured quantities by using the existing observations to the maximum extent. Adjoint method is an effective variational assimilation technique based on mathematics strictly. It takes ocean model equations, initial conditions, and boundary conditions as the constraint conditions and combines variational principle and the theory of optimal control. With the accessible oceanic elements, some inaccessible oceanic elements can be obtained by optimizing initial field and/or parameters (Sasaki [15]; Lu and Zhang [16]).

Adjoint method was widely used in both meteorology and oceanography. Zhang et al. [17] applied this method to study the similarities and the differences between the Ekman (linear) and the Quadratic (nonlinear) bottom friction parameterizations for a two-dimensional tidal model. The simulation results indicated that the nonlinear Quadratic parameterization is more accurate than the linear Ekman parameterization if the traditional constant boundary friction coefficient is used. However, when the spatially varying boundary friction coefficients were used, the differences between the Ekman and the Quadratic approaches diminished. In the study of Fan and Lv [18], SeaWiFS chlorophyll-a data were assimilated into a simple NPZD model by the adjoint method in a climatological physical environment provided by FOAM. The results showed that the values of the selected sensitive parameters were spatially variable and the application of spatial parameterizations could improve the assimilation results significantly. Many researches (Yu and O'Brien [19], Lawson et al. [20], Zhao et al. [21], Zhao and Lu [22], Qi et al. [23]) have proved the validity and rationality of the adjoint method. Therefore, in this paper, we apply this method to simulate the distribution of totalN in any time step, and the monthly mean distribution in May in the Bohai Sea can be obtained.

The contents of this paper are organized as follows. Section 2 describes the ecosystem model and the database. Section 3 illustrates the adjoint method and independent grids briefly. Section 4 describes the twin experiment to validate the model's capability of inversing pollutant initial field and finds the optimal strategy of setting independent grids. Based on twin experiments, practical experiment is performed in Section 5 in order to obtain the initial field of totalN in the Bohai Sea in May. The conclusions of our work are presented in Section 6.

2. Model and Data

2.1. Model Equations. Based on hydrodynamic model, the transporting diffusion process of pollution can be written as follows:

$$\frac{\partial C}{\partial t} + u\frac{\partial C}{\partial x} + v\frac{\partial C}{\partial y} + w\frac{\partial C}{\partial z} = \frac{\partial}{\partial x}\left(A_H\frac{\partial C}{\partial x}\right) + \frac{\partial}{\partial y}\left(A_H\frac{\partial C}{\partial y}\right) + \frac{\partial}{\partial z}\left(K_H\frac{\partial C}{\partial z}\right) - rC, \tag{1}$$

where A_H and K_H represent horizontal and vertical eddy diffusivities, respectively, C is the concentration of pollution, and r is the degradation coefficient of pollution. When the pollution is conservative substance, $r = 0$; otherwise, $r \neq 0$. In this paper, we treat the pollution as conservative substance, so $r = 0$; the finite-difference form can be seen in Appendix A.

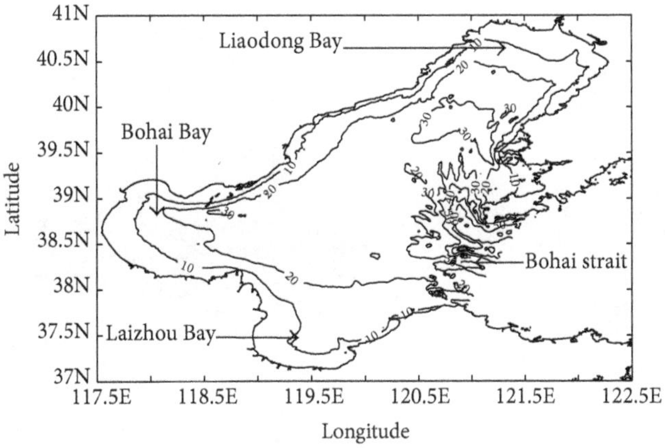

FIGURE 1: Location and morphology of the Bohai Sea. Values of the bathymetric isolines are in meters.

The three-dimensional Regional Ocean Model System (ROMS) is used to calculate the ambient physical velocities and temperature in the Bohai Sea (37°N~41°N, 122.5°E~127.5°E, Figure 1), and the horizontal resolution is 4 second in both latitude and longitude. The thickness of each layer from top to bottom is 10 m, 10 m, 10 m, 20 m, 25 m, and 25 m, respectively, and the integral time step is 6 hours.

Monthly mean horizontal currents in May in the Bohai Sea obtained by ROMS in 5 m depth, 15 m depth, and 25 m depth are shown in Figures 2(a), 2(b), and 2(c), respectively. On the whole, mean flow velocity decreases gradually from surface to bottom in the Bohai Sea. There is a clockwise vortex in the central Bohai Sea, and an anticlockwise vortex can be seen in the east part of Liaodong Bay. The circulation in Laizhou Bay is very weak, and there is a clockwise vortex in the mouth. Meanwhile, water flows out of Bohai Sea through north of Bohai Strait and into Bohai Sea through south of Bohai Strait.

2.2. Observations and Model-Generated Observations. The distribution of the conventional monitoring stations is depicted in Figure 3. We can see that most of the stations are located in the Bohai strait and the coastal areas while only a few of them are located in the central Bohai Sea. The observations are needed in practical experiments and some model-generated observations are needed in the twin experiments. We can choose some model-generated observations through the following methods: first, an initial pollutant distribution (initial field) in the Bohai Sea is assigned. Then, the forward model is run for 30 days, so the pollutant distribution at every time step can be obtained. As the sampling locations of the conventional monitoring stations have been known from Figure 3, we can pick up model-generated observations according to the following method: the sampling locations of the model-generated observations are the same as the conventional monitoring stations. The total number of the conventional monitoring stations is 121. Since the total calculating step is also 121, we prescribe that the number of model-generated stations are in one-to-one correspondence with the sampling time for the sake of simplicity. The observation numbers of each station depend on the depth of water. If the water depth is within three layers, the pollutant concentration of every layer is chosen as model-generated observations; otherwise, only the pollutant concentration of upper three layers is chosen as model-generated observations.

If a guess initial field of pollution is given, the initial field of the pollution can be optimized by using the observations (practical experiment) or the model-generated observations (twin experiment) through the adjoint method.

3. Method

3.1. Adjoint Assimilation Method. Adjoint assimilation method treats all the practical problems as minimum problems. It takes model equations, initial conditions, and boundary conditions as constraint conditions and minimizes a predetermined cost function which defines differences between model-derived quantities and measured quantities. The flowchart in Figure 4 summarizes the steps that make up the adjoint method.

Step 1. an initial distribution of pollution is given empirically (guess distribution);

Step 2. perform the simulation by running the forward model, and the simulation results are obtained;

Step 3. Calculate the cost function which defines the misfit between simulation results and observations. The cost function is defined by

$$J = \frac{1}{2} \sum K_C \left(C_{i,j,k} - \overline{C}_{i,j,k} \right)^2, \qquad (2)$$

where $C_{i,j,k}$ represents the simulation result, $\overline{C}_{i,j,k}$ is the observation, the index triplet (i, j, k) is a pointer to certain grid cell, and K_C is the weight of the observation, which is defined as follows:

$$K_C = \begin{cases} 1, & \text{if the observations are available} \\ 0, & \text{otherwise;} \end{cases} \qquad (3)$$

Step 4. the adjoint of the model (Appendix B) is run backward in time;

Step 5. calculate the gradient of the cost function with respect to initial field;

Step 6. update the initial pollutant distribution closer to the minimum of the cost function;

Step 7. return to Step 1; repeat the iteration with the updated pollutant distribution;

Step 8. end this procedure after a specific number of iterations or the cost function J is small enough to meet the criteria $J < \varepsilon$ (ε is a small real number, such as 0.01), and the optimized initial field of pollution is obtained. In this paper, we choose the former for easiness to compare the simulation results in twin experiment 1.

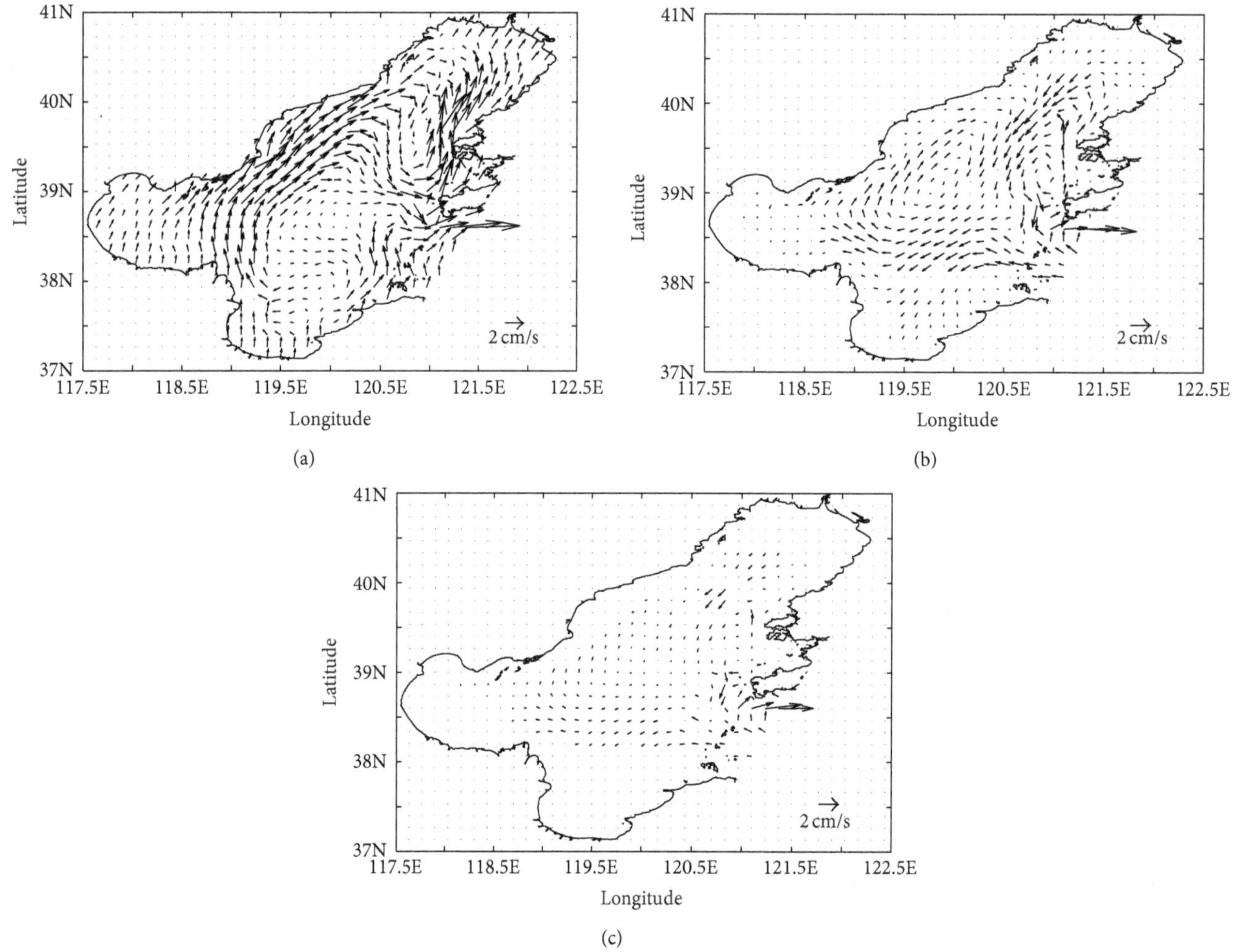

FIGURE 2: Monthly mean horizontal currents in May in the Bohai Sea in (a) 5 m depth, (b) 15 m depth, and (c) 25 m depth.

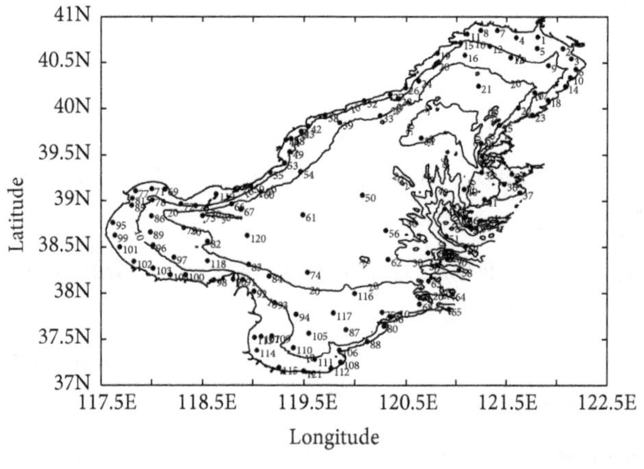

FIGURE 3: The regular monitoring stations in the Bohai Sea.

3.2. Independent Grids. If the pollutant concentration of each grid is optimized independently, then there are too many variables to constraint, and the pollutant distribution is not continuous, which is not reasonable. So, we can reduce variables that need to be optimized and guarantee that the simulation result coincides with the law of physics by using independent grids. Several grids are selected as independent grids and only pollutant concentrations of these independent grids need to be optimized while those of other grids are calculated by Cressman method [18]. Since cost function declines in the inverse direction of its gradient, the gradient is used to determine the direction to optimize the pollutant concentration. In this paper, the distributions of independent grids in longitude are the same as in latitude.

4. Numerical Experiments and Result Analysis

4.1. Twin Experiment 1: The Strategy of Independent Grids. The simulated results are affected by the number of independent grids, so we will discuss this factor in twin experiment 1. The experiment is designed as follows. Assume that the initial distribution of pollutant concentration shows a parabolic surface with upward convex, which means it is high in the centre and low in the surroundings. The independent grids are selected every 2 to 9 common grids. The influence radius is 1.2 times of the distance between adjacent independent

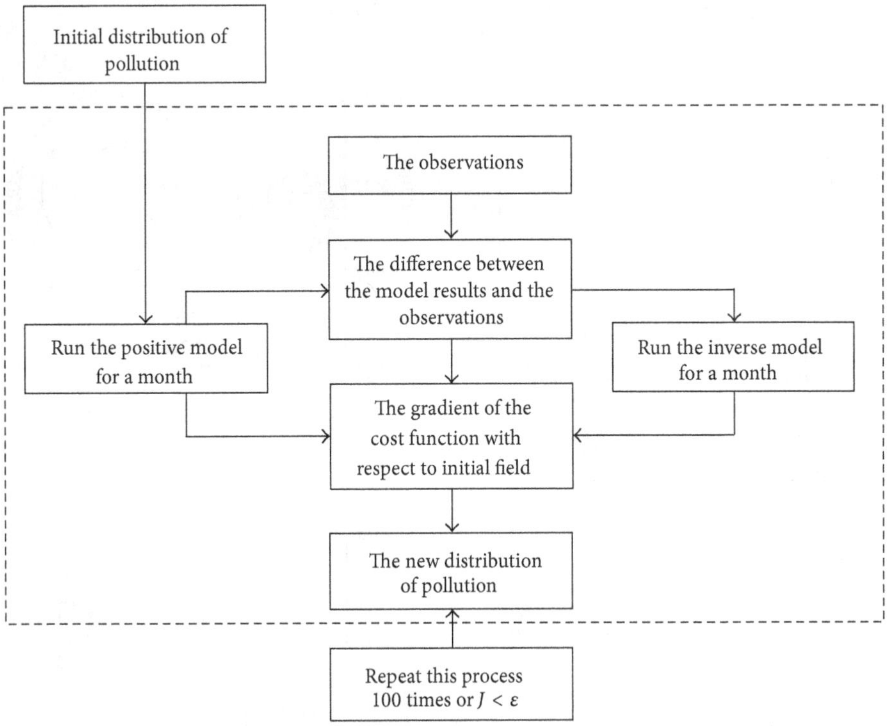

FIGURE 4: Flowchart of the adjoint assimilation method.

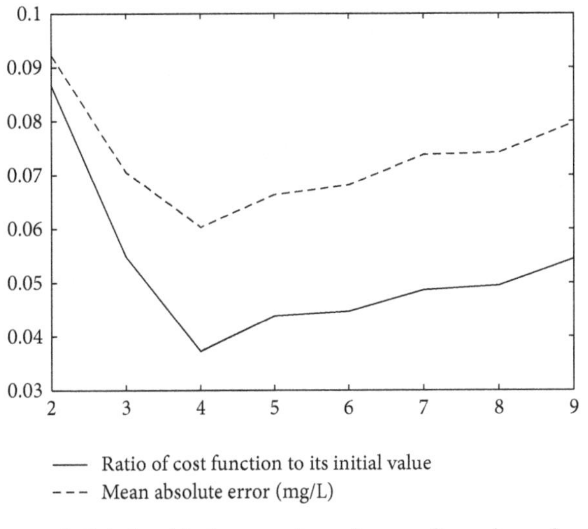

FIGURE 5: Relationship between inversion results and number of independent grid.

grids, and the number of iterations is set to 100 mainly based on the following considerations: (1) the misfit between "observations" and the simulated results is very small and approximately constant when it declines to a certain value after 100 iterations; (2) the cost functions are almost no longer falling after 100 iterations; and (3) the calculation amount is acceptable for the 100 iterations. The results are given in Figure 5.

In Figure 5, X-axis represents that there is one independent grid every 2 to 9 common grids. The solid line shows the ratio of cost function to its initial value while the dotted line indicates mean absolute error (MAE) of the model-generated observations. It shows that when independent grids get fewer, both the cost function and the MAE of the model-generated observations get smaller at first but larger after 1.2°. This means that choosing an independent grid every four grids is the best choice. Therefore, in the following experiments, we select an independent grid every four grids both in longitude and latitude.

4.2. *Twin Experiment 2: The Pollutant Concentration Shows a Parabolic Surface*

4.2.1. Parabolic Surface with Upward Convex. Assume that the initial distribution of pollutant concentration shows a parabolic surface with upward convex, which means it is high in the centre and low in the surroundings, and the pollutant concentration at any grid can be calculated by

$$C(i,j) = \left((\text{lon}(i) - 120.0)^2 + (\text{lat}(j) - 39.0)^2\right) \times 0.2 + 0.05, \tag{4}$$

where lon (i) and lat (j) indicate the longitude and latitude of grid (i, j), respectively. Equation (4) shows that the pollutant concentration varies from 0.05 to 2.10.

We guess that the initial pollutant concentration at any grid is 0.05 mg/L, and the number of iterations is the same as twin experiment 1.

As we can see from Figures 6 and 7, the cost function can reduce to 3.7 percent of its initial value. Table 1

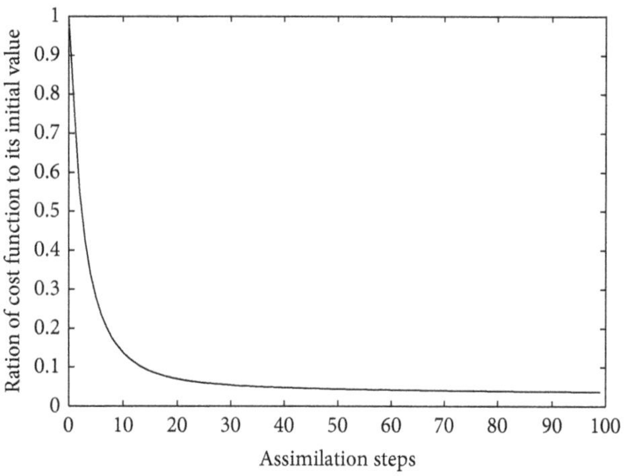

FIGURE 6: Ratio of cost function to its initial value versus assimilation step.

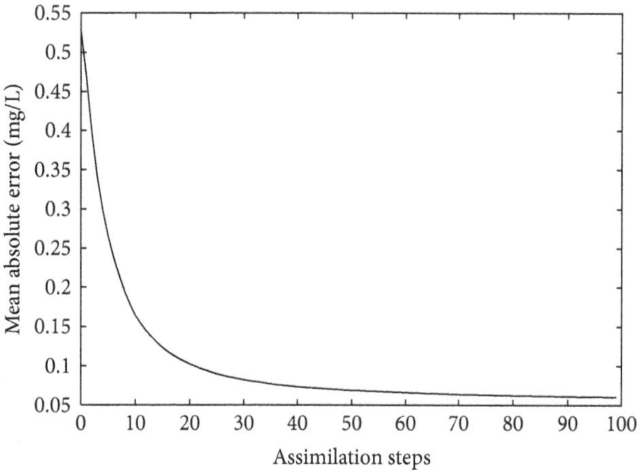

FIGURE 7: Mean absolute error versus assimilation step.

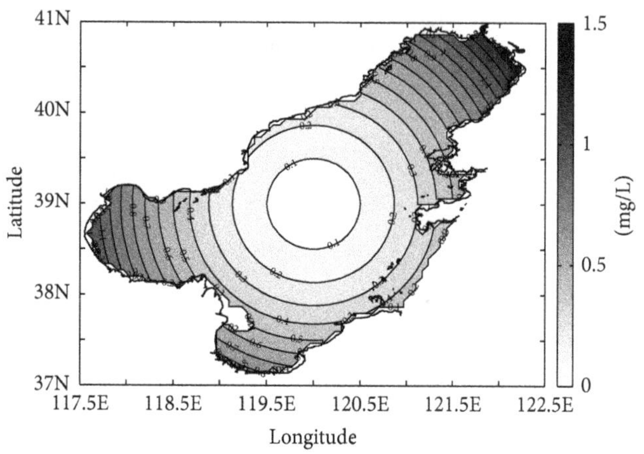

FIGURE 8: Prescribed initial distribution of pollutant concentration.

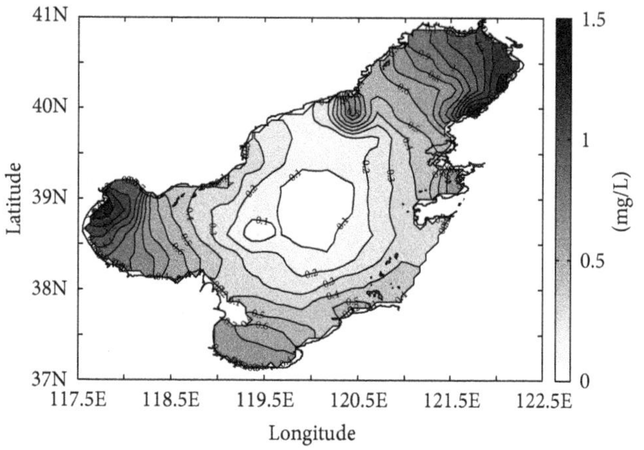

FIGURE 9: Inversion results of initial distribution of pollutant concentration.

shows that the MAE of the model-generated observations declines from 0.53 mg/L to 0.06 mg/L, which decreases by 88.7 percent, and Table 2 shows that the mean relative error (MAE) also declined obviously after adjoint assimilation. The results indicate that through adjoint assimilation, the model-generated observations have been effectively used, and the given distribution can be inverted successfully. The given distribution and inversion results can be seen in Figures 8 and 9. The inversion results of the central Bohai Sea and the Laizhou Bay are very satisfactory while those of the transition zones between the central Bohai Sea and the three bays are a little worse. The inversion results are basically the same as the prescribed distribution.

4.2.2. Parabolic Surface with Downward Convex. Assume that the initial distribution of pollutant concentration shows a parabolic surface with downward convex, which means it is low in the centre and high in the surroundings, and the pollutant concentration at any grid can be calculated by

$$C(i,j) = -\left((\mathrm{lon}(i) - 120.0)^2 + (\mathrm{lat}(j) - 39.0)^2\right) \times 0.2 + 2.10. \tag{5}$$

Repeat the process of Section 4.2.1. The cost function can reduce to 8.3 percent of its initial value, and the MAE of the model-generated observations declines from 1.5 mg/L to 0.2 mg/L, indicating that the model is stable and valid. No matter which convex is given, the initial distribution can be inverted successfully.

4.3. Twin Experiment 3: The Pollutant Concentration Shows a Conical Surface. Assume that the initial pollutant concentration shows a conical surface with upward convex, and the pollutant concentration at any grid can be calculated by

$$C(i,j) = \sqrt{0.41 \times \left((\mathrm{lon}(i) - 120.0)^2 + (\mathrm{lat}(j) - 39.0)^2\right)} + 0.05. \tag{6}$$

TABLE 1: The change of cost function and MAE after reversion.

Type of pollutant distribution	Cost function	Initial value of MAE (mg/L)	Initial value of MAE (mg/L)	Reduction ration of MAE (%)
Parabolic surface				
Up direction	3.7×10^{-2}	0.53	0.06	88.7
Down direction	8.3×10^{-2}	1.52	0.20	86.9
Conical surface				
Up direction	4.4×10^{-2}	0.98	0.11	92.5
Down direction	9.1×10^{-2}	1.07	0.15	92.1

TABLE 2: The change of cost function and MRE after reversion.

Type of pollutant distribution	Initial value of MRE (%)	Final value of MRE (%)	Reduction ration of MRE (%)
Parabolic surface			
Up direction	86.3	12.8	85.2
Down direction	96.7	12.2	87.4
Conical surface			
Up direction	93.9	12.7	86.5
Down direction	95.1	12.3	87.1

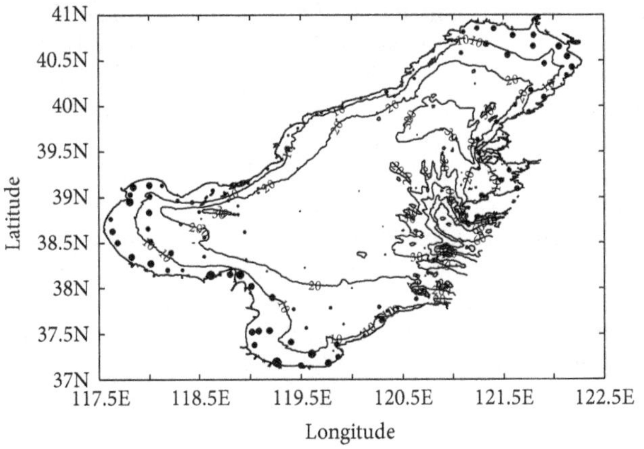

FIGURE 10: Distribution of totalN monitoring station in use in May, 2009.

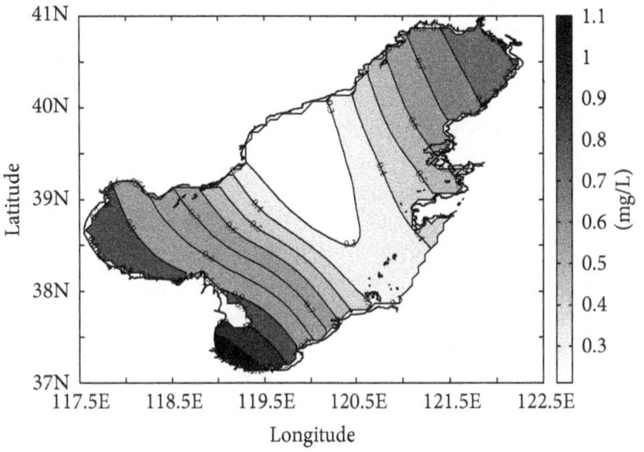

FIGURE 11: Monthly mean distribution of totalN in May in the Bohai Sea calculated by the Cressman method.

Assume that the initial pollutant concentration shows a conical surface with downward convex and the pollutant concentration at any grid can be calculated by

$$C(i,j) = -\sqrt{0.41 \times \left((\text{lon}(i) - 120.0)^2 + (\text{lat}(j) - 39.0)^2\right)} + 2.10. \quad (7)$$

Twin experiments can also be evaluated by the absolute and the relative difference between the prescribed and the inversion results.

In these circumstances, the initial pollutant concentration also varies from 0.05 to 2.10. Repeat the process of twin experiment 2, and the results are similar to those of twin experiment 2. The detailed information is presented in Table 1. These results once again verify the stability and validation of this model, indicating that the coupled model can be used in practical experiment. In other words, the initial field of the Bohai Sea can be inverted through adjoint method by using the regular monitoring observations.

5. Practical Experiment and Result Analysis

The distribution of totalN monitoring stations in May, 2009 is shown in Figure 10. The larger the dot is, the greater the observation it represents. According to the traditional method, the monthly mean distribution of totalN in surface layer is always obtained by interpolation method, such as Cressman and Kriging. The monthly mean distribution of totalN in surface layer calculated by Cressman method is depicted in Figure 11. When we make use of the coupled model mentioned in this paper, not only the initial totalN distribution in May, but also the distribution in any time step can be obtained. And then, we can get the monthly mean distribution of totalN in surface layer, which is depicted in Figure 12.

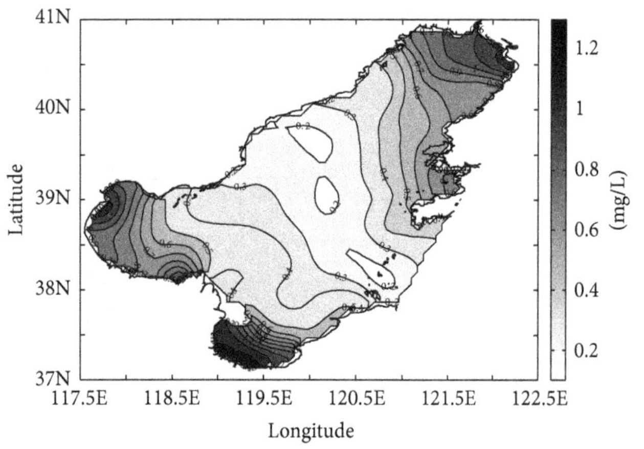

Figure 12: Monthly mean distribution of totalN in May in the Bohai Sea obtained by the adjoint method.

In the Bohai strait and the central Bohai Sea, the distribution of totalN calculated by the Cressman method is similar to the result obtained by the adjoint method. However, the latter is higher in the Laizhou Bay and lower in the Liaodong Bay than the former. The reason may be that the sampling time of each routine monitor station is not the same, and the observation cannot always reflect the mean value in a month. Moreover, the Cressman method only takes the spatial information of the stations into consideration and ignores the time information.

The average concentration of totalN in surface layer in the Bohai Sea is 0.49 mg/L by using the Cressman method while it is only 0.41 mg/L by using the adjoint method. The former is almost twenty percent greater than the latter. The reason is probably that most of routine monitoring stations are located in the Bohai strait and the coastal areas while only a few of those located in the central Bohai Sea. In the meantime, the observations in the Bohai strait and the coastal areas are of relativly high values while the observations in the central Bohai Sea are of relative low values. When we calculate the pollutant concentration between high-value observations and low-value observations by using the Cressman method, there may be more high-value observations involved in the calculation. Therefore, it will lead to a larger pollutant concentration than the real value in most areas. Moreover, although the interpolation results can guarantee the continuity of the totalN distribution, the totalN concentration at any grid cannot be larger or smaller than the observations we already have.

The adjoint method can make use of the observations to a maximum extent. By using this method, not only the values and locations of the observations, but also the sampling time is taken into consideration. The distribution characteristics of totalN in the Bohai Sea in May are depicted in Figure 12. The concentration of totalN is lowest in the central Bohai Sea and Bohai strait. In fact, the concentration is almost zero in these areas, and the highest concentration is no more than 0.5 mg/L. The totalN concentrations of three bays are higher than the central Bohai Sea, among them the concentration of Bohai Bay is the lowest, while Liaodong Bay takes the second place, and the concentration of most areas is less than 1.0 mg/L. The totalN concentration of the Laizhou is the highest. It is more than 1.0 mg in half of areas, and it even reaches 1.4 mg/L in some areas.

6. Conclusion

Recently, more and more numerical models for simulating pollutant dispersion have been actually developed since they can be used for decision-making purposes after releases of contaminants into the marine environment (Periáñez [10]). Regardless of which numerical model we employ, the initial field of pollution has dramatic influence on the simulated results. Therefore, in order to simulate the dispersion process of the pollution accurately, we need not only a good pollutant model and reasonable model parameters, but also an accurate initial distribution of pollution. Although some researchers have tried to inverse the pollutant location and strength, this is the first time to inverse the initial distribution of pollution as far as we know.

Based on the simulation of a marine ecosystem dynamical model in the Bohai Sea, routine monitoring data were assimilated to study the initial field of pollution in this paper. Firstly, in order to reduce variables that need to be optimized and make the simulation results more reasonable, an independent grid is selected every four grids both in longitude and latitude, and only the pollutant concentrations of these independent grids needed to be optimized while the other grids were calculated by interpolation method. Based on this method, the stability and reliability of this model were proved by a set of twin experiments. Therefore, this model could be applied in real experiment to simulate the initial field of the totalN in May, 2009. Furthermore, the distribution of totalN in any time step could be calculated by this model, and the monthly mean distribution in May in the Bohai Sea could be obtained. Compared with the Cressman method, the adjoint method can make use of the existing observations to the maximum extent. Not only the values and locations of the observations, but also the sampling time are taken into consideration by using this method. Therefore, it can reduce the misfit between inversion results and observations significantly and make the simulated results closer to reality. This method can also be used to simulate the distribution of other pollutions, such as total phosphorus, chemical oxygen demand, and petroleum hydrocarbon. So, it could most probably be used in solving environmental problems in the future.

Appendices

A. Finite-Difference Form of Equation (1)

Consider the following:

$$\frac{\left(C_{i,j,k}^{l+1} - C_{i,j,k}^{l}\right)}{\Delta t} - \left[\frac{K_H\left(C_{i,j,k+1}^{l+1} - C_{i,j,k}^{l+1}\right)}{\Delta z_{k+1/2}\cdot\Delta z_{k+1}} - \frac{K_H\left(C_{i,j,k}^{l+1} - C_{i,j,k-1}^{l+1}\right)}{\Delta z_{k+1/2}\cdot\Delta z_{k}}\right]$$

$$= -\frac{u^l_{i,j,k}\left(C^l_{i+1,j,k} - C^l_{i-1,j,k}\right)}{2\Delta x_j} - \frac{v^l_{i,j,k}\left(C^l_{i,j+1,k} - C^l_{i,j-1,k}\right)}{2\Delta y}$$

$$- \frac{w^l_{i,j,k}\left(C^l_{i,j,k+1} - C^l_{i,j,k-1}\right)}{2\Delta z_k}$$

$$+ \left[\frac{A_H\left(C^l_{i+1,j,k} - C^l_{i,j,k}\right)}{\Delta x_{j+1/2}\cdot\Delta x_{j+1}} - \frac{A_H\left(C^l_{i,j,k} - C^l_{i-1,j,k}\right)}{\Delta x_{j+1/2}\cdot\Delta x_j}\right]$$

$$+ \left[\frac{A_H\Delta t\cdot\left(C^l_{i,j+1,k} - C^l_{i,j,k}\right)}{(\Delta y)^2} - \frac{A_H\Delta t\cdot\left(C^l_{i,j,k} - C^l_{i,j-1,k}\right)}{(\Delta y)^2}\right], \tag{A.1}$$

where u and v are horizontal velocities and w is vertical velocity.

B. Adjoint Equation

Consider the following:

$$-\frac{\partial C^*}{\partial t} - \frac{\partial}{\partial z}\left(K_H\frac{\partial C^*}{\partial z}\right)$$

$$= \frac{\partial}{\partial x}(uC^*) + \frac{\partial}{\partial y}(vC^*) + \frac{\partial}{\partial z}(wC^*)$$

$$+ \frac{\partial}{\partial x}\left(A_H\frac{\partial C^*}{\partial x}\right) + \frac{\partial}{\partial y}\left(A_H\frac{\partial C^*}{\partial y}\right) - K_C\left(C - \overline{C}\right), \tag{B.1}$$

where $\overline{C}$, C represent observation and corresponding simulating result, respectively. C^* is the adjoint operator of C. The finite-difference form of adjoint equation is as follows:

$$\frac{\left(C^{*l-1}_{i,j,k} - C^{*l}_{i,j,k}\right)}{\Delta t}$$

$$- \left[\frac{K_H\left(C^{*l-1}_{i,j,k+1} - C^{*l-1}_{i,j,k}\right)}{\Delta zz_k\cdot\Delta z_{k+1}} - \frac{K_H\left(C^{*l-1}_{i,j,k} - C^{*l-1}_{i,j,k-1}\right)}{\Delta zz_k\cdot\Delta z_k}\right]$$

$$= \frac{\left(u^l_{i+1,j,k}C^{*l}_{i+1,j,k} - u^l_{i-1,j,k}C^{*l}_{i-1,j,k}\right)}{2\Delta x_j}$$

$$+ \frac{\left(v^l_{i,j+1,k}C^{*l}_{i,j+1,k} - v^l_{i,j-1,k}C^{*l}_{i,j-1,k}\right)}{2\Delta y}$$

$$+ \frac{\left(w^l_{i,j,k+1}C^{*l}_{i,j,k+1} - w^l_{i,j,k-1}C^{*l}_{i,j,k-1}\right)}{2\Delta z_k}$$

$$+ \left[\frac{A_H\left(C^{*l}_{i+1,j,k} - C^{*l}_{i,j,k}\right)}{\Delta x_{j+1/2}\cdot\Delta x_{j+1}} - \frac{A_H\left(C^{*l}_{i,j,k} - C^{*l}_{i-1,j,k}\right)}{\Delta x_{j+1/2}\cdot\Delta x_j}\right]$$

$$+ \left[\frac{A_H\left(C^{*l}_{i,j+1,k} - C^{*l}_{i,j,k}\right)}{(\Delta y)^2} - \frac{A_H\left(C^{*l}_{i,j,k} - C^{*l}_{i,j-1,k}\right)}{(\Delta y)^2}\right]$$

$$- K_C\left(C^l_{i,j,k} - \overline{C^l_{i,j,k}}\right). \tag{B.2}$$

Acknowledgments

The authors acknowledge the support of the Major State Basic Research Development Program of China through Grant 2013CB956500, the National Natural Science Foundation of China through Grants 41076006 and 41206001, the Natural Science Foundation of Jiangsu Province of China through Grant BK2012315, and the Fundamental Research Funds for the Central Universities Grants 201261006 and 201262007.

References

[1] Q. Chen, K. Tan, C. Zhu, and R. Li, "Development and application of a two-dimensional water quality model for the Daqinghe River Mouth of the Dianchi Lake," *Journal of Environmental Sciences*, vol. 21, no. 3, pp. 313–318, 2009.

[2] J. G. Duan and S. K. Nanda, "Two-dimensional depth-averaged model simulation of suspended sediment concentration distribution in a groyne field," *Journal of Hydrology*, vol. 327, no. 3-4, pp. 426–437, 2006.

[3] M. E. Lee and I. W. Seo, "Analysis of pollutant transport in the Han River with tidal current using a 2D finite element model," *Journal of Hydro-Environment Research*, vol. 1, no. 1, pp. 30–42, 2007.

[4] X.-B. LIU, W.-Q. PENG, G.-J. HE, J.-L. LIU, and Y.-C. WANG, "A Coupled Model of Hydrodynamics and Water Quality for Yuqiao Reservoir in Haihe River Basin," *Journal of Hydrodynamics*, vol. 20, no. 5, pp. 574–582, 2008.

[5] I. Gupta, S. Dhage, A. A. Chandorkar, and A. Srivastav, "Numerical modeling for Thane creek," *Environmental Modelling and Software*, vol. 19, no. 6, pp. 571–579, 2004.

[6] R. Periáñez, "GISPART: a numerical model to simulate the dispersion of contaminants in the Strait of Gibraltar," *Environmental Modelling and Software*, vol. 20, no. 6, pp. 797–802, 2005.

[7] R. Periáñez, "Modelling the transport of suspended particulate matter by the Rhone River plume (France). Implications for pollutant dispersion," *Environmental Pollution*, vol. 133, no. 2, pp. 351–364, 2005.

[8] R. Rajar, M. Cetina, and A. Sirca, "Hydrodynamic and water quality modelling: case studies," *Ecological Modelling*, vol. 101, no. 2-3, pp. 209–228, 1997.

[9] R. Rajar and M. Cetina, "Hydrodynamic and water quality modelling: an experience," *Ecological Modelling*, vol. 101, no. 2-3, pp. 195–207, 1997.

[10] R. Periáñez, "A particle-tracking model for simulating pollutant dispersion in the Strait of Gibraltar," *Marine Pollution Bulletin*, vol. 49, no. 7-8, pp. 613–623, 2004.

[11] H. Huang, G. Chen, and Q.-F. Zhang, "The distribution characteristics of pollutants released at different cross-sectional positions of a river," *Environmental Pollution*, vol. 158, no. 5, pp. 1327–1333, 2010.

[12] I. H. Harms, M. J. Karcher, and D. Dethleff, "Modelling Siberian river runoff—implications for contaminant transport in the Arctic Ocean," *Journal of Marine Systems*, vol. 27, no. 1–3, pp. 95–115, 2000.

[13] S.-Q. Peng and L. Xie, "Effect of determining initial conditions by four-dimensional variational data assimilation on storm surge forecasting," *Ocean Modelling*, vol. 14, no. 1-2, pp. 1–18, 2006.

[14] C. T. Allen, G. S. Young, and S. E. Haupt, "Improving pollutant source characterization by better estimating wind direction with a genetic algorithm," *Atmospheric Environment*, vol. 41, no. 11, pp. 2283–2289, 2007.

[15] Y. Sasaki, "Some basic formalisms in numerical variational analysis," *Monthly Weather Review*, vol. 98, pp. 875–883, 1970.

[16] X. Lu and J. Zhang, "Numerical study on spatially varying bottom friction coefficient of a 2D tidal model with adjoint method," *Continental Shelf Research*, vol. 26, no. 16, pp. 1905–1923, 2006.

[17] J. Zhang, X. Lu, P. Wang, and Y. P. Wang, "Study on linear and nonlinear bottom friction parameterizations for regional tidal models using data assimilation," *Continental Shelf Research*, vol. 31, no. 6, pp. 555–573, 2011.

[18] W. Fan and X. Lv, "Data assimilation in a simple marine ecosystem model based on spatial biological parameterizations," *Ecological Modelling*, vol. 220, no. 17, pp. 1997–2008, 2009.

[19] L. S. Yu and J. J. O'Brien, "On the initial condion in parameter estimation," *Journal of Physical Oceanography*, vol. 22, pp. 1361–1364, 1992.

[20] L. M. Lawson, Y. H. Spitz, E. E. Hofmann, and R. B. Long, "A data assimilation technique applied to a predator-prey model," *Bulletin of Mathematical Biology*, vol. 57, no. 4, pp. 593–617, 1995.

[21] Q. Zhao, X. Hu, X. Lü, X. Xiong, and B. Yang, "Study on the transport of COD in the sea area around Maidao off Qingdao coast using data assimilation," *Journal of Ocean University of China*, vol. 6, no. 4, pp. 339–344, 2007.

[22] Q. Zhao and X. Lu, "Parameter estimation in a three-dimensional marine ecosystem model using the adjoint technique," *Journal of Marine Systems*, vol. 74, no. 1-2, pp. 443–452, 2008.

[23] P. Qi, C. Wang, X. Li, and X. Lv, "Numerical study on spatially varying control parameters of a marine ecosystem dynamical model with adjoint method," *Acta Oceanologica Sinica*, vol. 30, no. 1, pp. 7–14, 2011.

6

New Jacobi Elliptic Function Solutions for the Kudryashov-Sinelshchikov Equation Using Improved F-Expansion Method

Yinghui He

Department of Mathematics, Honghe University, Mengzi, Yunnan 661100, China

Correspondence should be addressed to Yinghui He; heyinghui07@163.com

Academic Editor: Gradimir Milovanovic

Based on the F-expansion method with a new subequation, an improved F-expansion method is introduced. As illustrative examples, some new exact solutions expressed by the Jacobi elliptic function of the Kudryashov-Sinelshchikov equation are obtained. When the modulus m of the Jacobi elliptic function is driven to the limits 1 and 0, some exact solutions expressed by hyperbolic function and trigonometric function can also be obtained. The method is straightforward and concise and is promising and powerful for other nonlinear evolution equations in mathematical physics.

1. Introduction

It has recently become more interesting to obtain exact solutions of nonlinear partial differential equations. These equations are mathematical models of complex physical phenomena that arise in engineering, applied mathematics, chemistry, biology, mechanics, physics, and so forth. Thus, the investigation of the traveling wave solutions to nonlinear evolution equations (NLEEs) plays an important role in mathematical physics. A lot of physical models have supported a wide variety of solitary wave solutions.

In 2010, Kudryashov and Sinelshchikov [1] introduced the following equation:

$$u_t + \gamma u u_x + u_{xxx} - \varepsilon (u u_{xx})_x - \kappa u_x u_{xx} - \nu u_{xx} - \delta (u\, u_x)_x = 0, \quad (1)$$

where γ, ε, κ, ν, and δ are real parameters. Equation (1) describes the pressure waves in the liquid with gas bubbles taking into account the heat transfer and viscosity. When $\varepsilon = \kappa = \delta = 0$ and $\varepsilon = \kappa = \nu = \delta = 0$, (1) becomes the BKdV equation and the KdV equation, respectively. So, (1) can be considered as the generalization of KdV equation. Therefore, the study to (1) is more meaningful than KdV equation and BKdV equation. We call this equation the Kudryashov-Sinelshchikov equation.

Equation (1) was studied by many researchers in various methods. In the case of $\nu = \delta = 0$, it was studied by Ryabov, using a modification of the truncated expansion method [2], by Randrüüt in a more straightforward manner [3], by Li et al., using the bifurcation method of dynamical systems [4–6], by Nadjafikhah and Shirvani-Sh, using the Lie symmetry method [7], and by He, using G'/G-expansion method [8]. In the case of $\nu \neq 0$, $\delta \neq 0$, (1) was studied by Efimova using the modified simplest equation method [9], by Mirzazadeh and Eslami, using first integral method [10]. And they obtained some results when β took special values.

We noticed that the Jacobi elliptic function solutions of (1) are only reported in [8] with special case $\beta = -3$ and $\beta = -4$. Our aim is to find some new solutions expressed by the Jacobi elliptic function making use of improved F-expansion method.

The organization of the paper is as follows: in Section 2, a brief description of the improved F-expansion for finding traveling wave solutions of nonlinear equations is given. In Sections 3 and 4, we will study, respectively, the Kudryashov-Sinelshchikov equation with the situation $\nu = \delta = 0$ and

TABLE 1: Relations between values of r, p, q, and corresponding $F(\xi)$ in $F'^2 = qF^3 + pF^2 + rF$.

Case	r	p	q	$F(\xi)$
1	4	$-4(m^2+1)$	$4m^2$	$\mathrm{sn}^2(\xi)$
2	$4(1-m^2)$	$4(2m^2-1)$	$-4m^2$	$\mathrm{cn}^2(\xi)$
3	$-4(1-m^2)$	$4(2-m^2)$	-4	$\mathrm{dn}^2(\xi)$
4	m^2	$2(1+m^2)$	m^2	$(\mathrm{sn}(\xi) \pm i\mathrm{cn}(\xi))^2$
5	$-(1-m^2)^2$	$2(m^2-2)$	-1	$(m\mathrm{cn}(\xi) \pm \mathrm{dn}(\xi))^2$
6	$1-m^2$	$2(1+m^2)$	$1-m^2$	$\left(\dfrac{\mathrm{cn}(\xi)}{1+\mathrm{sn}(\xi)}\right)^2$
7	m^2-1	$2(1+m^2)$	m^2-1	$\left(\dfrac{\mathrm{dn}(\xi)}{1+m\mathrm{sn}(\xi)}\right)^2$
8	m^2	$2(m^2-1)$	m^2	$\left(\dfrac{m\mathrm{sn}(\xi)}{1+\mathrm{dn}(\xi)}\right)^2$
9	1	$2(1-2m^2)$	1	$\left(\dfrac{\mathrm{sn}(\xi)}{1+\mathrm{cn}(\xi)}\right)^2$

$\nu \neq 0$, $\delta \neq 0$ by the improved F-expansion methods. Finally conclusions are given in Section 5.

2. Description of the Improved Methods

Based on F-expansion method [11–13], the main procedures of the improved F-expansion method are as follows.

Step 1. Consider a general nonlinear PDE in the form

$$F(u, u_x, u_t, u_{xx}, u_{xt}, \ldots) = 0. \tag{2}$$

Using $u(x,t) = U(\xi)$, $\xi = x - ct$, we can rewrite (2) as the following nonlinear ODE:

$$F(U, U', U'', \ldots) = 0, \tag{3}$$

where the prime denotes differentiation with respect to ξ.

Step 2. Suppose that the solution of ODE (3) can be written as follows:

$$U(\xi) = A_0 + \sum_{i=1}^{n} \left(A_i F^i(\xi) + B_i F^{-i}(\xi)\right) \tag{4}$$

or

$$U(\xi) = A_0 + \sum_{i=1}^{n} \left(A_i F^i(\xi) + B_i F^{i-1} F'(\xi)\right), \tag{5}$$

where A_i, B_i $(i = 1, 2, \ldots, n)$ are constants to be determined later and n is a positive integer that is given by the homogeneous balance principle. And $F(\xi)$ satisfies the following equation

$$\left(F'(\xi)\right)^2 = qF^3(\xi) + pF^2(\xi) + rF(\xi), \tag{6}$$

where r, p, and q are constant.

Step 3. Substituting (4) or (5) along with (6) into (3) and then setting all the coefficients of $F^j(\xi)F'^k(\xi)$ $(j = 1, 2, \ldots, k = 0, 1)$ of the resulting system to zero yield a set of overdetermined nonlinear algebraic equations for A_0, A_i, and B_i $(i = 1, 2, \ldots n)$.

Step 4. Assuming that the constants A_0, A_i, and B_i $(i = 1, 2, \ldots n)$ can be obtained by solving the algebraic equations in Step 3, then by substituting these constants and the solutions of (6) that can be found in Table 1 into (4), we can obtain the explicit solutions of (2) immediately.

3. Exact Solutions of the Kudryashov-Sinelshchikov Equation in the Case of $\delta = \nu = 0$

Using scale transformation

$$x = x', \qquad t = t', \qquad u = \frac{u'}{\varepsilon}, \tag{7}$$

we can write the Kudryashov-Sinelshchikov equation (1) in the form

$$u_t + \alpha u u_x + u_{xxx} - (u u_{xx})_x - \beta u_x u_{xx} - \nu u_{xx} - \mu (u\, u_x)_x = 0, \tag{8}$$

where $\alpha = \gamma/\varepsilon$, $\beta = \kappa/\varepsilon$, and $\mu = \delta/\varepsilon$ (primes are omitted). When $\delta = \nu = 0$, (8) becomes

$$u_t + \alpha u u_x + u_{xxx} - (u u_{xx})_x - \beta u_x u_{xx} = 0. \tag{9}$$

Let

$$u(x,t) = 1 - \varphi(\xi), \qquad \xi = x - ct, \tag{10}$$

where c is the wave speed. Under this transformation, (9) can be reduced to the following ordinary differential equation (ODE):

$$c\varphi' - \alpha(1-\varphi)\varphi' - \varphi''' + \left((1-\varphi)\varphi''\right)' - \beta\varphi'\varphi'' = 0. \tag{11}$$

Integrating (11) once with respect to ξ and setting the constant of integration to R, we have

$$\frac{1}{2}\alpha\varphi^2 + (c-\alpha)\varphi - \varphi\varphi'' - \frac{1}{2}\beta(\varphi')^2 + R = 0. \tag{12}$$

Balancing $\varphi\varphi''$ with $(\varphi')^2$ in (12) we find that $n + n + 2 = 2(n + 1)$, so n is an arbitrary positive integer. For simplicity, we take $n = 2$. Suppose that (12) owns the solutions in the form

$$\varphi(\xi) = A_0 + A_1 F(\xi) + A_2 F^2(\xi) + \frac{B_1}{F(\xi)} + \frac{B_2}{F^2(\xi)}. \quad (13)$$

Substituting (13) and (6) into (12) and then setting all the coefficients of F^k $(k = -5, \ldots, 5)$ of the resulting system to zero, we can obtain the following results:

$$A_1 = A_2 = B_2 = 0,$$
$$c = \frac{2A_0B_1p - B_1^2q + 3rA_0 - B_1p - 3rA_0^2}{B_1}, \quad (14)$$
$$\alpha = \frac{3rA_0 - B_1p}{B_1}, \qquad \beta = -3,$$

$$A_2 = B_1 = B_2 = 0,$$
$$c = \frac{2A_0A_1p - A_1p + 3qA_0 - A_1^2r - 3qA_0^2}{A_1}, \quad (15)$$
$$\alpha = \frac{3qA_0 - A_1p}{A_1}, \qquad \beta = -3,$$

$$A_2 = 0, \qquad B_1 = \frac{A_1r}{q}, \qquad B_2 = 0,$$
$$c = \frac{4A_1^2r + 2A_0A_1p - A_1p + 3qA_0 - 3qA_0^2}{A_1}, \quad (16)$$
$$\alpha = \frac{3qA_0 - A_1p}{A_1}, \qquad \beta = -3,$$

$$A_0 = \frac{A_2(2wp - r)}{3q}, \qquad A_1 = 2wA_2, \qquad B_1 = B_2 = 0,$$
$$c = -\frac{-18q^2w + 2A_2p^2w - 6wA_2rq + 6pq - A_2pr}{3q},$$
$$\alpha = -2p + 6wq, \qquad \beta = -\frac{5}{2}, \quad (17)$$

where $w = (p \pm \sqrt{p^2 - 3rq})/3q$, $q \neq 0$. Consider the following:

$$A_0 = \frac{2A_2(2wp + 5r)}{3q}, \qquad A_1 = 4A_2w,$$
$$B_1 = \frac{4A_2wr}{q}, \qquad B_2 = \frac{A_2r^2}{q^2},$$
$$c = -\left(2\left(-18wq^2 + 2A_2p^2w + 24A_2wrq + 3pq - 16A_2pr\right)\right) \times (3q)^{-1}, \quad (18)$$
$$\alpha = -2p + 12wq, \qquad \beta = -\frac{5}{2},$$

where $w = (p \pm \sqrt{p^2 - 3rq})/3q$, $q \neq 0$.

$$A_0 = \frac{B_2(2w_1p - q)}{3r},$$
$$A_1 = A_2 = 0, \qquad B_1 = 2w_1B_2,$$
$$c = -\frac{-18w_1r^2 + 2B_2p^2w_1 - 6w_1B_2qr + 6pr - B_2pq}{3r}, \quad (19)$$
$$\alpha = -2p + 6w_1r, \qquad \beta = -\frac{5}{2},$$

where $w_1 = (p \pm \sqrt{p^2 - 3rq})/3r$, $r \neq 0$.

Substituting (14)–(19) into (13) with (10), we obtain, respectively, the following formal solution of (9):

$$u(x,t) = 1 - A_0 - \frac{B_1}{F(\xi)}, \quad (20)$$

where $\xi = x - ((2A_0B_1p - B_1^2q + 3rA_0 - B_1p - 3rA_0^2)/B_1)t$, $\alpha = (3rA_0 - B_1p)/B_1$, and $\beta = -3$. Consider the following:

$$u(x,t) = 1 - A_0 - A_1F(\xi), \quad (21)$$

where $\xi = x - ((2A_0A_1p - A_1p + 3qA_0 - A_1^2r - 3qA_0^2)/A_1)t$, $\alpha = (3qA_0 - A_1p)/A_1$, and $\beta = -3$. Consider the following:

$$u(x,t) = 1 - A_0 - A_1F(\xi) - \frac{A_1r}{qF(\xi)}, \quad (22)$$

where $\xi = x - ((4A_1^2r + 2A_0A_1p - A_1p + 3qA_0 - 3qA_0^2)/A_1)t$, $\alpha = (3qA_0 - A_1p)/A_1$, and $\beta = -3$. Consider the following:

$$u(x,t) = 1 - \frac{A_2(2wp - r)}{3q} - 2wA_2F(\xi) - A_2F^2(\xi), \quad (23)$$

where $\xi = x + ((-18q^2w + 2A_2p^2w - 6wA_2rq + 6pq - A_2pr)/3q)t$, $\alpha = -2p + 6wq$, and $\beta = -5/2$. Consider the following:

$$u(x,t) = 1 - \frac{2A_2(2wp + 5r)}{3q} - 4wA_2F(\xi) - A_2(F(\xi))^2 - \frac{4wA_2r}{qF(\xi)} - \frac{A_2r^2}{q^2F^2(\xi)}, \quad (24)$$

where $\xi = x + (2(-18wq^2 + 2A_2p^2w + 24A_2wrq + 3pq - 16A_2pr)/3q)t$, $\alpha = -2p + 12wq$, and $\beta = -5/2$. Consider the following:

$$u(x,t) = 1 - \frac{B_2w_1}{3r} - \frac{2B_2w_1}{F(\xi)} - \frac{B_2}{F^2(\xi)}, \quad (25)$$

where $\xi = x + ((-18w_1r^2 + 2B_2p^2w_1 - 6w_1B_2qr + 6pr - B_2pq)/3r)t$, $\alpha = -2p + 6w_1r$, and $\beta = -5/2$.

Combining (20)–(25) with Table 1, many exact solutions of (9) can be obtained. For simplicity, we just give out case 1 of Table 1, the other cases can be discussed similarly.

When $r = 4$, $p = -4(m^2+1)$, and $q = 4m^2$, the solution of elliptic Equation (6) is $F(\xi) = \mathrm{sn}^2(\xi, m)$. Substituting it into (20)–(24), we can obtain the following solutions of (9).

From (20), one has

$$u_1(x,t) = 1 - A_0 - B_1 \mathrm{ns}^2(\xi, m), \quad (26)$$

where $\xi = x + (4(2A_0B_1m^2 + 2A_0B_1 + B_1^2m^2 - 3A_0 - B_1m^2 - B_1 + 3A_0^2)/B_1)t$, $\alpha = (4(3A_0 + B_1m^2 + B_1)/B_1)$, and $\beta = -3$.

When $m \to 1$ and $\mathrm{sn}(\xi, m) \to \tanh(\xi)$, (26) becomes

$$u(x,t) = 1 - A_0 - B_1 \coth^2(\xi), \quad (27)$$

where $\xi = x + (4(4A_0B_1 - 3A_0 - 2B_1 + B_1^2 + 3A_0^2)/B_1)t$, $\alpha = 4(3A_0 + 2B_1)/B_1$, and $\beta = -3$.

When $m \to 0$ and $\mathrm{sn}(\xi, m) \to \sin(\xi)$, (26) becomes

$$u(x,t) = 1 - A_0 - B_1 \sec^2(\xi), \quad (28)$$

where $\xi = x + (4(2A_0B_1 - 3A_0 - B_1 + 3A_0^2)/B_1)t$, $\alpha = (4(3A_0 + B_1)/B_1)$, and $\beta = -3$.

From (21), we have

$$u(x,t) = 1 - A_0 - A_1 \mathrm{sn}^2(\xi, m), \quad (29)$$

where $\xi = x + 4((2A_0A_1m^2 + 2A_0A_1 + 3m^2A_0^2 + A_1^2 - A_1m^2 - A_1 - 3m^2A_0)/A_1)t$, $\alpha = 4((A_1m^2 + A_1 + 3m^2A_0)/A_1)$, and $\beta = -3$.

When $m \to 1$, (29) becomes a hyperbolic function solution,

$$u(x,t) = 1 - A_0 - B_1 \tanh^2(\xi), \quad (30)$$

where $\xi = x + (4(4A_0B_1 - 3A_0 - 2B_1 + B_1^2 + 3A_0^2)/B_1)t$, $\alpha = (4(3A_0 + 2B_1)/B_1)$, and $\beta = -3$.

When $m \to 0$, (29) becomes a trigonometric function solution,

$$u(x,t) = 1 - A_0 - B_1 \sin^2(\xi), \quad (31)$$

where $\xi = x + 4(2A_0 + A_1 - 1)t$, $\alpha = 4$, and $\beta = -3$.

From (22), we have

$$u(x,t) = 1 - A_0 - A_1 \mathrm{sn}^2(\xi, m) - \frac{A_1 r}{q} \mathrm{ns}^2(\xi, m), \quad (32)$$

where $\xi = x + (4(2A_0A_1m^2 + 2A_0A_1 + 3m^2A_0^2 - 4A_1^2 - A_1m^2 - A_1 - 3m^2A_0)/A_1)t$, $\alpha = (4(A_1m^2 + A_1 + 3m^2A_0)/A_1)$, and $\beta = -3$.

When $m \to 1$, (32) becomes a hyperbolic function solution,

$$u(x,t) = 1 - A_0 - A_1 \tanh^2(\xi) - \frac{A_1 r}{q} \coth^2(\xi), \quad (33)$$

where $\xi = x + (4(4A_0A_1 + 3A_0^2 - 4A_1^2 - 2A_1 - 3A_0)/A_1)t$, $\alpha = (4(2A_1 + 3A_0)/A_1)$, and $\beta = -3$.

From (23), we have

$$u(x,t) = 1 + \frac{A_2\left(2w\left(m^2+1\right)+1\right)}{3m^2} - 2wA_2\mathrm{sn}^2(\xi, m) - A_2\mathrm{sn}^4(\xi, m), \quad (34)$$

where $w = (-m^2 - 1 \pm \sqrt{m^4 - m^2 + 1})/3m^2$, $\xi = x + (4(-2wA_2m^2 + 2wA_2m^4 + 2wA_2 - 18wm^4 + A_2m^2 + A_2 - 6m^4 - 6m^2)/3m^2)t$, $\alpha = 8(3wm^2 + m^2 + 1)$, and $\beta = -5/2$.

When $m \to 1$, (34) becomes a hyperbolic function solution,

$$u(x,t) = 1 - \frac{1}{12}A_2 w - 2wA_2\tanh^2(\xi) - A_2\tanh^4(\xi, x), \quad (35)$$

where $w = (-2 \pm 1)/3$, $\xi = 8(3w - (1/3)A_2w + 2 - (1/3)A_1)t$, $\alpha = 16 + 24w$, and $\beta = -5/2$.

From (24), we have

$$u(x,t) = 1 + \frac{2A_2\left(2w\left(m^2+1\right)-5\right)}{3m^2} - 4wA_2\mathrm{sn}^2(\xi, m) - A_2\mathrm{sn}^4(\xi, m) - \frac{4wA_2}{m^2}\mathrm{ns}^2(\xi, m) - \frac{A_2}{m^4}\mathrm{ns}^4(\xi, m), \quad (36)$$

where $w = (-m^2 - 1 \pm \sqrt{m^4 + 14m^2 + 1})/6m^2$, $\xi = x + (8(-18wm^4 + 2wA_2m^4 + 28wA_2m^2 + 2wA_2 - 3m^4 - 3m^2 + 16A_2m^2 + 16A_2)/3m^2)t$, $\alpha = 8(m^2 + 1 + 6wm^2)$, and $\beta = -5/2$.

When $m \to 1$, (36) becomes a hyperbolic function solution,

$$u(x,t) = 1 + \frac{2}{3}A_2(4w - 5) - 4wA_2\tanh^2(\xi) - A_2\tanh^4(\xi) - 4wA_2\coth^2(\xi) - A_2\coth^4(\xi), \quad (37)$$

where $w = (-1 \pm 2)/3$, $\xi = x - (48w - (256/3)wA_2 + 16 - (256/3)A_2)\, t$, $\alpha = 16 + 48w$, and $\beta = -5/2$.

From (25), we have

$$u(x,t) = 1 + \frac{1}{3}B_2\left(2w\left(m^2+1\right)+m^2\right) - 2wB_2\mathrm{ns}^2(\xi, m) - B_2\mathrm{ns}^4(\xi, m), \quad (38)$$

where $w = (-m^2 - 1 \pm \sqrt{m^4 - m^2 + 1})/3$, $\xi = x + ((8/3)wB_2m^4 + (8/3)wB_2m^2 - (8/3)wB_2 + 24w + 8m^2 + 8 - (4/3)B_2m^4 - (4/3)B_2m^2)t$, $\alpha = 8(3w + m^2 + 1)$, and $\beta = -5/2$.

When $m \to 1$, (38) becomes a hyperbolic function solution,

$$u(x,t) = 1 + \frac{1}{3}B_2(4w + 1) - 2wB_2\coth^2(\xi) - B_2\coth^4(\xi), \quad (39)$$

where $w = (-2 \pm 1)/3$, $\xi = x - (24w - (8/3)wB_2 + 16 - (8/3)B_2)t$, $\alpha = 16 + 24w$, and $\beta = -5/2$.

When $m \to 0$, (38) becomes a triangle function solution,

$$u(x,t) = 1 + \frac{2}{3}wB_2 - 2wB_2\csc^2(\xi) - B_2\csc^4(\xi), \quad (40)$$

where $w = (-1 \pm 1)/3$, $\xi = x - (24w + 8 - (8/3)B_2w)t$, $\alpha = 24w + 8$, and $\beta = -5/2$.

We notice that when $\beta = -3$, some Jacobi elliptic function solutions have been given in [8]. Part of our results may be the same with them. However, when $\beta = -(5/2)$, the Jacobi elliptic function solutions of (9) have not been reported in the related literatures, so we believe that our solutions (34), (36), and (38) are new.

4. Exact Solutions of the Kudryashov-Sinelshchikov Equation in the Case of $\delta \neq 0, \mu \neq 0$

In this case, by similar process, (1) can be changed into the following PDE:

$$\begin{aligned}&\frac{1}{2}\alpha\varphi^2 + (c-\alpha)\varphi - \varphi\varphi''\\&\quad - \frac{1}{2}\beta(\varphi')^2 + \nu\varphi' + \mu(1-\varphi)\varphi' + R = 0,\end{aligned} \tag{41}$$

where $\varphi' = d\varphi/d\xi$, $\xi = x - ct$, and R is an integral constant.

Suppose that (41) owns the solutions in the form

$$\begin{aligned}\varphi(\xi) &= A_0 + A_1F(\xi) + A_2F^2(\xi)\\&\quad + B_1F(\xi)' + B_2F(\xi)F(\xi)'.\end{aligned} \tag{42}$$

Substituting (42) and (6) into (41) and then setting all the coefficients of $F^j(\xi)F'^k(\xi)$ $(j = 0,1,\ldots,6, k = 0,1)$ of the resulting system to zero, we can obtain the following results:

$$\begin{aligned}&A_0 = 0, \quad A_1 = wB_1, \quad A_2 = B_2 = 0,\\&c = -\frac{5}{3}wB_1r - \frac{4}{3}p, \quad \alpha = -\frac{4}{3}p,\\&\beta = -\frac{8}{3}, \quad \lambda = \frac{2}{3}B_1r + \frac{1}{3}w, \quad \mu = -\frac{1}{3}w,\end{aligned} \tag{43}$$

where $w = \pm\sqrt{p}$. Consider the following:

$$\begin{aligned}&A_0 = \frac{B_2w(2pw^2 - 2p^2 + 3rq)}{9q^2},\\&A_1 = \frac{2B_2w(w-p)}{3q},\\&A_2 = -B_2w, \quad B_1 = -\frac{B_2(w^2-p)}{3q},\\&c = -w(14w^2B_2p^2 - 42w^2B_2rq + 63B_2rpq\\&\quad +108wq^2 - 14B_2p^3)\times(45q^2)^{-1},\\&\alpha = -\frac{12}{5}w^2, \quad \beta = -\frac{12}{5},\\&\lambda = -(2w^2B_2p^2 - 2B_2p^3 + 9B_2rpq\\&\quad -6w^2B_2rq + 9wq^2)\times(45q^2)^{-1},\\&\mu = \frac{1}{5}w\end{aligned} \tag{44}$$

where $w = \pm\sqrt[4]{p^2 - 3rq}$.

Substituting (43) and (44) along with (10) into (42), we obtain, respectively, the following formal solution of (8):

$$u(x,t) = 1 - wB_1F(\xi) - B_1F'(\xi), \tag{45}$$

where $w = \pm\sqrt{p}$, $\xi = x + ((5/3)wB_1r + (4/3)p)t$, $\alpha = -(4/3)p$, $\beta = -8/3$, $\lambda = (2/3)B_1r + (1/3)w$, and $\mu = -(1/3)w$. Consider the following:

$$\begin{aligned}u(x,t) &= 1 - \frac{B_2w(2pw^2 - 2p^2 + 3rq)}{9q^2}\\&\quad - \frac{2B_2w(w^2-p)F(\xi)}{3q} + wB_2F^2(\xi)\\&\quad + \frac{B_2(w^2-p)F'(\xi)}{3q} - B_2F(\xi)F'(\xi),\end{aligned} \tag{46}$$

where $w = \pm\sqrt[4]{p^2 - 3rq}$, $\xi = x + (w(14w^2B_2p^2 - 42w^2B_2rq + 63B_2rpq + 108wq^2 - 14B_2p^3)/45q^2)t$, $\alpha = -(12/5)w^2$, $\beta = -12/5$, $\lambda = -((2w^2B_2p^2 - 2B_2p^3 + 9B_2rpq - 6w^2B_2rq + 9wq^2)/45q^2)$, and $\mu = (1/5)w$.

Combining (45) and (46) with Table 1, many exact solutions of (8) can be obtained. For simplicity, we just give out case 3 of Table 1, and the other cases can be discussed similarly.

When $r = -4(1-m^2)$, $p = 4(2-m^2)$, $q = -4$, and the solution of elliptic Equation (6) is $F(\xi) = \mathrm{dn}^2(\xi,m)$. Substituting it into (45) and (46), we can obtain the following Jacobi elliptic function solutions of (8):

$$\begin{aligned}u(x,t) &= 1 - B_1(w\,\mathrm{dn}^2(\xi,m)\\&\quad -2m^2\mathrm{sn}(\xi,m)\,\mathrm{cn}(\xi,m)\,\mathrm{dn}(\xi,m)),\end{aligned} \tag{47}$$

where $w = \pm2\sqrt{2-m^2}$, $\xi = x - ((20/3)wB_1 - (20/3)wB_1m^2 - (32/3) + (16/3)m^2)t$, $\alpha = (16/3)(m^2-2)$, $\beta = -8/3$, $\nu = (1/3)(-8B_1 + 8B_1m^2 + w)$, $\mu = -(1/3)w$, and

$$\begin{aligned}u(x,t) &= 1 - \frac{B_2w(2pw^2 - 2p^2 + 3rq)}{9q^2}\\&\quad - \frac{2B_2w(w^2-p)\,\mathrm{dn}^2(\xi,m)}{3q} + wB_2\mathrm{dn}^4(\xi,m)\\&\quad - \frac{2m^2B_2(w^2-p)\,\mathrm{dn}(\xi,m)\,\mathrm{cn}(\xi,m)\,\mathrm{sn}(\xi,m)}{3q}\\&\quad + 2m^2B_2\,\mathrm{dn}^3(\xi,m)\,\mathrm{cn}(\xi,m)\,\mathrm{sn}(\xi,m),\end{aligned} \tag{48}$$

where $w = \pm2\sqrt[4]{1-m^2-m^4}$, $\xi = x + (2w/45)(7w^2B_2 - 7w^2B_2m^2 + 28B_2 - 42B_2m^2 - 42B_2m^4 + 7w^2B_2m^4 + 28B_2m^6 + 54w)t$, $\alpha = -(12/5)w^2$, $\beta = -(12/5)$, $\nu = -(2/45)w^2B_2 + (2/45)w^2B_2m^2 - (2/45)w^2B_2m^4 - (8/45)B_2 + (4/15)B_2m^2 + (4/15)B_2m^4 - (8/45)B_2m^6 - (1/5)w$, and $\mu = (1/5)w$.

Similarly, when $m \rightarrow 1$ we can obtain hyperbolic function solutions of (8). Here, we omit them.

We indicate that these results with $\beta = -12/5$ and $\beta = -8/3$ are new.

5. Conclusions

The Jacobi elliptic function solutions of (1) are only reported in [8] with special case of $\beta = -3$ and $\beta = -4$. In the present work, we successfully obtained some new Jacobi elliptic function solutions of the Kudryashov-Sinelshchikov equation with $\beta = -3$, $\beta = -5/2$, $\beta = -8/3$, and $\beta = -12/5$ using the improved F-expansion method. When the modulus m of the Jacobi elliptic function is driven to the limits 1 and 0, some exact solutions expressed by hyperbolic function and trigonometric function can also be obtained. All the results we obtained have been verified. The related results are enriched.

Acknowledgments

This research is supported by the Natural Science Foundation of China (11161020), the Natural Science Foundation of Yunnan (2013FZ117), and Research Foundation of Honghe University (10XJY120).

References

[1] N. A. Kudryashov and D. I. Sinelshchikov, "Nonlinear waves in bubbly liquids with consideration for viscosity and heat transfer," *Physics Letters A*, vol. 374, no. 19-20, pp. 2011–2016, 2010.

[2] P. N. Ryabov, "Exact solutions of the Kudryashov-Sinelshchikov equation," *Applied Mathematics and Computation*, vol. 217, no. 7, pp. 3585–3590, 2010.

[3] M. Randrüüt, "On the Kudryashov-Sinelshchikov equation for waves in bubbly liquids," *Physics Letters A*, vol. 375, pp. 3687–3692, 2011.

[4] J. Li, "Exact travelling wave solutions and their bifurcations for Kudryashov-Sinelshchikovequation," *International Journal of Bifurcation and Chaos*, vol. 5, Article ID 125011, 19 pages, 2012.

[5] B. He, Q. Meng, and Y. Long, "The bifurcation and exact peakons, solitary and periodic wave solutions for the Kudryashov-Sinelshchikov equation," *Communications in Nonlinear Science and Numerical Simulation*, vol. 17, no. 11, pp. 4137–4148, 2012.

[6] B. He, Q. Meng, J. Zhang, and Y. Long, "Periodic loop solutions and their limit forms for the Kudryashov-Sinelshchikov equation," *Mathematical Problems in Engineering*, vol. 2012, Article ID 320163, 10 pages, 2012.

[7] M. Nadjafikhah and V. Shirvani-Sh, "Lie symmetry analysis of Kudryashov-Sinelshchikov equation," *Mathematical Problems in Engineering*, vol. 2011, Article ID 457697, 9 pages, 2011.

[8] Y. He, "Exact solutions of the Kudryashov-Sinelshchikov equation using the multiple G'/G-expansion method," *Mathematical Problems in Engineering*, vol. 2013, Article ID 708049, 7 pages, 2013.

[9] O. Y. Efimova, "The modified simplest equation method to look for exact solutions of nonlinear partial differential equations," http://arxiv.org/abs/1011.4606, 2010.

[10] M. Mirzazadeh and M. Eslami, "Exact solutions of the Kudryashov-Sinelshchikov equation and nonlinear telegraph equation via the first integral method," *Nonlinear Analysis. Modelling and Control*, vol. 17, no. 4, pp. 481–488, 2012.

[11] N. A. Kudryashov, "Exact solutions of the generalized Kuramoto-Sivashinsky equation," *Physics Letters A*, vol. 147, no. 5-6, pp. 287–291, 1990.

[12] N. A. Kudryashov, "Simplest equation method to look for exact solutions of nonlinear differential equations," *Chaos, Solitons and Fractals*, vol. 24, no. 5, pp. 1217–1231, 2005.

[13] M. Wang and X. Li, "Applications of F-expansion to periodic wave solutions for a new Hamiltonian amplitude equation," *Chaos, Solitons and Fractals*, vol. 24, no. 5, pp. 1257–1268, 2005.

7

Markov Chain Models for the Stochastic Modeling of Pitting Corrosion

A. Valor,[1] F. Caleyo,[1] L. Alfonso,[2] J. C. Velázquez,[3] and J. M. Hallen[1]

[1] *Departamento de Ingeniería Metalúrgica, IPN-ESIQIE, UPALM s/n, Edificio 7, Zacatenco, 07738 México, DF, Mexico*
[2] *Universidad Autónoma de la Ciudad de México, 09790 México, DF, Mexico*
[3] *Departamento de Ingeniería Química Industrial, ESIQIE-IPN, UPALM Edificio 7, Zacatenco, 07738 México, DF, Mexico*

Correspondence should be addressed to A. Valor; almavalor@gmail.com

Academic Editor: Wuquan Li

The stochastic nature of pitting corrosion of metallic structures has been widely recognized. It is assumed that this kind of deterioration retains no memory of the past, so only the current state of the damage influences its future development. This characteristic allows pitting corrosion to be categorized as a Markov process. In this paper, two different models of pitting corrosion, developed using Markov chains, are presented. Firstly, a continuous-time, nonhomogeneous linear growth (pure birth) Markov process is used to model external pitting corrosion in underground pipelines. A closed-form solution of the system of Kolmogorov's forward equations is used to describe the transition probability function in a discrete pit depth space. The transition probability function is identified by correlating the stochastic pit depth mean with the empirical deterministic mean. In the second model, the distribution of maximum pit depths in a pitting experiment is successfully modeled after the combination of two stochastic processes: pit initiation and pit growth. Pit generation is modeled as a nonhomogeneous Poisson process, in which induction time is simulated as the realization of a Weibull process. Pit growth is simulated using a nonhomogeneous Markov process. An analytical solution of Kolmogorov's system of equations is also found for the transition probabilities from the first Markov state. Extreme value statistics is employed to find the distribution of maximum pit depths.

1. Introduction

Localized corrosion, specifically pitting corrosion of metals and alloys, constitutes one of the main failure mechanisms of corroding structures such as pressurized containers and pipes. Pits cause failure through perforation of the component wall. In other cases, pits become nucleation centers for cracks [1]. In the oil and gas industry, pitting corrosion is a major problem, especially for transporting pipelines [2].

Pitting corrosion comprises two main processes: pit initiation and stable pit growth (pit repassivation is not considered in this paper). It is accepted that pit initiation can be a consequence of the breakdown of the passive layer caused by random fluctuations in local conditions. This process takes some time, usually called induction (nucleation or initiation) period [3]. Passive layer breakdown, followed by localized metal dissolution, is the most common mechanism of pitting corrosion. However, pitting can also occur as the result of the active dissolution of certain regions of the material at its surface, such as nonmetallic inclusions, which are susceptible and dissolve faster than the rest of the surface [4]; in this case, the pitting induction time is typically shorter.

After a pit has nucleated, it can repassivate (stop growing) immediately or grow and then repassivate. This process is regarded as metastable pitting. If a pit is able to grow indefinitely, it becomes a stable growing pit [5]. Those pits that reach the stable growth regime become part of a pit population that shows a remarkable stochastic behavior [6, 7].

The time evolution of pit depth due to corrosion is often expressed as a power function of time [7–11]: $D(t) = \alpha(t - t_{\text{ini}})^{\beta}$, where t is the exposure time, α and β are parameters related to the corrosion process, and t_{ini} stands for pit initiation time. This same function has been found to govern the pitting corrosion growth in stainless and mild steels and aluminum alloys [11, 12].

Pitting corrosion occurs in a wide range of metals and environments. This fact points out to the universality of this phenomenon and suggests that randomness is an inherent and unavoidable characteristic of this damage over time, so that stochastic models are better suited to describe pitting corrosion than deterministic ones. Localized corrosion cannot be explained without assuming stochastic points of view due to the large scatter in the measurable parameters such as corrosion rate, maximum pit depth, and time to perforation [13]. Many variables of the metal-environment system such as alloy composition and microstructure, and composition of the surrounding media and temperature, are all involved in the pitting process [4]. Such complexity imposes the development of theoretical models and simulation tools for a better understanding of the outcome of the pitting corrosion process. These tools help predict more accurately the time evolution of pit depth in corroding structures as the key factor in structural reliability assessment.

Another important characteristic of the pitting corrosion process that is worth noting is the time and pit-depth dependence of the corrosion rate [6, 7]. It has been established that, for a given pit, the growth rate decreases with time, while for pits with equal lifetimes, the corrosion rate is larger for deeper ones.

Provan and Rodriguez [14] are amongst the first authors to use a nonhomogenous Markov process to model pit depth growth. In their model, the authors divided the space of possible pit depths into discrete, non overlapping states and numerically solved the system of Kolmogorov's forward equations (1) for the transition probabilities $p_{i,j}(t)$ between damage states i and j. However, Provan and Rodriguez modeled pitting without taking into account the pit generation process and proposed an expression for the intensities $\lambda_i(t)$ of the process that conveyed no physical meaning. They did not discuss the method used to solve the system of equations either. This has made it impossible to reproduce their results (deeper discussion on this topic can be found in [15])

$$\frac{dp_{i,j}(t)}{dt} = \begin{cases} \lambda_{j-1}(t)\, p_{i,j-1}(t) - \lambda_j(t)\, p_{i,j}(t), & j \geq i+1 \\ \lambda_i(t)\, p_{i,i}(t). \end{cases} \tag{1}$$

Other authors who made use of Markov chains to model corrosion were Morrison and Worthingham [16]. Making use of a continuous time birth process with linear intensity λ, they determined the reliability of high-pressure corroding pipelines. For their purpose, these authors divided the space of the load-resistance ratio into discrete states and numerically solved the Kolmogorov's equations in order to find the intensities of transition between damage states. Afterwards, probability distribution function of the load-resistance ratio was estimated and compared to the distribution obtained from field measurements. Worthingham's model was further improved by Hong [17], who used an analytical solution to the system of Kolmogorov's equations for the same homogeneous continuous type of Markov process and derived the process probability transition matrix in order to evaluate the probability of failure. Hong also investigated the influence of pit depth on the load-resistance ratio. The merits and limitations of these two models are discussed in detail in [15, 18].

In recent years, modeling of pitting corrosion with Markov chains has shown new advances. For example, Bolzoni et al. [19] have modeled the first stages of localized corrosion using a continuous-time, three-state Markov process. The Markov states of the metal surface are passivity, meta stability, and stable pit growth. On the other hand, Timashev and coworkers [20] formulated a model based on the use of a continuous-time, discrete-state pure birth homogenous Markov process for stochastically describing the growth of corrosion-caused metal loss. The goal was to assess the conditional probability of pipeline failure and to optimize the maintenance of operating pipelines. In their model, the intensities of the process were calculated by iteratively solving the proposed system of Kolmogorov's forward equations. The drawbacks of Timashev's model are discussed in detail in [18].

In the present paper, a review of the Markov models developed by the authors in an attempt to describe pitting corrosion is presented. The first model intends to solve a problem that is crucial in reliability-based pipeline integrity management: the accurate estimation of future pit depth and growth rate distributions from a (single) measured or assumed pit depth distribution. It has been recognized [21] that such estimation can be carried out only if oversimplifications are made, or if additional information, besides the pit depth distribution, is available. In the developed model, it is postulated that in the case of external pitting corrosion in underground pipelines, this additional information can be attained from the available predictive models for pit growth as a function of the soil characteristics [8, 22]. A model for pit growth previously developed by the authors has been used to perform Monte Carlo simulations in order to predict the distribution of maximum pit depths as a function of the pipeline age and the physicochemical characteristics of the soil [9].

A nonhomogenous linear growth pure birth Markov process, with discrete states in continuous time, is used to model external pitting corrosion in underground pipelines. The system of forward Kolmogorov's equations (1) is analytically solved using the binomial closed-form solution for the transition probabilities between Markov states in a given time interval. This Markov framework is used to predict the time evolution of the pitting depth and rate distributions. The analytical solution becomes available under the assumption that Markov-derived stochastic mean of the pit depth distribution is equal to the deterministic mean of the distribution obtained through Monte Carlo simulations. This assumption is made for different exposure times and different soil classes defined according to soil physicochemical characteristics that are easy to measure in the field. In this way, the transition probabilities are obtained as a function of soil properties for a given time point, and the corrosion rate and future pit depth distributions are predicted. Real-life case studies are presented to illustrate the proposed Markov model. The main advantage of this model resides in its capability of correctly predicting the time evolution of pit depth and rate distributions over time.

One of the main goals of reliability analysis is to estimate the risk of perforation of in-service components produced by the deepest pit. Extreme value statistics is commonly used together with pit depth growth modeling to predict the risk of failure of in-service components and structures [23, 24].

It is recognized [6, 7] that the deepest pits grow at higher rates than the rest of the defects right from the beginning of the corrosion process, so that the maximum pit depths are commonly sampled from an exponential parent distribution that constitutes the right tail of the pit depth distribution of the whole pit population [7]. Besides that, in the corrosion literature [13, 14, 24, 25], it is a well-established fact that the Gumbel extreme value distribution [26, 27] is a good fit to the maximum pit depth distribution obtained by measuring the maximum pit depth on several areas. The cumulative function of the Gumbel distribution, with location parameter μ and scale parameter σ, is

$$G(x) = \exp\left[-\exp\left(-\left(\frac{x-\mu}{\sigma}\right)\right)\right], \quad -\infty < x < \infty. \quad (2)$$

It is important to underline that the previously mentioned Markov model has proved successful in modeling the time evolution of the entire pit depth population but failed to correctly describe the evolution of pit-depth extremes. The second model presented in this work focuses on the simulation of the time evolution of extreme pit depths. The model is based on the stochastic description of pitting corrosion, taking into account pit initiation and growth. A nonhomogeneous Poisson process is used to model pit initiation. The distribution of pit nucleation times is simulated using a Weibull process. Pit depth growth is also modeled as a nonhomogeneous Markov process. The system of forward Kolmogorov's equations (1) is solved analytically for the transition probability from the first state to any jth state during a given time interval [28]. From this solution, the cumulative distribution function of pit depth for the one-pit case can be found. This distribution function has an exponential character and corresponds to the parent distribution from which the extremes can be readily drawn. The extreme depths distribution for the m-pit case is obtained by the multiplication of the m parent cumulative distribution functions of the pits population. This stochastic model is able to predict the time evolution of the probability distribution of maximum pit depths.

2. Stochastic Models of Pitting Corrosion

In this section, two different Markov chain models are presented to describe pitting corrosion. The first one is focused on the description of time evolution of pit depth and rate distributions in operating underground pipelines. The second model deals with the description of maximum pit depths when multiple corrosion pits are taken into account in a laboratory (controlled) pitting corrosion experiment.

In the following, only the definitions relevant to the focus of the presented models are given. The reader is referred to [28, 29] for the formalism theory of Markov processes.

In both models, the possible pit depths constitute the Markov space. The material thickness (along which pits propagate) is divided into N discrete Markov states. The corrosion damage (pit) depth, at time t, is represented by a discrete random variable $D(t)$ such that $P\{D(t) = i\} = p_i(t)$, with $i = 1, 2, \ldots, N$. Furthermore, it is postulated that the probability that the damage that is at the ith state at the moment t increases by one state in a very small interval of time δt is expressed as $\lambda_i(t)\delta t + o(\delta t)$. For a continuous-time, nonhomogenous linear growth Markov process with intensities $\lambda_i(t) = i\lambda(t)$, the probability $p_{i,j}(t)$ that the process in state i will be in state j $(j \geq i)$ at some later time obeys the system of Kolmogorov's forward equations presented in (1). In this infinitesimal transition scheme, $\lambda(t)$ can be interpreted as the jump frequency between the ith to the $(i+1)$th states during the time interval $[t, t+\delta t]$. This means that the number of states transited by the corrosion pit in a short time interval $[0, t]$ can be written as

$$\rho(t) = \int_0^t \lambda(\tau)\, d\tau. \quad (3)$$

It is not difficult to find out that the functions $\lambda(t)$ and $\rho(t)$ have direct physical meaning when Markov processes are used to model pitting corrosion. They are related to the pit growth rate and pit depth, respectively.

2.1. Markov Chain Modeling of Pitting Corrosion Depth and Rate in Underground Pipelines. In [28, page 304], it is shown that for a Markov process defined by the system of (1), the conditional probability $p_{m,n}(t_0, t) = P\{D(t) = n \mid D(t_0) = m\}$ of transition from the mth state to the nth state $(n \geq m)$ in the interval (t_0, t) can be obtained analytically and has the form

$$p_{m,n}(t_0, t) = \binom{n-1}{n-m} e^{-\{\rho(t)-\rho(t_0)\}m}\left(1 - e^{-\{\rho(t)-\rho(t_0)\}}\right)^{n-m}, \quad (4)$$

where $\rho(t)$ is defined by (3).

Equation (4) shows that the increase in pit depth over $\Delta t = t - t_0$ follows a negative binomial distribution NegBin(r, p) with parameters $r = m$ and $p = p_s = e^{-\{\rho(t)-\rho(t_0)\}}$. From the transition probability $p_{m,n}(t_0, t)$ it is possible to estimate the probability distribution function $f(v)$ of the pitting corrosion rate v over the time interval Δt, when the pit depth is at the mth state as

$$f(v; m, t_0, t) = p_m(t_0)\, p_{m,m+v\Delta t}(t_0, t)\, \Delta t. \quad (5)$$

From the distribution function $f(v; m, t_0, t)$, it is straightforward to derive the pitting rate probability distribution associated with the entire pit population (all possible depths) as

$$f(v; t_0, t) = \sum_{m=1}^{N} f(v; m, t_0, t). \quad (6)$$

Until this point, we have the transition probabilities from any state m to the state n in the interval (t_0, t), given by (4). The corrosion rate distribution can also be derived through (6). In principle, if the probability distribution of the corrosion depth at t_0 is known, that is, $P\{D(t_0) = m\} = p_m(t_0)$, the pit depth distribution at any future moment in time can be estimated using

$$p_n(t) = \sum_{m=1}^{n} p_m(t_0)\, p_{m,n}(t_0, t). \tag{7}$$

For the case of buried pipelines, the initial probability distribution of pit depths $p_m(t_0)$ can be obtained if the corrosion damage in the pipeline is monitored using in-line inspection (ILI), being t_0 the time of the inspection. The values of the probabilities p_m can be estimated from the ratio of the number of corrosion pits with depths in the mth state to the total number of observed pits.

It is evident from (4) that the transition probabilities $p_{m,n}(t_0, t)$ depend on the functions $\lambda(t)$ and $\rho(t)$. At this point, physically sound expression for $\rho(t)$ and $\lambda(t)$ should be proposed in order to estimate the evolution of the pit depth and corrosion rate distributions. For that, the knowledge about the pitting corrosion process in soils must be used. It is postulated that the stochastic mean pit depth $M(t)$ can be assumed to be equal to the deterministic mean $Đ(t)$ of the pitting process, which can be estimated from the observed evolution of the average pit depth over time as

$$M(t) = Đ(t). \tag{8}$$

This equality is true under certain simple assumptions that are explained in [29]. In summary, a sufficient condition for the equality of the stochastic and deterministic means is that for any positive integer q, the structure of the process starting from n individuals is identical to that of the sum of q separated systems each starting from n.

Cox and Miller [29] have shown that if the initial damage state is n_i at $t = t_i$, so that $D(t_i) = n_i$, then the time-dependent stochastic mean $M(t) = E[D(t)]$ of the linear growth Markov process can be expressed as

$$M(t) = n_i e^{\rho(t - t_i)}. \tag{9}$$

If we consider that a power function is an accurate deterministic representation of the pit growth process, one can postulate that the deterministic mean pit depth at time t is [8]

$$Đ(t) = \kappa (t - t_{sd})^{\nu}, \tag{10}$$

where κ and ν are the pitting proportionality and exponent parameters, respectively, and t_{sd} is the starting time of the pitting corrosion process. In systems where passivity breakdown and/or inclusion dissolution are the prevalent mechanisms for pit initiation, t_{sd} would represent the initiation time of stable pit growth. In the case of underground pipelines, this parameter corresponds to the total elapsed time from pipeline commissioning to coating damage plus the time period when the cathodic protection is still effective preventing or attenuating external pitting corrosion after coating damage.

The increase of the deterministic mean pit depth in the time interval Δt is

$$\Delta Đ(t) = ƛ(t)\, Đ(t)\, \Delta t, \tag{11}$$

where $ƛ$ can be interpreted as the deterministic intensity (rate) of the process. The pitting rate, obtained by taking the time derivative of $Đ(t)$ (10), obeys the functional form of (11) with $ƛ = \nu/\tau$ and $\tau = t - t_{sd}$ as

$$\frac{dĐ(t)}{dt} = \frac{\nu}{\tau} Đ(t). \tag{12}$$

If δt represents the stay time of the corrosion pit in the first state of the chain, then $n_i = 1$ during the time interval between t_{sd} and $t_{sd} + \delta t$. If δt is significantly less than the simulation time span, it is easy to show from (8)–(10) that the value of the function $\rho(t)$ can be approximated as

$$\rho(t) = \ln\left[\kappa (t - t_{sd})^{\nu}\right]. \tag{13}$$

From this, taking into account (3), it follows that

$$\lambda(t) = \frac{\nu}{t - t_{sd}}. \tag{14}$$

Note that the intensity of the Markov process $\lambda(t)$ is inversely proportional to the exposure time τ, as it is the case of the deterministic intensity $ƛ(t)$ according to (11) and (12).

One can substitute the expression for $\rho(t)$ from (13) into (4) and to show that the probability parameter $p_s = e^{-\{\rho(t)-\rho(t_0)\}}$ can be expressed as

$$p_s = \left(\frac{t_0 - t_{sd}}{t - t_{sd}}\right)^{\nu}, \quad t \geq t_0 \geq t_{sd}. \tag{15}$$

Suppose that a pit is in the state m at t_0. Let $E[n-m]/\Delta t$ be the average damage rate in the time interval $(t_0, t_0 + \Delta t)$. It can be shown [19] that the instantaneous pitting rate v predicted by the stochastic model is

$$v(m, t_0) = \frac{E[n - m]}{\Delta t} \xrightarrow[\Delta t \to 0]{} \nu \frac{m}{t_0}. \tag{16}$$

The stochastically predicted instantaneous damage rate agrees with the deterministic rate given by (12). This coincidence is a sign of the adequacy of the proposed Markov pitting-corrosion model.

For the case of underground pipelines, the pitting corrosion damage evolution can be undertaken as follows. The measured or assumed pit depth probability distribution at t_0 is used as the initial corrosion damage distribution $p_m(t_0)$, where t_0 is the time of the pit depth measurement. The value of the probabilities p_m is estimated from the ratio of the number of corrosion pits with depths in the mth state to the total number of observed pits. The transition probability function $p_{m,n}(t_0, t)$ can be identified from (4) and (15) if the parameters t_{sd} and ν are known.

The estimation of the pitting parameters is possible thanks to the previously developed [8, 9] predictive model for localized corrosion in buried pipelines, which relates the physicochemical conditions of the pipe and soil to parameters t_{sd}, κ and ν. The model is based on a multivariate nonlinear regression analysis using (10), with the field-measured maximum pit depths as the dependent variable to be predicted and

TABLE 1: Estimated pit initiation times for underground pipelines (from [8]).

Soil class	t_{sd} (years)
Clay	3.0
Clay loam	3.1
Sandy clay loam	2.6
All	2.9

the pipe age and soil-pipe characteristics as the independent ones. The experimental details and corrosion data used to produce the model can be found elsewhere [8, 9]. From this model, the pitting initiation time was estimated to have the values displayed in Table 1 for different soil classes. In Table 1, class "All" corresponds to a general class that includes all the soils found in the field survey carried out in South Mexico [8]. The pitting parameters κ and ν in (10) were estimated as linear combinations of soil and pipe characteristics [8] for the different soil classes.

Later on, Caleyo et al. [9] used a Monte Carlo simulation based on (10) and on the obtained linear combinations of environmental variables for the pitting parameters to predict the time evolution of the maximum pit depth probability distribution in underground pipelines. The measured distributions of the considered soil-pipe independent variables were used as inputs in Monte Carlo simulations of the pitting process for each soil class to produce the simulated extreme pit depth distribution for each soil category and exposure time considered. The reader is referred to [9] for details about the used pipe-soil properties distributions and the Monte Carlo simulation framework. The Monte Carlo produced extreme pit depth values for each soil class at different time points were fitted to the generalized extreme value (GEV) distribution [30]:

$$\text{GEV}(x) = \exp\left\{-\left[1+\zeta\left(\frac{x-\mu}{\sigma}\right)\right]^{-1/\zeta}\right\}, \quad (17)$$

where ζ, σ, and μ are the shape, scale, and location parameters of the distribution, respectively.

Figure 1(a) shows the time evolution of the GEV distribution fitted to the Monte Carlo predicted extreme pit depths for a general soil class found in southern Mexico [8]. The inset of Figure 1(a) shows the evolution of the mean and variance of such distributions with time. These average values are used to estimate the pitting proportionality and exponent factors (κ and ν) associated with the typical (average) values of the predictor variables in this soil category. The time evolution of the mean of the simulated maximum pit depth distributions was fitted to (10) using the corresponding value of t_{sd} given in Table 1. Typical values κ_{Typ} and ν_{Typ}, derived following this method, are displayed in Table 2 for the analyzed soil textural classes.

From the estimated values of κ and ν for the general soil class and the value of t_{sd} shown in the last column of Table 1, it is possible to obtain the function $\rho(t)$ (13) and p_s (15) for typical conditions of the "All" soil category [8, 9].

Figure 1(b) shows the time evolution of $\rho(t)$ and p_s predicted from the parameters associated with the pitting

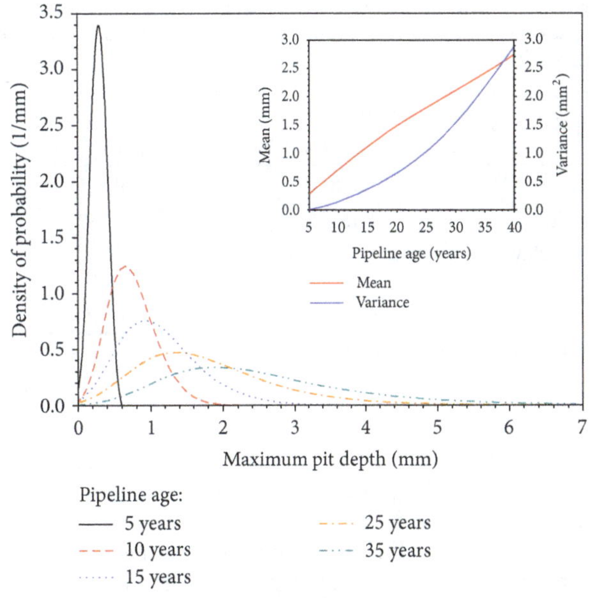

(a)

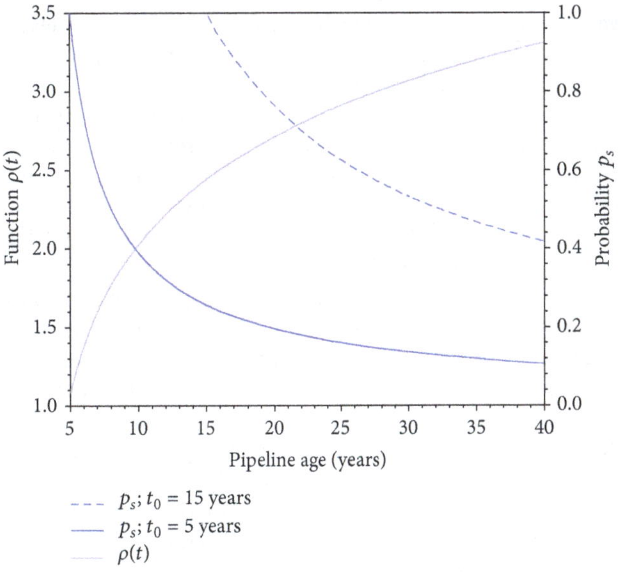

(b)

FIGURE 1: (a) Evolution of the Generalized Extreme Value distribution fitted to the extreme pit depths predicted in a Monte Carlo framework [7] for a general soil class. The time evolution of the mean value and variance of such distributions are shown in the inset. (b) Time evolution of the functions $\rho(t)$ (13) and p_s (15), calculated using the predicted pitting parameters κ, ν, and t_{sd}. The two curves of p_s correspond to two different values of t_0 in (15).

process in soils. Given that $\rho(t)$ is completely determined by the extent of the damage [15], its value is unique for each soil class. Also, the probability p_s is unique in each soil type. However, in Figure 1(b), two different curves of p_s are displayed to illustrate the dependence of p_s on time. They correspond to two distinct values of t_0 (see (15)), the time when the initial pit depth distribution is measured. The value of p_s at a given moment in time $t \geq t_0$ increases with increasing t_0 and

Table 2: Estimated pitting parameters (10) associated with the typical values of the physicochemical pipe-soil variables in each soil category [9].

Soil class	Typical pitting coefficient κ_{Typ} (mm/years$^{\nu_{\text{Typ}}}$)[a]	Typical pitting exponent ν_{Typ} (years)[a]
Clay	0.178	0.829
Clay loam	0.163	0.793
Sandy clay loam	0.144	0.734
All	0.164	0.780

[a]Correspond to parameters in (10).

decreases when the length of the interval (t_0, t) increases. The increase in p_s with t_0 indicates that the mean and variance of the pitting rate decrease as the lifetime of the pitting damage increases, which is in agreement with experimental observations. Also, the form of $\rho(t)$ in Figure 1(b) suggests that, for pits with equal lifetimes, the deeper the pit, the smaller the value of p_s and, therefore, the larger the mean and variance of the pitting rate. This characteristic of the developed model will permit accurately predicting the corrosion rate distribution in a pitting corrosion experiment.

To estimate the evolution of pit depth and pitting rate distributions using the proposed Markov model one can use (7) and (6), respectively. As has been already stated, from the results of an ILI of the pipe and the knowledge of the local soil characteristics, it would be possible to estimate the pitting corrosion damage evolution.

To illustrate the application of the proposed Markov chain model, it is employed in the analysis of an 82 km long operating pipeline, used to transport sweet gas since its commissioning in 1981. This pipeline is coal-tar coated and has a wall thickness of 9.52 mm (0.375″). It was inspected in 2002 and 2007 using magnetic flux leakage (MFL) ILI. The pit depth distributions present in the pipeline were obtained by calibrating the ILI tools using a methodology described by Caleyo et al. [31]. These distributions are shown in Figure 2 in the form of histograms. The first, grey-shadowed histogram represents the depth distribution of $N_{02} = 3577$ pits located and measured by ILI in 2002, while the hatched histogram represents the depth distribution of $N_{07} = 3851$ pits measured by ILI in mid-2007.

In order to apply the proposed model, the pipe wall thickness was divided in 0.1 mm thick non overlapping Markov states, so that the pitting damage is represented through Markov chains with states ranging from $m = 1$ to $m = N = 100$. Unless otherwise specified, this scheme of discretization of the pipe wall thickness is used hereafter to represent the pitting damage penetration.

The empirical depth distribution observed in 2002 was used as the initial distribution so that $t_0 = 21$ years, $\Delta t = 5.5$ years, and $p_m(t_0 = 21) = N_m/N_{02}$, N_m being the number of pits with measured depth in the mth state. The soil characteristics along the pipeline were taken as those of the "All" (generic) soil class and the value of t_{sd} was taken equal to 2.9 years, as indicated in Table 1. From Table 2, a value of

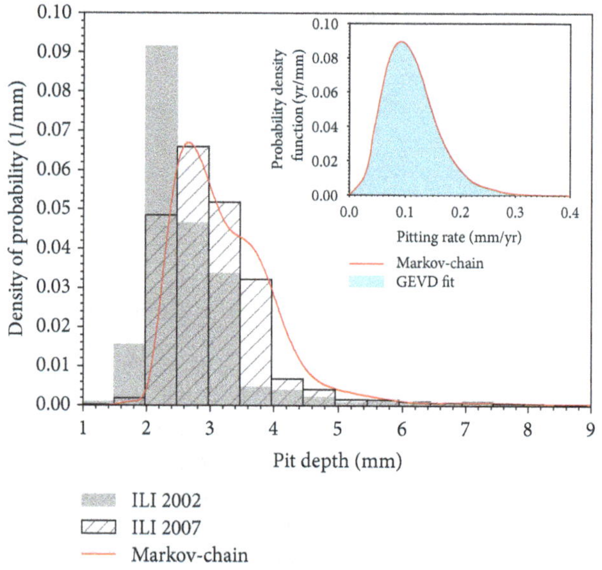

Figure 2: Histograms of the pipeline pit depth distributions measured by ILI in 2002 (grey) and in 2007 (hatched). The solid line is the Markov-estimated pit depth distribution for 2007. In the inset, the Markov-derived pitting rate distribution for the period 2002–2007 is presented.

0.780 was assigned to ν in order to obtain the predicted pit depth distribution p_n $(t = 26$ years) for 2007. The values of $p_n(t)$ can be calculated using the following expression, which is deducted from (4), (7), and (15):

$$p_n(t) = \sum_{m=1}^{n} p_m(t_0) \binom{n-1}{n-m} \left(\frac{t_0 - t_{sd}}{t_0 - t_{sd}}\right)^{\nu m} \left[1 - \left(\frac{t_0 - t_{sd}}{t_0 - t_{sd}}\right)^{\nu}\right]^{n-m}. \tag{18}$$

In Figure 2, the Markov chain-predicted pit depth distribution for 2007 is represented with a thick line.

In order to test the degree of coincidence between the empirical distribution (for 2007) and the Markov-model predicted distribution, the two-sample Kolmogorov Smirnov (K-S) and the two-sample Anderson-Darling (A-D) tests were used. This was done by means of a Monte Carlo simulation in which 1000 samples of 50 depth values were generated for each one of the distributions under comparison and then used in pairs for the tests. In the case of the Markov-predicted distribution the data samples were generated as (pseudo) random variates with probability density function (pdf) as displayed in Figure 2 and given by (18). The samples from the empirical data were produced by sampling the experimental 2007 dataset with replacement. The K-S test gave an average P value of 0.42, with only 7% of cases where the test failed at 5% significance. By its part, the A-D test produced an average P value of 0.39 with an 11% fraction of failed tests. This is an evidence of the validity of the null hypothesis about both samples, modeled and empirical, coming from the same distribution. The good agreement between the empirical pit depth distribution observed in 2007 and the Markov chain-modeled distribution also points to the accuracy of the proposed model.

The corrosion rate distribution $f(\upsilon; t_0, t)$ associated with the entire population of pits in the 2002–2007 period was also estimated using (19), which was derived from (4), (5), (6), and (15). Consider

$$f(\upsilon; t_0, t) = \sum_{m=1}^{N} p_m(t_0) \binom{m + \upsilon(t - t_0) - 1}{\upsilon(t - t_0)} \left(\frac{t_0 - t_{sd}}{t - t_{sd}}\right)^{\nu m} \times \left[1 - \left(\frac{t_0 - t_{sd}}{t - t_{sd}}\right)^{\nu}\right]^{\upsilon(t - t_0)}. \quad (19)$$

The estimated rate distribution is shown in the inset of Figure 2; it can be fitted to a GEV distribution (17) with negative shape parameter ζ, although it is very close to zero. This means that the Weibull and Gumbel subfamilies [27] of the GEV distribution are appropriate to describe the pitting rate in the pipeline over the selected estimation period.

It has to be noticed that the proposed Markov model can be used to predict the progression of other pitting processes. For this to be done, the values of the pitting exponent and starting time must be known for the process. These can be obtained, for example, from the analysis of repeated in-line inspections of the pipeline, from the study of corrosion coupons, or from the analysis of laboratory tests. To show this capability of the model, another example is given, using this time a different experimental data set gathered from two successive MFL-ILIs. The analysis involves a 28 km long pipeline used to transport natural gas since its commissioning in 1985. This pipeline is coal-tar coated, made of API 5L grade X52 steel [32], with a diameter of 457.2 mm (18″) and a wall thickness of 8.74 mm (0.344″). The first inspection was conducted in 1996 and the other in 2006. The ILIs were calibrated following procedures described elsewhere [31]. Before using the ILI data, the corrosion defects from both inspections were matched using the following criteria: (i) the matched defects should agree in location, and (ii) the depth of a defect in the second inspection must be equal to or greater than its depth in the first inspection. Only the matched defects were used in the analyses that followed to ensure that only the actual defect depth progression with time was considered (one-by-one if necessary) without including defects that might have nucleated in the interval between the two inspections. The depth distribution of the matched defects of the 1996 inspection was taken as the initial depth distribution fed into the proposed Markov model. The resulting depth distribution was subsequently compared to that of the defects observed in the 2006 inspection.

From the knowledge of the mean depth values from both inspections, an empirical value of the pitting exponent ν was estimated instead of using the value recommended in Table 2. It was calculated for the specific type of soil in which the pipe under analysis was buried by using the ratio between the empirical pit depth means $Đ_{96}$ and $Đ_{06}$, from the ILIs in 1996 and 2006, respectively. Assuming that the pit depth mean complies with (10) and that parameters κ, ν, and t_{sd} (which depend exclusively on the soil and pipe properties) are the same at the times of both inspections, it is possible to obtain the pitting exponential parameter ν from

$$\frac{Đ_{06}}{Đ_{96}} = \frac{\kappa(t_{06} - t_{sd})^{\nu}}{\kappa(t_{96} - t_{sd})^{\nu}}, \quad (20)$$

where $t_{96} = 11$ years and $t_{06} = 21$ years are the times of the inspections, and t_{sd} is the incubation time of the corrosion metal losses in the pipeline. The value of t_{sd} was taken equal to 2.9 years, the average pitting initiation time for a generic soil class (Table 1).

As in the previous example, the wall thickness was divided into N (0.1 mm thick) equally spaced Markov states. The observed defect depth was converted to Markov-state units, and the depth distribution was given in terms of the probability $p_m(t)$ for the depth in a state equal or less than m at time t. The proposed Markov model (18) was applied to estimate the pit depth distribution after a time interval $\delta t = t - t_0$ of 10 years, with $t_{sd} = 2.9$ years, $t_0 = 11$ years, and $t = 21$ years.

In Figure 3, the final (2006) pit depth distribution of the by-ILI measured and matched 179 defects in this pipeline is shown on a probability density scale in the form of a histogram, together with the pit depth distribution predicted by the Markov model for 2006. Again, the Monte Carlo framework described in the previous example was followed in order to apply the two-sample K-S test. The sample size was 50, and the sampling process was repeated 1000 times. The resulting average P value was 0.07, which is enough for not rejecting the null hypothesis that both the empirical and modeled distributions come from the same distribution. From the results of the test and from what it is shown in Figure 3, it can be concluded that the Markov model predicts a pit depth distribution that is close to the experimental one. More important, the model is also capable to correctly describe the form (shape) of the defect depth distribution, as seen also from this figure.

To further explore the capabilities of the model in estimating the time evolution of pit depths, the defects in the initial (1996) inspection were grouped into 6-state (0.6 mm) depth intervals in order to test the model efficacy in predicting the pit depth evolution by depth interval. An example of these grouping and modeling approach is shown in the inset of Figure 3. The chosen initial depth interval is that between 9 and 15 states (0.9 and 1.5 mm). In the inset, a histogram of the matched defect depths in the second ILI together with the distribution (solid line) predicted by the Markov model starting from the chosen initial depth subpopulation is depicted. Here as well, the model is capable of reproducing the experimental results. The same was proven true for the rest of depth intervals in the initial (1996) pit depth population. Therefore, it can be stated that the Markov model is not only accurate in estimating the time evolution of the entire defect depth distribution, but also in estimating the time evolution of defect subpopulations that differ in depths.

Additionally, to check the ability of the proposed Markov model to predict the corrosion rate (CR) distribution, an empirical CR distribution was determined using the data from both inspections based on the observed change in depth

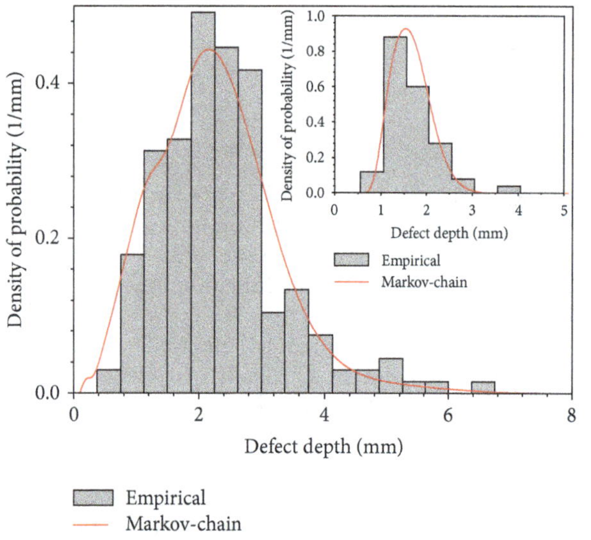

FIGURE 3: Histogram of the 179 matched defect depths, measured in 2006, together with the predicted pit depth distribution. The inset shows the histogram of the observed in the second inspection defect depths matched to 0.9–1.5 mm depths of first inspection, along with the Markov chain-predicted distribution (solid line).

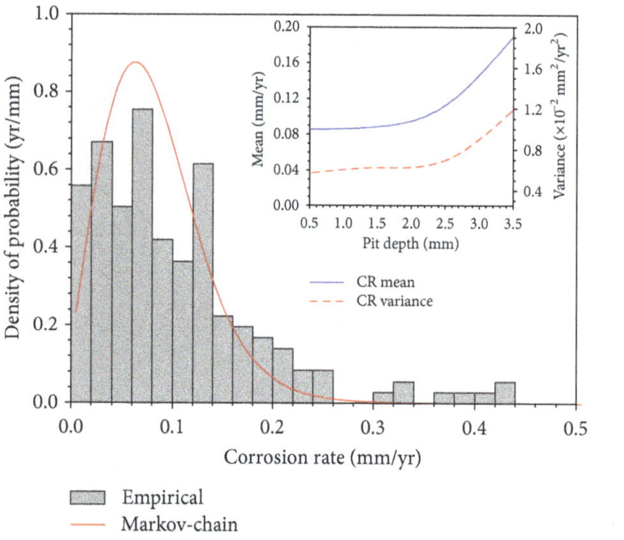

FIGURE 4: Comparison of the empirical corrosion rate (CR) distribution as determined from repeated ILIs (1996 and 2006) with the CR distribution predicted by the model. The inset is a plot of the experimental CR mean and variance against defect depth.

of the matched defects over the interval δt. The empirical rate distribution was then compared with the corrosion rate distribution predicted by the model (19). This comparison can be observed in Figure 4. The two-sample K-S test for 1000 pairs of 50-point CR samples yielded an average P value of 0.27 with only 12% of rejections at 5% significance level. Again, the Markov model satisfactorily reproduces the empirical CR distribution.

The reasons behind the foregoing results lie in the ability of the Markov chain model to capture the influence of both the depth and age of the corrosion defects on the deterministic pit depth growth together with its ability to reproduce the stochastic nature of the process, also as a function of pit depth and age. The inset of Figure 4 is a proof of this last statement. It shows the experimental corrosion rate mean and variance dependence on pit depth. To obtain these results, pit depths were grouped into depth intervals of ten states (1 mm) and the corrosion rate values, derived from the comparison of the two ILIs, were averaged within each group. From this inset, it can be concluded that both the corrosion rate mean and variance increase with defect depth, as was previously noted from different experimental evidence [7, 18]. This result is critical to support the criteria that a good corrosion rate model should consider both the age and depth of the to-be-evolved corrosion defects together with the actual shape of the function describing the observed dependence of the corrosion defect depth with time.

2.2. Markov Chain Modeling of Extreme Corrosion Pit Depths. The novelty in this second stochastic model is the consideration of multiple pits in a given corrosion area (called coupon hereafter). Pit initiation is modeled using a nonhomogeneous Poisson process, while the distribution of pit nucleation times is simulated using a Weibull process. The initiation time of each one of the m pits produced in a coupon is described as the time to the first failure of a part of the system [33]. So, if pit initiation time is understood as the time to first failure (passive layer breakdown or inclusion dissolution) of an individual part [33], one can assume that it follows a Weibull process with distribution

$$F(t) = 1 - \exp\left[-\left(\frac{t}{\varepsilon}\right)^{\xi}\right]. \tag{21}$$

The pit initiation times t_k are computed as the realizations of a Weibull probability distribution described by (21) with both the scale parameter ε and the time t expressed in days.

After a pit has been generated at time t_k, it is assumed that it starts to grow. The time evolution of pit depth $D(t)$ is then modeled using a nonhomogeneous Markov process. The transition probabilities $p_{i,j}$ from state i to state j satisfy the system of forward Kolmogorov's equations given in (1).

If it is assumed that the pitting damage (depth) is in state 1 at the initial time t_k (after initiation), then, for the Markov process defined by the system of (1), the transition probability from the first state to any jth state during the interval (t_k, t), that is, $p_{1,j}(t) = P\{D(t) = j \mid D(t_k) = 1\}$, can be found analytically [28] as

$$p_{1,j}(t) = \exp\left[-\rho(t) + \rho(t_k)\right]\left\{1 - \exp\left[-\rho(t) + \rho(t_k)\right]\right\}^{j-1}, \tag{22}$$

where $\rho(t)$ is defined by (3).

From (22) it follows that, for a single pit, the probability that its depth is equal or less than state i, after a time increment (t_k, t), is

$$F(i,t) = \sum_{j=1}^{i} p_{1,j}(t) = \frac{e^{-\rho(t)+\rho(t_k)}}{1-e^{-\rho(t)+\rho(t_k)}} \sum_{j=1}^{i} \left(1-e^{-\rho(t)+\rho(t_k)}\right)^j, \quad i = 1 \ldots N, \tag{23}$$

where N is the total number of states in the Markov chain.

It is easy to show [15] that expression (23) can be rewritten as

$$F(i,t) = 1 - \left\{1 - \exp\left[-\rho(t) + \rho(t_k)\right]\right\}^i. \tag{24}$$

Equation (24) represents the probability of finding a single pit in a state less than or equal to state i after a time interval (t_k, t). It is worth noting that in (22)–(24) the pit initiation and growth processes are combined by taking into account the pitting initiation time t_k.

As it was stated before, a direct physical meaning can be given to the functions $\lambda(t)$ and $\rho(t)$, which are related to the pit growth rate and pit depth, respectively. Taking into account that the dependence of pit depth on time is a power function (10), the functional dependence of ρ and λ on time was assumed to be

$$\rho(t) = \chi(t)^{\omega}, \tag{25}$$

$$\lambda(t) = \chi\omega(t-t_k)^{\omega-1}, \tag{26}$$

where χ has dimensions of distance over the ωth power of time and ω is less than 1.0.

In order to generalize from the single pit case to the m-pit case, the probability that the maximum damage state is less than or equal than a given value for a time interval is estimated under the assumption that all the pits are described by parent distributions $F(i; t_k, t)$ of the type given in (24). First, let us consider the simplest situation in which it is assumed that all the pits are generated at the same time $t_k = t_{\text{ini}}$. In this case, function (24) represents the cumulative distribution function of a parent population of m pits. To find the extreme depth value distribution, the distribution of (24) should be raised to the mth power [27]. It can be shown [15, 27] that for a large number of pits ($m \to \infty$) and under a suitable variable transformation, it follows that

$$\Phi(i,t) = [F(i,t)]^m \xrightarrow{m\to\infty} G(i,t). \tag{27}$$

Substituting (24) and (25) in expression (27), and after some transformations [15], $G(i,t)$ converges to a Gumbel function given by (2). The involved mathematical formalisms that led to this result can be found in Appendix A of [15].

Summarizing, when the pit initiation times are very short (smaller than the observation time) and/or when it can be considered that they are the same for all the pits, the distribution function for maximum pit depths is obtained by raising (24) to the power of the number of pits in the area of interest. The parameters of the model in this case are the number of pits m and the parameters of the $\rho(t)$ function: χ and ω (see expression (25)).

Consider now an alternative case, when m pits are generated at different times t_k, and expression (24) holds for each one of them. The m cumulative distribution functions, $F_k(i, t-t_k)$, $k = 1, \ldots, m$, must be combined in order to estimate the distribution of the deepest pit, under the assumption that pits nucleate and grow independently. In such a case, the probability that the deepest pit is in a state less than or equal to state i, at time t, can be estimated using the expression

$$\theta(i,t) = \prod_{k=1}^{m} \left\{1 - \left[1 - \exp\left(-\rho(t) + \rho(t_k)\right)\right]^i\right\}. \tag{28}$$

It can be shown that this cumulative function also follows a Gumbel distribution for large m [15].

In (28), expression (25) for $\rho(t)$ must be substituted together with the initiation times t_k, which are assumed to follow a Weibull distribution (21). Therefore, function $\theta(i,t)$ in (28) includes the model parameters ε, ξ, and m to simulate pit initiation, together with parameters χ and ω to model pit growth through the function $\rho(t)$.

At this point, owing to the fact that functions $\Phi(i,t)$ (from (27)) and $\theta(i,t)$ (from (28)) are extreme value distributions of the Gumbel type, it can be stated that function $F(i,t)$ (24) is the parent distribution that lies in the domain of attraction of the Gumbel distribution for maxima [27]. From (24), it can be observed that $F(i,t)$ is of exponential type; therefore, the parent distribution for pit depth extremes is actually an exponential function. This is in complete agreement with the findings in [7], where it was concluded, after measuring and analyzing all the pits in corroded low carbon steel coupons, that the exponential pit depth distribution fitted to the right tail of the pit depth distribution (of the Normal type, adjusted to the depths of the whole population of pits) is actually the extreme's parent distribution, which lies in the domain of attraction of the Gumbel extreme value distribution for maxima.

Following this idea, it is soundly to shift the initial Markov state to the depth value that constitutes the starting point of the exponential distribution tail. If we consider that this exponential distribution starts at some depth value u, then the Markov state i that appears in (24) should be changed to a new variable

$$i' = i + u. \tag{29}$$

Because the pits that contribute to the extreme pit depths values are only those whose depths exceed the threshold u, the number m of pits that should be taken into account when applying (27) or (28) can be equated to the average number of pits whose depths exceed the threshold value u. A more detailed justification of this procedure of variable change for the state i in (24) can be found elsewhere [34].

The empirical average number λ_e of exceedances over threshold u per coupon (to be compared to the model parameter m) can be estimated using (30), which relates λ_e with the threshold value u and the parameters σ and μ of the Gumbel distribution fitted to the observed extreme pit depths. Consider

$$\lambda_e = \exp\left(\frac{\mu - u}{\sigma}\right). \tag{30}$$

This expression is readily obtained from the existing relation between the Gumbel and exponential parameters, as has been shown in [7].

To run the proposed Markov model for extremes, (27) (pits initiate all at the same time) or (28) (pits initiate at different times) is computed and fitted to a Gumbel distribution for several times t (2). The model parameters (three for the instantaneous pit initiation: χ, ω, and m, and five for the generation of pits at different times: ε, ξ, χ, ω, and m) are adjusted considering the experimental distributions for the deepest pits. A series of corrosion experiments is considered, each one of them consisting in the exposition of n_c coupons of a corrodible material to a corrosion environment for a given time. After exposure, the depth of the deepest pit in each coupon is measured [6, 7], and the distribution of maximum depths is fitted with a Gumbel distribution (2). If the experiment is carried out for N_t exposure times, and the observed behavior of depth extremes is to be described with the proposed model, the model parameters are adjusted by minimizing a total error function E_T defined as

$$E_T = \sum_{l=1}^{N_t}\left(\sqrt{\left(\mu_e^l - \mu_p^l\right)^2} + \sqrt{\left(\sigma_e^{2l} - \sigma_p^{2l}\right)^2}\right), \tag{31}$$

where (μ_e^l, σ_e^{2l}) and (μ_p^l, σ_p^{2l}) are the mean and variance values of the lth experimental and estimated extreme value distributions, respectively.

The model parameters are adjusted through minimization of the average value of the error function E_T computed during N_{MC} Monte Carlo simulations, with $N_{MC} = 1000$.

2.2.1. Application of Markov Model for Extremes to Low Carbon Steel Corrosion Experiment. The described Markov model for extreme pit depths was applied to experimental data obtained in pitting corrosion experiments for low-carbon steel reported by Rivas et al. [7]. The details of the experiments can be found elsewhere [7].

In these experiments, groups of 20 coupons of API-5L X52 [32] pipeline steel, with $1 \times 1\ \text{cm}^2$ of exposed area, were immersed in a corroding solution for 1, 3, 7, 15, 21, and 30 days. After the immersion, all the pits with depths greater than 5 μm were measured. The maximum pit depth in each coupon was also recorded. The Gumbel distribution (2) was fitted to the resulting maxima data sets using the Maximum Likelihood Estimator (MLE) [27].

The Gumbel probability density functions fitted to the experimental pit depth maxima for the six exposure times are displayed in Figure 5. They are represented with red solid lines. Table 3 shows the values of the Gumbel location (μ_E) and scale (σ_E) parameters for the fitted distributions.

In their work, Rivas et al. [7] concluded that, as has been previously recognized [4, 35], in low carbon steel pits initiate

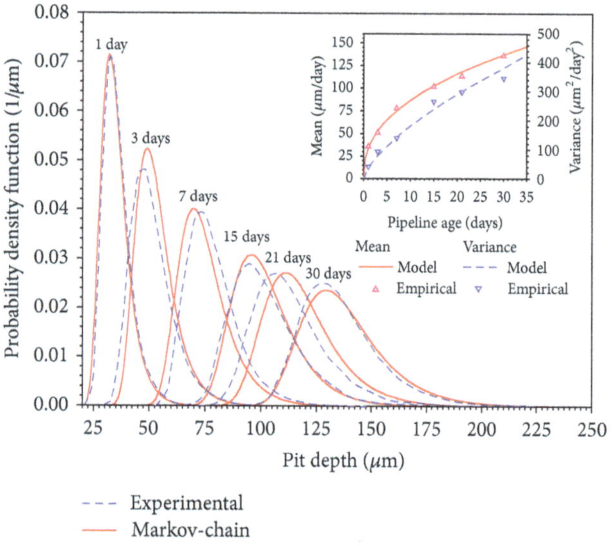

FIGURE 5: Comparison of the Gumbel probability density functions fitted to the experimental pit depth maxima with the model-predicted density functions for the six exposure times.

at the site of MnS inclusions. In the present experiment, the dissolution of these inclusions occurs in a matter of few hours after immersion, and then pits start propagating. The authors also established that the deepest pits of the entire pit population are the oldest ones, meaning that they are the defects that initiate first in time [7]. This conclusion leads to the use of (27) to model the extreme pits using Markov chains. Since the pit initiation time for this experiment has been considered to be very short, the nucleation time t_k in (27) was set to zero.

In order to apply the Markov model, the value of the threshold parameter u of (29) should be determined from the empirical corrosion data. Given that u represents the depth value from which the exponential tail of the whole pit depth distribution starts, it should coincide with the previously determined [7] threshold pit depth value u_P for the tail of the whole pit depth distribution.

When applying the so-called Peak over Threshold (POT) approach in pitting corrosion experiments, Rivas et al. proposed [7] a simplified method for determining the threshold depth value without the need of measuring the entire pit depth population. The method includes a proposition of the *a priori* determination of the threshold pit depth value u_P from the fitted Gumbel distribution to the pit depth extremes. It was suggested to take u_P as a depth value for which the Gumbel distribution shows a cumulative probability P in the range from 0.00005 to 0.005. This proposition was based on the empirical fact that in the analyzed pitting corrosion experiments [7], the Gumbel distribution for maximum depth starts to rise precisely at the beginning of the exponential tail of the normal distribution of the entire pit depth population. Therefore, as the starting point of the Gumbel distribution, it was proposed to use the 0.5%, 0.05%, or even the 0.005% quantile of the experimental pit depth distributions.

Following this suggestion, and taking into consideration the determined values of the location and scale from the

TABLE 3: Estimated location and scale parameters of the Gumbel distributions fitted to the experimental pit depth maxima and to the Markov model. The fourth and fifth columns correspond to the calculated pit depth threshold and average exceedances values, respectively. The last column displays the average P value of 1000 two-sample Kolmogorov-Smirnov tests.

Exp. time (days)	Estimated from the experimental data		$u_{0.005}$[a]	λ_e[b]	Estimated from the modeled distribution		P value
	μ_E (μm)	σ_E (μm)			μ_M (μm)	σ_M (μm)	
1	32.8	5.15	21.2	10	32.0	5.15	0.50
3	47.0	7.58	34.49059	5	49.1	7.02	0.36
7	72.9	9.29	50.20122	12	69.7	9.17	0.31
15	94.9	12.75	70.36299	9	95.6	11.94	0.56
21	105.9	13.52	81.67309	6	111.0	13.52	0.29
30	127.9	14.54	95.65344	9	129.4	15.52	0.53

[a]Estimated from (32).
[b]Estimated from (30).

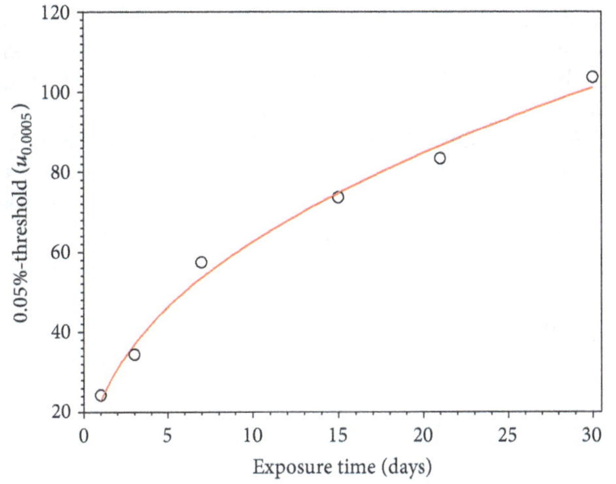

FIGURE 6: Plot of the calculated threshold depth values $u_{0.005}$ versus exposure time. The solid line is a power function fitted to the determined threshold values.

fitted Gumbel distribution, it is possible to determine the X_P quantile, being $P = 0.005, 0.0005$, or 0.00005. To achieve this, the following equation, obtained through the inversion of expression (1), is used:

$$X_P = \mu - \sigma \ln\left[-\ln(P)\right]. \quad (32)$$

Thus, the value of X_P can be used as the threshold u_P ($u_P = X_P$) [7]. For the analyzed experiment we are fixing the value of P to 0.0005. Substituting this value in (32) and using the estimated location (μ_E) and scale (σ_E) parameters for the empirical Gumbel distributions for each one of the exposure times (Table 3), six values of $u_{0.0005}$ are obtained. These values are plotted in Figure 6 against the exposure times. In this figure, a power function fitted to the empirical threshold values is also displayed. From this curve, the values of u to be substituted in (29) are taken. A more detailed proof of the correctness of this choice of u can be found in [34].

Once the value of the threshold is determined, the function $F(i,t)$ of (24) is raised to the mth power, and the resulting function $\Phi(i,t)$ (27) is fitted to a Gumbel distribution (2). The fitted function is compared with the empirical Gumbel distribution in order to adjust the model parameters. For this particular case, the parameters to take into account are the number of pits m, which should approximately match the average experimental number of exceedances over the threshold u per coupon, and the parameters χ and ω, which define the function $\Phi(i,t)$. The model parameters are refined by minimizing the error function of (31).

The minimization process gives an adjusted parameter $m = 7.6$. This value can be approximated to 8 pits. It represents the average number of pits per coupon whose depths exceed the threshold value u. This fact can be confirmed by comparing the adjusted value for m with the values of the empirical exceedances λ_e displayed in the fifth column of Table 3. The exceedances for each exposure period were calculated by substituting the corresponding estimated Gumbel parameters μ_E and σ_E and the threshold value $u_{0.0005}$ (all given in Table 3) into (30). The mean value of λ_e (from Table 3) equals 8, which is in good agreement with the model prediction.

The probability density functions corresponding to the distributions $\Phi(i,t)$ obtained after raising $F(i,t)$ to the power of the adjusted parameter $m = 7.6$ are shown in Figure 5 for the six exposure times. In this figure, the modeled distributions can be compared with the empirical Gumbel distributions. One can see a good agreement between the modeled and experimental Gumbel distributions. Even for this case, in which the number of pits m is small (see Section 2.1 and (27)), both the experimental and modeled distributions agree with the Gumbel model for maxima.

The Gumbel distribution (2) was also fitted to the functions predicted by the model using the Maximum Likelihood Estimator (MLE) [27]. In Table 3, the estimated parameters μ_M and σ_M for the fitted Gumbel distributions (to the Markov-predicted functions) are displayed together with the corresponding parameters of the empirical Gumbel distributions. The differences between the empirical and modeled parameters do not exceed 8%. The last column of Table 3 shows the average P values of the two-sample K-S test performed on 1000 pairs of 50-depth-point samples (one for the empirical and one for the Markov-modeled distributions) for the six exposure periods. The smaller P value among them is 0.29, which leads to not reject the null hypothesis that the two samples belong to the same distribution.

The good agreement between the modeled and the empirical Gumbel distributions can also be established from the inset of Figure 5, where the time evolution of the Gumbel mean and variance for the observed data and for the result of the proposed model are compared. These results demonstrate the suitability of the proposed model to describe the initiation and evolution of the maximum pit depths in a pitting corrosion experiment, which is of great importance in many applications such as reliability assessment and risk management.

The advantages of the proposed model compared with previous Markov models (developed by other authors) can also be established. The details regarding this topic can be found in [15].

3. Concluding Remarks

Two Markov chain models have been presented to simulate pitting corrosion. They have been developed and validated using experimental pitting corrosion data. Both models are attractive due to the existence of analytical solutions of the system of Kolmogorov's forward equations.

The first model describes the time evolution of pit depths that correspond to the general population of defects in underground pipelines. It has been developed using a continuous-time, nonhomogenous linear growth (pure birth) Markov process, under the assumption that the Markov-chain-derived stochastic mean of the pitting damage equals the deterministic, empirical mean of the defect depths. Such an assumption allows the transition probability function to be identified only from the pitting starting time and exponent parameter. Moreover, this assumption requires that the functional form of the stochastic and deterministic instantaneous pitting rates also agree. This supports the idea that the intensities of the transitions in the Markov process are closely related to the pitting damage rate. This model permits predicting the time evolution of pit depth distribution, which is of paramount importance for reliability estimations. It is also able to capture the dependence of the pitting rate on the pit depth and lifetime. This is an advantage of the Markov chain approach over deterministic and other stochastic models for pitting corrosion. It could also be extended to investigate pitting corrosion in environments other than soils, for example, in laboratory experiments.

The second Markov chain pitting corrosion model gives account for the maximum pit depths. In it, pitting corrosion is modeled as the combination of two independent nonhomogeneous in time physical processes, one for pit initiation and one for pit growth. Both processes are combined using well-suited physical and statistical methodologies, such as extreme value statistics, to produce a unified stochastic model of pitting corrosion.

Pit initiation is described by means of a non-homogeneous Poisson process so that a set of multiple pit nucleation events can be modeled as a Weibull process. Pit growth is modeled as a nonhomogeneous Markov process. Given that the intensity of the process is related to the corrosion rate, its functional form can be proposed from the results of the experimental tests.

Taking into account the experimental evidence that the pit depth parent distribution that leads to the distribution for extremes is the exponential tail of the pit depth population distribution, the solution of the Markov chain is shifted to the threshold pit depth value, where the exponential tail of the pit depth distribution starts. The threshold value is determined from the pit depth value from which the empirical Gumbel distribution for maximum pit depth becomes significant.

Extreme value statistics has been used to show the accuracy of the model describing the experimental observations. The model is capable of predicting not only the correct Gumbel distributions for pitting corrosion maxima in low-carbon steel, but also of estimating the number of extreme pits that has physical sense and that matches the experimental findings.

In order to simulate the whole pitting process, five model parameters are necessary. Two parameters are required to simulate pit initiation as a Weibull process; another two parameters are required to simulate pit growth with the nonhomogeneous Markov process; finally, the number of pits is necessary to combine these two processes. If the assumption that all pits nucleate instantaneously holds, as is the case of the presented experimental example, only three parameters will be necessary to fit the model to the experimental data.

Given the fact that the model parameters and assumptions do not depend on the corroded material, nor on the corrosion environment, the model is suited for different corrosion systems. The model can be easily adapted to describe situations in which the distribution of pit initiation times and the functional form of the time dependence for pit growth differ from those considered in this work.

References

[1] G. S. Chen, K. C. Wan, G. Gao, R. P. Wei, and T. H. Flournoy, "Transition from pitting to fatigue crack growth—modeling of corrosion fatigue crack nucleation in a 2024-T3 aluminum alloy," *Materials Science and Engineering A*, vol. 219, no. 1-2, pp. 126–132, 1996.

[2] M. Nessin, "Estimating the risk of pipeline failure due to corrosion," in *Uhlig's Corrosion Handbook*, W. Revie, Ed., p. 85, John Wiley & Sons, 2nd edition, 2000.

[3] T. Shibata and T. Takeyama, "Pitting corrosion as a stochastic process," *Nature*, vol. 260, no. 5549, pp. 315–316, 1976.

[4] G. S. Frankel, "Pitting corrosion of metals: a review of the critical factors," *Journal of the Electrochemical Society*, vol. 145, no. 6, pp. 2186–2198, 1998.

[5] G. T. Burstein, C. Liu, R. M. Souto, and S. P. Vines, "Origins of pitting corrosion," *Corrosion Engineering Science and Technology*, vol. 39, no. 1, pp. 25–30, 2004.

[6] P. M. Aziz, "Application of the statistical theory of extreme values to the analysis of maximum pit depth data for aluminum," *Corrosion*, vol. 12, pp. 495t–506t, 1956.

[7] D. Rivas, F. Caleyo, A. Valor, and J. M. Hallen, "Extreme value analysis applied to pitting corrosion experiments in low carbon steel: comparison of block maxima and peak over threshold approaches," *Corrosion Science*, vol. 50, no. 11, pp. 3193–3204, 2008.

[8] J. C. Velázquez, F. Caleyo, A. Valor, and J. M. Hallen, "Predictive model for pitting corrosion in buried oil and gas pipelines," *Corrosion*, vol. 65, no. 5, pp. 332–342, 2009.

[9] F. Caleyo, J. C. Velázquez, A. Valor, and J. M. Hallen, "Probability distribution of pitting corrosion depth and rate in underground pipelines: A Monte Carlo Study," *Corrosion Science*, vol. 51, no. 9, pp. 1925–1934, 2009.

[10] Z. Szklarska-Smialowska, "Pitting corrosion of aluminum," *Corrosion Science*, vol. 41, no. 9, pp. 1743–1767, 1999.

[11] W. Zhang and G. S. Frankel, "Localized corrosion growth kinetics in AA2024 alloys," *Journal of the Electrochemical Society*, vol. 149, no. 11, pp. B510–B519, 2002.

[12] F. Hunkeler and H. Bohni, "Mechanism of pit growth on aluminum under open circuit conditions," *Corrosion*, vol. 40, no. 10, pp. 534–540, 1984.

[13] T. Shibata, "1996 W.R. Whitney Award Lecture: statistical and stochastic approaches to localized corrosion," *Corrosion*, vol. 52, no. 11, pp. 813–830, 1996.

[14] J. W. Provan and E. S. Rodriguez, "Part I: Development of a Markov description of pitting corrosion," *Corrosion*, vol. 45, no. 3, pp. 178–192, 1989.

[15] A. Valor, F. Caleyo, L. Alfonso, D. Rivas, and J. M. Hallen, "Stochastic modeling of pitting corrosion: a new model for initiation and growth of multiple corrosion pits," *Corrosion Science*, vol. 49, no. 2, pp. 559–579, 2007.

[16] T. B. Morrison and R. G. Worthingham, "Reliability of high pressure line pipe under external corrosion," in *Proceedings of the 11th ASME International Conference Offshore Mechanics and Arctic Engineering*, vol. 5, part B, pp. 401–408, Calgary, Canada, June 1992.

[17] H. P. Hong, "Inspection and maintenance planning of pipeline under external corrosion considering generation of new defects," *Structural Safety*, vol. 21, no. 3, pp. 203–222, 1999.

[18] F. Caleyo, J. C. Velázquez, A. Valor, and J. M. Hallen, "Markov chain modelling of pitting corrosion in underground pipelines," *Corrosion Science*, vol. 51, no. 9, pp. 2197–2207, 2009.

[19] F. Bolzoni, P. Fassina, G. Fumagalli, L. Lazzari, and E. Mazzola, "Application of probabilistic models to localised corrosion study," *Metallurgia Italiana*, vol. 98, no. 6, pp. 9–15, 2006.

[20] S. A. Timashev, M. G. Malyukova, L. V. Poluian, and A. V. Bushiskaya, "Markov description of corrosion defects growth and its application to reliability based inspection and maintenance of pipelines," in *Proceedings of the 7th ASME International Pipeline Conference (IPC '08)*, Calgary, Canada, September 2008, Paper IPC2008-64546.

[21] T. B. Morrison and G. Desjardin, "Determination of corrosion rates from single inline inspection of a pipeline," in *Proceedings of the NACE One Day Seminar*, Houston, Tex, USA, December 1998.

[22] Y. Katano, K. Miyata, H. Shimizu, and T. Isogai, "Predictive model for pit growth on underground pipes," *Corrosion*, vol. 59, no. 2, pp. 155–161, 2003.

[23] A. K. Sheikh, J. K. Boah, and D. A. Hansen, "Statistical modelling of pitting corrosion and pipeline reliability," *Corrosion*, vol. 46, no. 3, pp. 190–196, 1990.

[24] M. Kowaka, Ed., *Introduction to Life Prediction of Industrial Plant Materials: Application of the Extreme Value Statistical Method for Corrosion Analysis*, Allerton Press, New York, NY, USA, 1994.

[25] A. Valor, D. Rivas, F. Caleyo, and J. M. Hallen, "Discussion: statistical characterization of pitting corrosion—Part 1: Data analysis and part 2: Probabilistic modeling for maximum pit depth," *Corrosion*, vol. 63, no. 2, pp. 107–113, 2007.

[26] E. J. Gumbel, *Statistics of Extremes*, Columbia University Press, New York, NY, USA, 2004.

[27] E. Castillo, *Extreme Value Theory in Engineering*, Academic Press, San Diego, Calif, USA, 1988.

[28] E. Parzen, *Stochastic Processes*, vol. 24, SIAM, Philadelphia, Pa, USA, 1999.

[29] D. R. Cox and H. D. Miller, *The Theory of Stochastic Processes*, Chapman & Hall/CRC, Boca Raton, Fla, USA, 1st edition, 1965.

[30] S. Coles, *An Introduction to Statistical Modeling of Extreme Values*, Springer Series in Statistics, Springer, London, UK, 2001.

[31] F. Caleyo, L. Alfonso, J. H. Espina-Hernández, and J. M. Hallen, "Criteria for performance assessment and calibration of in-line inspections of oil and gas pipelines," *Measurement Science and Technology*, vol. 18, no. 7, pp. 1787–1799, 2007.

[32] *Specification for Line Pipe, API Specification 5L*, American Petroleum Institute, Washington, DC, USA, 42nd edition, 2001.

[33] H. Ascher and H. Feingold, *Repairable Systems Reliability. Modeling, Inference, Misconceptions and Their Causes*, vol. 7, Marcel Dekker, New York, NY, USA, 1984.

[34] A. Valor, F. Caleyo, D. Rivas, and J. M. Hallen, "Stochastic approach to pitting-corrosion-extreme modelling in low-carbon steel," *Corrosion Science*, vol. 52, no. 3, pp. 910–915, 2010.

[35] G. Wranglen, "Pitting and sulphide inclusions in steel," *Corrosion Science*, vol. 14, no. 5, pp. 331–349, 1974.

8

A Mollification Regularization Method for a Fractional-Diffusion Inverse Heat Conduction Problem

Zhi-Liang Deng,[1] Xiao-Mei Yang,[2] and Xiao-Li Feng[3]

[1] *School of Mathematical Sciences, University of Electronic Science and Technology of China, Chengdu 610054, China*
[2] *School of Mathematics, Southwest Jiaotong University, Chengdu 610031, China*
[3] *Department of Mathematical Sciences, Xidian University, Xi'an 710071, China*

Correspondence should be addressed to Zhi-Liang Deng; zhiliangdeng@gmail.com

Academic Editor: Fatih Yaman

The ill-posed problem of attempting to recover the temperature functions from one measured transient data temperature at some interior point of a one-dimensional semi-infinite conductor when the governing linear diffusion equation is of fractional type is discussed. A simple regularization method based on Dirichlet kernel mollification techniques is introduced. We also propose *a priori* and *a posteriori* parameter choice rules and get the corresponding error estimate between the exact solution and its regularized approximation. Moreover, a numerical example is provided to verify our theoretical results.

1. Introduction

It is well known that the classical diffusion equation requires second spatial derivatives and first time derivatives. However, people are shifting their partial focus to fractional-order differential equations with the realization that the use of fractional-order derivatives and integrals leads to formulas of certain physical processes which is more economical and useful than the classical approach in terms of Fick's laws of diffusion [1–4]. Some time fractional diffusion equations involving only a first-order spatial derivative and a half-order time derivative are successfully used for modeling some anomalous diffusion physical phenomena (see, e.g., [5–7] and the references therein). One can also find some applications of such fractional diffusion equations in several scenarios in [2, 7], such as relaxation to equilibrium in systems (such as polymers chains and membranes) with long temporal memory, anomalous transport in disordered systems, diffusion on fractals, and to model non-Markovian dynamical processes in protein folding. Fractional diffusion equations have been extensively investigated both in theory itself and numerical computation based on the broad applications in many application fields, especially in describing phenomena related to anomalous diffusion processes. The following is a partial list of articles which contain theoretical results and numerical tests. Some fundamental solutions and Green functions of fractional differential equations are given in [8, 9]. Finite difference techniques [10–12] and finite element methods [13, 14] provide some efficiency numerical results for several kinds of fractional differential equations. In [15], the decomposition method is used to construct analytical approximate solutions of time-fractional wave equation subject to specified boundary conditions. An analytical solution of a fractional diffusion equation by Adomian decomposition method is presented in [16]. One can also refer [17–19] to the regularity of the solution of fractional diffusion equations and some *a priori* estimates.

In this paper, we consider the following one-dimensional fractional diffusion problem on a semi-infinite slab: suppose the temperature $f(t)$ at some interior point $x = x_0 > 0$ is approximately measurable. For convenience, we set $x_0 = 1$. The temperature $u(x,t)$ at the $0 < x < 1$ is desired and

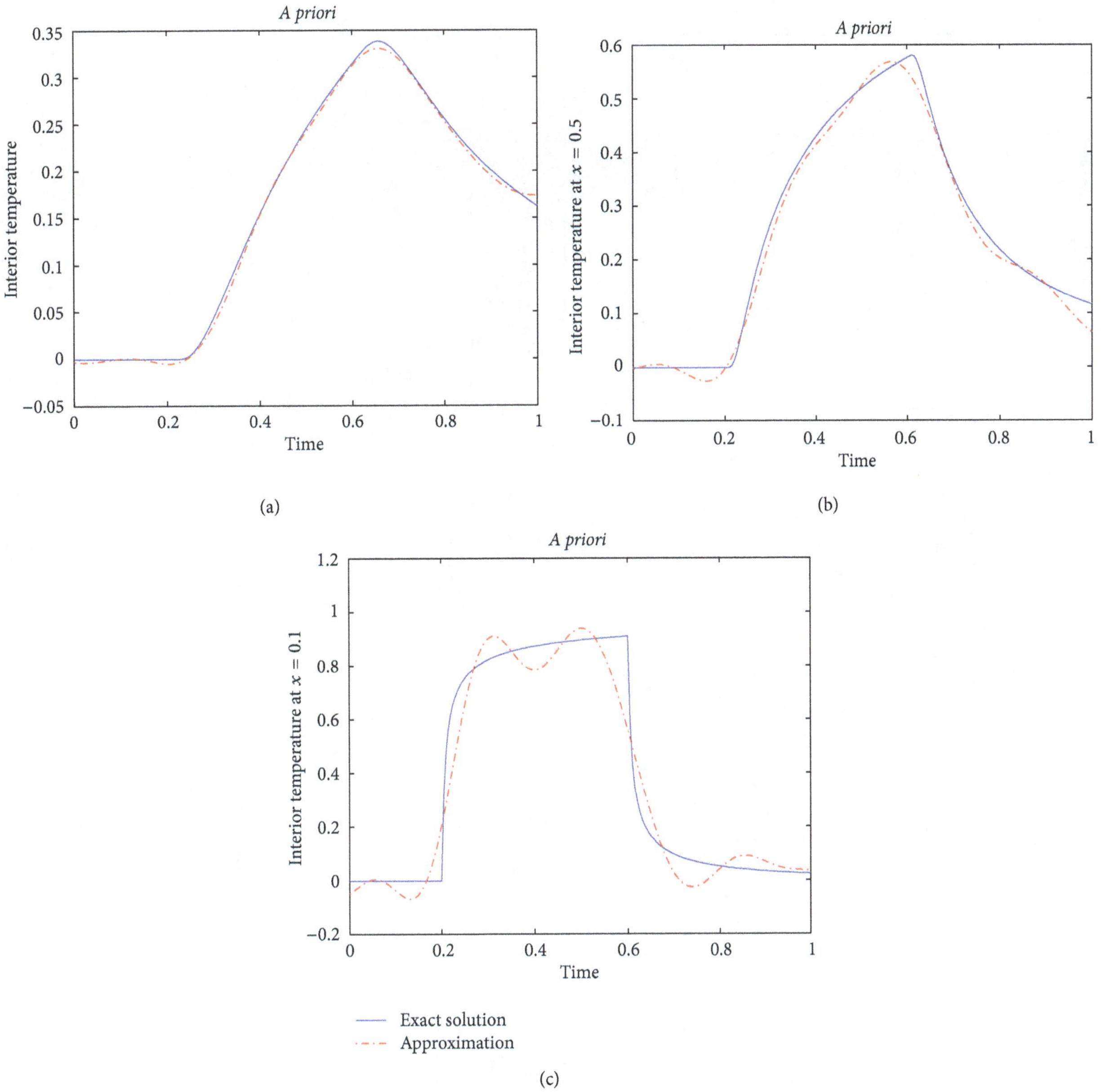

FIGURE 1: The comparisons of exact solution and regularization solution at (a) $x = 0.9$, (b) $x = 0.5$, (c) $x = 0.1$, $\nu = 34.3956$, $\delta = 0.01$.

unknown. The mathematical description of the fractional diffusion problem is listed in the following. The unknown temperature $u(x,t)$ satisfies

$$\frac{\partial}{\partial x}u(x,t) = -\frac{1}{\sqrt{\kappa}}\frac{\partial^{1/2}}{\partial t^{1/2}}u(x,t) + \frac{u_0}{\sqrt{\pi\kappa t}}, \quad x > 0, \ t \geq 0,$$
$$u(1,t) = f(t), \tag{1}$$

where κ is the constant diffusivity coefficient, $u_0 = u(x,0)$, a constant. The half-time differentiation $(\partial^{1/2}/\partial t^{1/2})u(x,t)$ indicates the Riemann-Liouville fractional derivative with order $\alpha = 1/2$ which is given by the convolution integral

$$(\partial^{\alpha}u)(t) = \frac{1}{\Gamma(1-\alpha)}\frac{d}{dt}\int_0^t u(s)(t-s)^{-\alpha}ds, \quad 0 \leq t \leq T, \ 0 < \alpha < 1, \tag{2}$$

where $\Gamma(\cdot)$ is the Gamma function. More detailed information on fractional derivatives and a general historical perspective may be found in [7, 20, 21]. The situations we will treat are those in which the system is initially at equilibrium, so that

$$u(x,t) = u_0, \quad t < 0, \ x \geq 0. \tag{3}$$

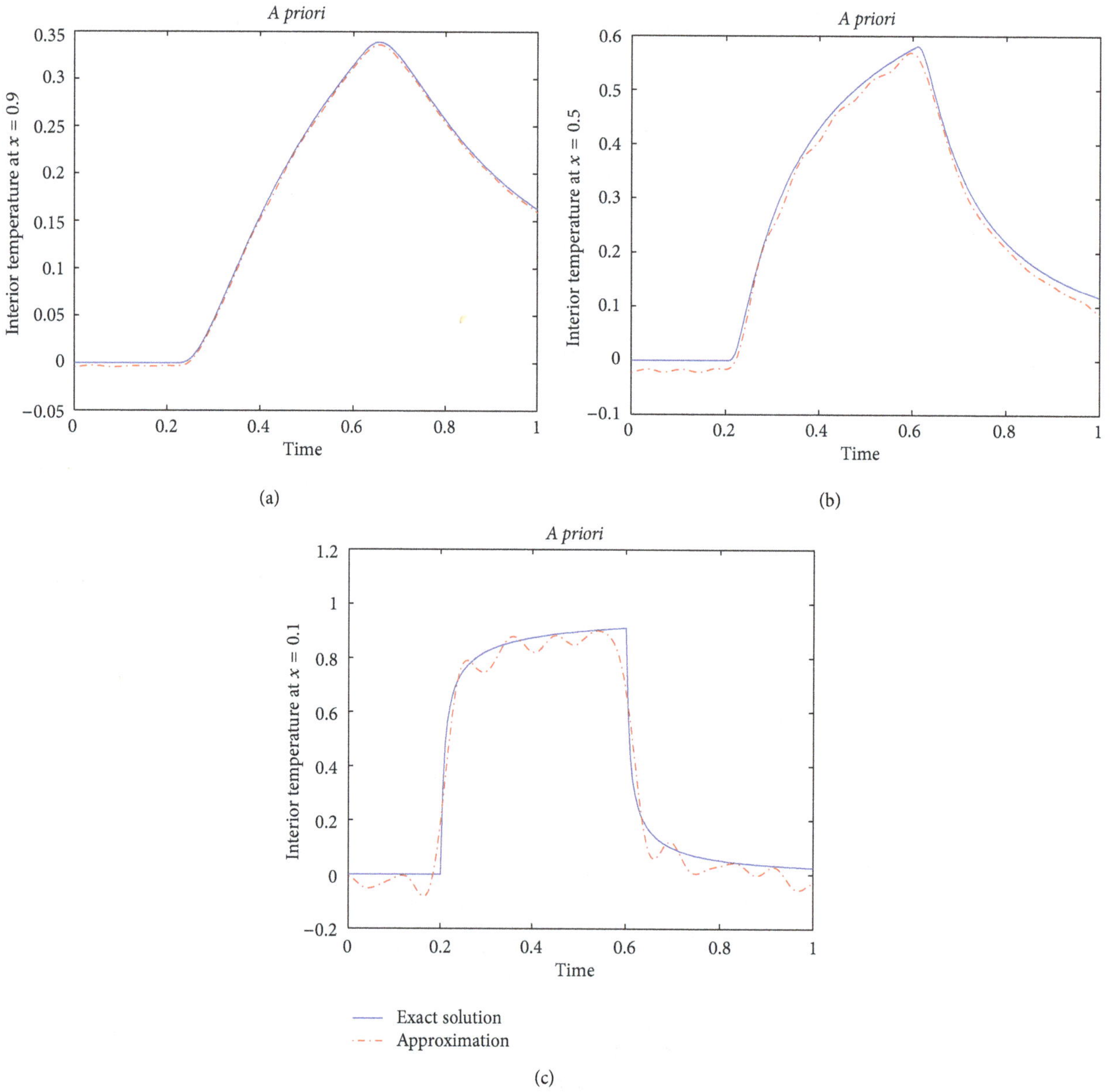

FIGURE 2: The comparisons of exact solution and regularization solution at (a) $x = 0.9$, (b) $x = 0.5$, (c) $x = 0.1$, $\nu = 83.1949$, $\delta = 0.001$.

At $t = 0$ a perturbation of the system commences by some unspecified process occurring at the boundary. During times of interest T, this perturbation does not affect regions remote from the $x = 0$ boundary, so that the relationship

$$\lim_{x\to+\infty} u(x,t) = u_0, \quad t \le T \tag{4}$$

applies. In [22], Murio discussed the similar problems using mollification method with Gauss kernel. The idea used in this current work is a development of the ideas in [22]. However, there is only formal stability and discrete error discussion in the cited paper. Here, we give some analysis of error estimates under *a priori* and *a posteriori* regularization parameter, and a comparison of those two choice methods in our regularization method. In the following, the regularization method is outlined.

In order to simplify the Fourier analysis of fractional diffusion problem (1), and in the rest of the paper, we assume without loss of generality, $\kappa = 1$, $u_0 = 0$. We also assume that all the functions involved are $L^2(\mathbb{R})$ and use the corresponding L^2 norm, as defined in the following, to measure errors:

$$\|y\| = \left(\int_{\mathbb{R}} |y(t)|^2 dt\right)^{1/2}. \tag{5}$$

If the Fourier transform of a function $y(t)$ is written as

$$\hat{y}(\xi) = \frac{1}{\sqrt{2\pi}} \int_{-\infty}^{+\infty} y(t)\, e^{-i\xi t} dt, \quad i = \sqrt{-1}, \tag{6}$$

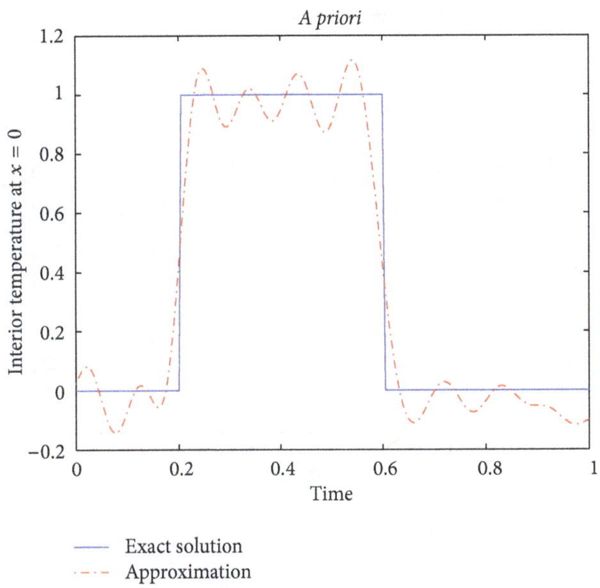

FIGURE 3: The comparisons of exact solution and regularization solution at $x = 0$, $\nu = 83.1949$, $\delta = 0.001$.

then the Fourier inversion formula reads

$$y(t) = \frac{1}{\sqrt{2\pi}} \int_{-\infty}^{+\infty} \hat{y}(\xi) e^{i\xi t} d\xi. \quad (7)$$

Fourier transforming (1) with respect to variable t, it follows that

$$\frac{\partial}{\partial x} \hat{u}(x,\xi) = -\sqrt{i\xi}\hat{u}(x,\xi), \quad x > 0, \ \xi \in \mathbb{R} \quad (8)$$

according to the fact that [7]

$$\widehat{\frac{\partial^{1/2}}{\partial t^{1/2}} h(t)} = \sqrt{i\xi}\hat{h}(\xi). \quad (9)$$

The first-order ordinary differential equation (8) has the general solution

$$\hat{u}(x,\xi) = \hat{u}(1,\xi) e^{\sqrt{i\xi}(1-x)} = \hat{f}(\xi) e^{\sqrt{i\xi}(1-x)}, \quad (10)$$

or equivalently,

$$e^{-\sqrt{i\xi}(1-x)} \hat{u}(x,\xi) = \hat{f}(\xi). \quad (11)$$

Now, we can write the problem (1) as

$$\widehat{A}(\xi,x) \hat{u}(x,\xi) = \hat{f}(\xi), \quad (12)$$

where $\widehat{A}(\xi,x) := e^{-\sqrt{i\xi}(1-x)}$ is a multiplication operator.

Since we require $\hat{u}(x,\cdot) \in L^2(\mathbb{R})$ and $u_0 = 0$, we see from (4) and (10) that the real part of $\sqrt{i\xi}$ is positive. Thereby, the multiplication operator $\widehat{A}(\xi,x)$ is a bounded linear operator for $0 \le x < 1$. Then its inverse operator is unbounded. Therefore problem (12) is linear ill posed. For the general theory of linear ill posed problems, we can refer to [23]. It is worth pointing out that ill posed problems of a large number of diffusion equations, both fractional-order as well as integral order, have been discussed by many authors. Yang et al. [24–28] discuss the identification of source terms for some integral-order diffusion equations using some regularization strategies. Hon et al. [29, 30] apply some meshless methods to the ill posed problems of heat conduction equations. In [17, 22, 29, 31, 32], some uniqueness results and numerical methods are given for some fractional diffusion ill posed problems. Here, we apply a simple stabilizing method, namely, the mollification method with Dirichlet kernel [33], to stabilize the problem (1). Suppose that the measured data function $f_\delta(t)$ satisfies

$$\|f - f_\delta\| \le \delta, \quad (13)$$

where δ is noise level. Take the Dirichlet function

$$D_\nu(t) := \sqrt{\frac{2}{\pi}} \frac{\sin \nu t}{t} \quad (14)$$

as the mollifier kernel, where ν is a positive constant. Define operator M_ν as

$$M_\nu u(x,t) := \int_{-\infty}^{+\infty} D_\nu(\tau) u(x, t-\tau) d\tau. \quad (15)$$

Then, we have the following associated problem: for some $\delta > 0$, find $M_\nu u_\delta(x,t)$ that satisfies

$$\begin{aligned} \frac{\partial}{\partial x} M_\nu u_\delta(x,t) &= -\frac{\partial^{1/2}}{\partial t^{1/2}} M_\nu u_\delta(x,t), \quad x > 0, \ t \ge 0, \\ M_\nu u_\delta(1,t) &= M_\nu f_\delta(t). \end{aligned} \quad (16)$$

In fact, ν plays the role of regularization parameter.

In Section 2 an *a priori* parameter choice rule and the corresponding error estimate are discussed. In Section 3, we propose an *a posteriori* parameter choice rule and get the error bound. Numerical tests are given in Section 4.

2. The Error Estimate with *a Priori* Parameter Choice

In this section, the error estimate of the mollification regularization method will be derived under the *a priori* parameter choice rule. Suppose that the following source condition holds:

$$\|u(0,\cdot)\| \le E. \quad (17)$$

The source condition (17) is indispensable, otherwise, there can be no uniform convergence rates for any regularization method, that is, the convergence is arbitrarily slow

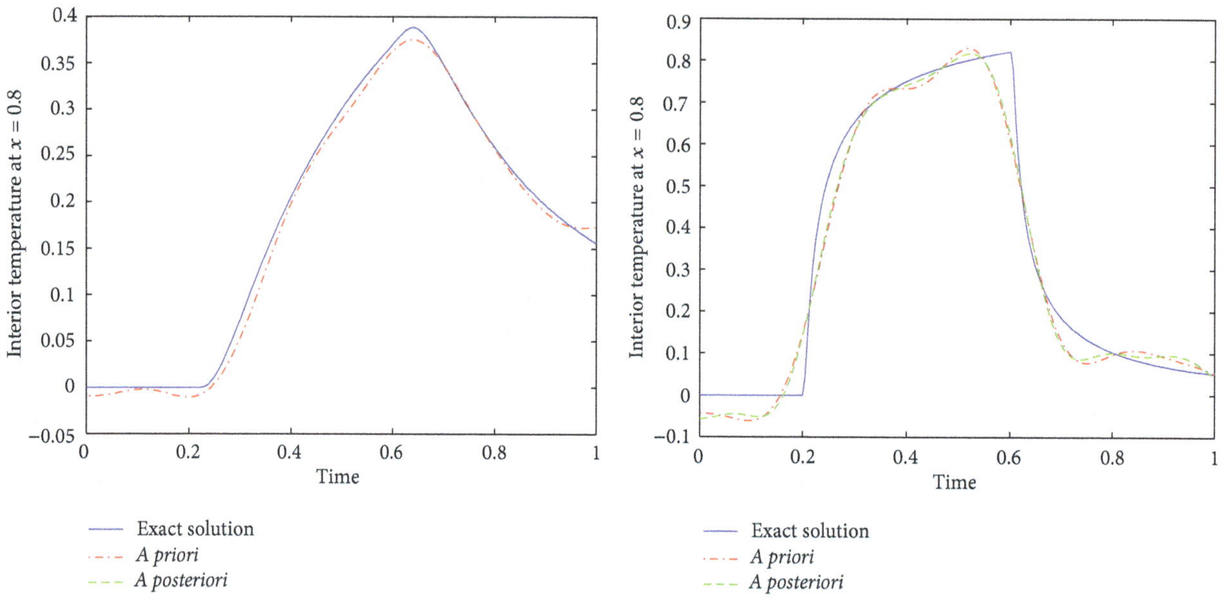

(a) (b)

FIGURE 4: The comparisons of exact solution and regularization solution at (a) $x = 0.8$, (b) $x = 0.2$ with $\delta = 0.01$.

(see Proposition 3.11 in [23]). For $0 < x < 1$, by Parseval formula and triangle inequality, we know that

$$\begin{aligned}\|M_\nu u_\delta(x,\cdot)-u(x,\cdot)\| &= \|\widehat{M_\nu u_\delta}(x,\cdot)-\hat{u}(x,\cdot)\| \\ &= \left\|e^{(1-x)\sqrt{i\xi}}\widehat{D_\nu}\widehat{f_\delta}(\xi)-e^{(1-x)\sqrt{i\xi}}\hat{f}(\xi)\right\| \\ &\le \left\|e^{(1-x)\sqrt{i\xi}}\widehat{D_\nu}\widehat{f_\delta}(\xi)-e^{(1-x)\sqrt{i\xi}}\widehat{D_\nu}\hat{f}(\xi)\right\| \\ &\quad+\left\|e^{(1-x)\sqrt{i\xi}}\widehat{D_\nu}\hat{f}(\xi)-e^{(1-x)\sqrt{i\xi}}\hat{f}(\xi)\right\| \\ &:= I_1+I_2.\end{aligned}\tag{18}$$

Since $|e^{\sqrt{i\xi}}| = e^{\sqrt{|\xi|/2}}$, for I_1, we get by (13)

$$\begin{aligned}I_1 &= \left\|e^{(1-x)\sqrt{i\xi}}\widehat{D_\nu}\widehat{f_\delta}(\xi)-e^{(1-x)\sqrt{i\xi}}\widehat{D_\nu}\hat{f}(\xi)\right\| \\ &= \left(\int_{-\nu}^{\nu}\left|e^{(1-x)\sqrt{i\xi}}\left(\widehat{f_\delta}-\hat{f}\right)\right|^2 d\xi\right)^{1/2} \\ &\le e^{(1-x)\sqrt{\nu/2}}\delta.\end{aligned}\tag{19}$$

For I_2, we use the source condition (17) and obtain

$$\begin{aligned}I_2 &= \left\|e^{(1-x)\sqrt{i\xi}}\widehat{D_\nu}\hat{f}(\xi)-e^{(1-x)\sqrt{i\xi}}\hat{f}(\xi)\right\| \\ &= \left(\int_{|\xi|\ge\nu}\left|e^{(1-x)\sqrt{i\xi}}\hat{f}(\xi)\right|^2 d\xi\right)^{1/2} \\ &= \left(\int_{|\xi|\ge\nu}\left|e^{(1-x)\sqrt{i\xi}}e^{-\sqrt{i\xi}}\hat{u}(0,\xi)\right|^2 d\xi\right)^{1/2} \\ &\le e^{-\sqrt{\nu/2}x}E.\end{aligned}\tag{20}$$

Therefore,

$$\|M_\nu u_\delta(x,\cdot)-u(x,\cdot)\| \le e^{(1-x)\sqrt{\nu/2}}\delta + e^{-\sqrt{\nu/2}x}E. \tag{21}$$

Minimizing the right-hand side of (21), we set $e^{(1-x)\sqrt{\nu/2}}\delta = e^{-\sqrt{\nu/2}x}E$ and get

$$\nu = 2\left(\log\frac{E}{\delta}\right)^2, \tag{22}$$

which also implies

$$\|M_\nu u_\delta(x,\cdot)-u(x,\cdot)\| \le 2E^{1-x}\delta^x, \quad \text{for } 0<x<1. \tag{23}$$

Therefore, we get the following theorem.

Theorem 1. *Assume conditions* (13), (17) *hold. If the regularization parameter* ν *is taken by* (22), *then for* $0 < x < 1$, *there holds the error estimate* (23).

Remark 2. The error estimate (23) is order optimal in the sense of Tautenhahn [34]. In our application $\|u(0,\cdot)\|$ is usually not known, therefore we have no exact *a priori* bound E and cannot choose the parameter ν according to (22). However, if selecting $\nu = 2(\log(1/\delta))^2$, we can obtain the convergence rate

$$\left\|M_\nu u_\delta(x,\cdot) - u(x,\cdot)\right\| \le 2\delta^x, \quad \text{for } 0 < x < 1. \tag{24}$$

Theorem 1 provides no information about the convergence and convergence rates of $M_\nu u_\delta(x,\cdot)$ at $x = 0$. The question is settled by our next result. We now give the error estimate at $x = 0$ under a stronger *a priori* assumption

$$\|u(0,\cdot)\|_p \le E, \quad p > 0, \tag{25}$$

where $\|\cdot\|_p$ denotes the norm on Sobolev space $H_p(\mathbb{R})$ defined by

$$\|f\|_p^2 := \int_{-\infty}^{\infty} \left(1+\xi^2\right)^p \left|\hat{f}(\xi)\right|^2 d\xi \tag{26}$$

for $f \in H^p(\mathbb{R})$. We only need to reestimate I_2 for $x = 0$. Under the stronger *a priori* bound (25), it is not hard to get

$$\begin{aligned} I_2 &= \left(\int_{|\xi|\ge\nu} \left|\hat{u}(0,\xi)\right|^2 d\xi\right)^{1/2} \\ &= \left(\int_{|\xi|\ge\nu} \left(1+\xi^2\right)^{-p}\left(1+\xi^2\right)^p \left|\hat{u}(0,\xi)\right|^2 d\xi\right)^{1/2} \\ &\le \frac{1}{(1+\nu^2)^p} E \le \frac{E}{\nu^{2p}}. \end{aligned} \tag{27}$$

Then, (18), (19), and (27) lead to the following error bound:

$$\left\|M_\nu u_\delta(0,\cdot) - u(0,\cdot)\right\| \le e^{\sqrt{\nu/2}}\delta + \frac{E}{\nu^{2p}}. \tag{28}$$

The error bound (28) does not provide the convergence as $\delta \to 0$ obviously. Hence, we need to choose a proper parameter ν. For this reason, some proper lower bound of the right-hand side of (28) should be given. In order to minimize the right-hand side of (28), it is necessary to introduce the following lemma and its proof can be found in [34].

Lemma 3 (see [34]). *Let the function $f(\lambda) : (0,a] \to \mathbb{R}$ be given by*

$$f(\lambda) = \lambda^b \left[d \log \frac{1}{\lambda}\right]^{-c} \tag{29}$$

with a constant $c \in \mathbb{R}$ and positive constants $a < 1$, b and d, then for the inverse function $f^{-1}(\lambda)$ one has

$$f^{-1}(\lambda) = \lambda^{1/b}\left[\frac{d}{b}\log\frac{1}{\lambda}\right]^{c/b}(1+o(1)) \quad \textit{for } \lambda \longrightarrow 0. \tag{30}$$

Minimizing the right-hand side of (28), we let $e^{\sqrt{\nu/2}}\delta = E/\nu^{2p}$. Denote $e^{-\sqrt{\nu/2}} := \lambda$. Simple computation shows $\lambda(\log(1/\lambda))^{-4p} = \delta 2^{2p}/E$. By using (30) in Lemma 3, we obtain

$$\lambda = \frac{\delta 2^{2p}}{E}\left(\log\frac{E}{\delta 2^{2p}}\right)^{4p}(1+o(1)), \quad \text{for } \delta \longrightarrow 0, \tag{31}$$

which also implies that

$$\nu = 2\log^2\left(\frac{\delta 2^{2p}}{E}\left(\log\frac{E}{\delta 2^{2p}}\right)^{4p}\right)(1+o(1)), \quad \text{for } \delta \longrightarrow 0. \tag{32}$$

Therefore, the following estimate holds

$$\begin{aligned} &\left\|M_\nu u_\delta(0,\cdot) - u(0,\cdot)\right\| \\ &\quad \le CE\left(2\log^2\left(\frac{\delta 2^{2p}}{E}\left(\log\frac{E}{\delta 2^{2p}}\right)^{4p}\right)\right)^{(-2p)}, \end{aligned} \tag{33}$$

where C is a constant, which also shows that the convergence rate at $x = 0$ is logarithmic.

Theorem 4. *Assume conditions* (13), (25) *hold. If the regularization parameter ν is taken as given by* (32)*, then the error estimate* (33) *holds.*

Remark 5. From the error estimate (23), as $x \to 0$, we see that the accuracy of regularization solution becomes progressively lower and even cannot get convergence for $x = 0$. This is common in the theory of ill posed problems. Nevertheless, if a stronger *a priori* assumption (25) is imposed, the regularization solution converges to the exact solution at $x = 0$, but only in a slower way.

3. The Error Estimate with *a Posteriori* Parameter Choice

In this section, we consider the *a posteriori* regularization parameter choice rule. Choose the regularization parameter ν as the solution of the equation

$$d(\nu) := \left\|\widehat{D_\nu}\widehat{f_\delta}(\cdot) - \widehat{f_\delta}(\cdot)\right\| = \tau\delta, \tag{34}$$

where $\tau > 1$ is a constant. To establish existence and uniqueness of solution of (34), we need the following lemma.

Lemma 6. *If $\delta > 0$, then there hold the following:*

(a) *$d(\nu)$ is a continuous function;*

(b) $\lim_{\nu\to 0} d(\nu) = \|\widehat{f_\delta}\|$;

(c) $\lim_{\nu\to+\infty} d(\nu) = 0$;

(d) *$d(\nu)$ is a strictly decreasing function.*

The proof is very easy and we omit it here.

Denote

$$\zeta(x,t) := M_\nu u_\delta(x,t) - u(x,t). \tag{35}$$

We give the main result of this section as follows.

Theorem 7. *Assume the conditions* (11) *and* (17) *hold and $\tau > 1$. Take the solution ν of* (34) *as the regularization parameter, then there holds the error estimate for $0 < x < 1$:*

$$\left\|M_\nu u_\delta(x,t) - u(x,t)\right\| \le C(E,\tau)E^{1-x}\delta^x, \tag{36}$$

where $C(E,\tau)$ is a constant depending on E, τ.

TABLE 1: Errors with *a priori* rule.

x	$\delta = 0.01$		$\delta = 0.001$	
	e_r	e_i	e_r	e_i
0.1	0.1590	1.2154	0.1000	0.7643
0.5	0.0596	0.2729	0.0517	0.2370
0.9	0.0173	0.0497	0.0154	0.0443

Proof. By (35), the Parseval formula, triangle inequality, and (17), we have

$$\begin{aligned}\|\zeta(0,\cdot)\| &= \|M_\nu u_\delta(0,\cdot) - u(0,\cdot)\| = \left\|\widehat{M_\nu u_\delta}(0,\cdot) - \hat{u}(0,\cdot)\right\| \\ &= \left\|\widehat{D_\nu}\widehat{u_\delta}(0,\cdot) - \hat{u}(0,\cdot)\right\| = \left\|\widehat{D_\nu}e^{\sqrt{i\xi}}\widehat{f_\delta} - e^{\sqrt{i\xi}}\hat{f}\right\| \\ &= \left\|\widehat{D_\nu}e^{\sqrt{i\xi}}\left(\widehat{f_\delta} - \hat{f}\right) + \left(\widehat{D_\nu} - 1\right)e^{\sqrt{i\xi}}\hat{f}\right\| \\ &\le \left\|\widehat{D_\nu}e^{\sqrt{i\xi}}\left(\widehat{f_\delta} - \hat{f}\right)\right\| + E \\ &\le e^{\sqrt{\nu/2}}\delta + E.\end{aligned} \tag{37}$$

By virtue of (34), we know that

$$\begin{aligned}\tau\delta &= \left\|\widehat{D_\nu}\widehat{f_\delta} - \widehat{f_\delta}\right\| = \left\|\left(\widehat{D_\nu} - 1\right)\left(\widehat{f_\delta} - \hat{f} + \hat{f}\right)\right\| \\ &\le \left\|\left(\widehat{D_\nu} - 1\right)\left(\widehat{f_\delta} - \hat{f}\right)\right\| + \left\|\left(\widehat{D_\nu} - 1\right)\hat{f}\right\| \\ &\le \delta + \left\|\left(\widehat{D_\nu} - 1\right)e^{-\sqrt{i\xi}}\hat{u}(0,\xi)\right\| \\ &\le \delta + e^{-\sqrt{\nu/2}}E,\end{aligned} \tag{38}$$

which means

$$(\tau - 1)\delta \le e^{-\sqrt{\nu/2}}E. \tag{39}$$

Thus it is obvious that

$$e^{\sqrt{\nu/2}}\delta \le \frac{E}{\tau - 1}. \tag{40}$$

Moreover, inserting (40) into (37), we get the following inequality:

$$\|\zeta(0,\cdot)\| \le \frac{\tau}{\tau - 1}E. \tag{41}$$

In addition,

$$\begin{aligned}\|\zeta(1,\cdot)\| &= \|M_\nu u_\delta(1,\cdot) - u(1,\cdot)\| = \left\|\widehat{M_\nu u_\delta}(1,\cdot) - \hat{u}(1,\cdot)\right\| \\ &\le \left\|\widehat{M_\nu u_\delta}(1,\cdot) - \widehat{u_\delta}(1,\cdot)\right\| + \left\|\widehat{u_\delta}(1,\cdot) - \hat{u}(1,\cdot)\right\| \\ &\le \tau\delta + \delta = (\tau + 1)\delta.\end{aligned} \tag{42}$$

It is easy to see

$$\begin{aligned}&\|\zeta(x,\cdot)\|^2 \\ &\quad = \int_{-\infty}^{+\infty}\left|\widehat{D_\nu}\widehat{u_\delta}(x,\xi) - \hat{u}(x,\xi)\right|^2 d\xi \\ &\quad = \int_{-\infty}^{+\infty}\left|\widehat{D_\nu}e^{\sqrt{i\xi}(1-x)}\widehat{f_\delta}(\xi) - e^{\sqrt{i\xi}(1-x)}\hat{f}(\xi)\right|^2 d\xi \\ &\quad = \int_{-\infty}^{+\infty}\left|e^{\sqrt{i\xi}(1-x)}\left(\widehat{D_\nu}\widehat{f_\delta}(\xi) - \hat{f}(\xi)\right)\right|^2 d\xi.\end{aligned} \tag{43}$$

Denote $\theta_f =: \widehat{D_\nu}\widehat{f_\delta}(\xi) - \hat{f}(\xi)$. By using the Hölder inequality, we know that

$$\begin{aligned}&\|\zeta(x,\cdot)\|^2 \\ &\quad = \int_{-\infty}^{+\infty}\left|e^{\sqrt{i\xi}(1-x)}\right|^2\left|\theta_f\right|^{2x}\left|\theta_f\right|^{2(1-x)} d\xi \\ &\quad \le \left[\int_{-\infty}^{+\infty}\left|\left(e^{\sqrt{i\xi}}\theta_f\right)^{2(1-x)}\right|^{1/(1-x)} d\xi\right]^{1-x}\left[\int_{-\infty}^{+\infty}\left(\theta_f^{2x}\right)^{1/x} d\xi\right]^x \\ &\quad = \left[\int_{-\infty}^{+\infty}\left|\left(e^{\sqrt{i\xi}}\theta_f\right)^2\right| d\xi\right]^{1-x}\left[\int_{-\infty}^{+\infty}\left(\theta_f^2\right) d\xi\right]^x.\end{aligned} \tag{44}$$

Therefore, we obtain

$$\|\zeta(x,\cdot)\| \le \|\zeta(0,\cdot)\|^{1-x}\|\zeta(1,\cdot)\|^x, \tag{45}$$

which also implies that

$$\begin{aligned}\|M_\nu u_\delta(x,\cdot) - u(x,\cdot)\| &\le \|\zeta(0,\cdot)\|^{1-x}\|\zeta(1,\cdot)\|^x \\ &\le \left(\frac{\tau}{\tau - 1}E\right)^{1-x}((\tau + 1)\delta)^x.\end{aligned} \tag{46}$$

Therefore, we complete the proof. □

4. Numerical Examples

For linear heat diffusion, analytic solutions for the temperature distribution $u(x,t)$, $x > 0$, $t > 0$ in a semi-infinite solid with zero initial temperature, and $u(0,t) := H(t)$ at the surface, are obtained using the integral equation

$$u(x,t) = \left(H * \frac{\partial}{\partial t}K\right)(x,t) = \int_0^t H(s)\frac{\partial}{\partial t}K(x,t-s)\,ds, \tag{47}$$

TABLE 2: Errors with *a priori* rule at $x = 0$.

x	$\delta = 0.001$	
	e_r	e_i
0	0.1873	1.6748

TABLE 3: The comparison between *a priori* and *a posteriori* rules with $\delta = 0.01$.

x	*A priori*		*A posteriori*	
0.8	$e_r = 0.0321$	$e_i = 0.1063$	$e_r = 0.0313$	$e_i = 0.1010$
0.2	$e_r = 0.1175$	$e_i = 0.7817$	$e_r = 0.1085$	$e_i = 0.7214$

where the kernel function

$$K(x,t) = \operatorname{erfc}\left(\frac{x}{2\sqrt{t}}\right) \tag{48}$$

is the temperature distribution corresponding to a unit step boundary temperature, $H(t) = 1, t > 0$. Here, $\operatorname{erfc}(\cdot)$ denotes the complementary error function defined by

$$\operatorname{erfc}(z) := \frac{2}{\sqrt{\pi}} \int_z^{\infty} e^{-t^2} dt. \tag{49}$$

We take the example from [22]. As an interesting and challenging test for the numerical method, we proposed the surface temperature function $H(t)$, which is 1 between 0.2 and 0.6 and zero otherwise. The exact solutions for the FICHP, in this example, are the functions

$$u(x,t) = \begin{cases} 0, & 0 < t \le 0.2, \\ \operatorname{erfc}\left(\dfrac{x}{2\sqrt{t-0.2}}\right), & 0.2 < t \le 0.6, \\ \operatorname{erfc}\left(\dfrac{x}{2\sqrt{t-0.2}}\right) - \operatorname{erfc}\left(\dfrac{x}{2\sqrt{t-0.6}}\right), & 0.6 < t < \infty. \end{cases} \tag{50}$$

Consequently, the exact interior data temperature is given by

$$f(t) = \begin{cases} 0, & 0 < t \le 0.2, \\ \operatorname{erfc}\left(\dfrac{1}{2\sqrt{t-0.2}}\right), & 0.2 < t \le 0.6, \\ \operatorname{erfc}\left(\dfrac{1}{2\sqrt{t-0.2}}\right) - \operatorname{erfc}\left(\dfrac{1}{2\sqrt{t-0.6}}\right), & 0.6 < t < \infty. \end{cases} \tag{51}$$

Suppose the vector F represents samples from the function $f(t)$. The noisy discrete data function is generated by adding random errors to the exact data function, at every grid point, that is,

$$F_\delta(n) = F(n) + \epsilon_n, \qquad |\epsilon_n| \le \delta, \quad n = 0, 1, \ldots, N_t, \tag{52}$$

where the (ϵ_n)'s are independent random Gaussian variables with variance $\sigma^2 = \epsilon^2$. The absolute and relative weighted l^2 errors for the recovered interior temperatures are calculated as

$$e_i := \left[\frac{1}{N_t + 1} \sum_{n=0}^{N_t} \left|M_\nu u_\delta(x,n) - u(x,n)\right|\right]^{1/2}, \tag{53}$$

$$e_r := \frac{\left[\left(1/(N_t+1)\right) \sum_{n=0}^{N_t} \left|M_\nu u_\delta(x,n) - u(x,n)\right|\right]^{1/2}}{\left[\left(1/(N_t+1)\right) \sum_{n=0}^{N_t} |u(x,n)|^2\right]^{1/2}}, \tag{54}$$

respectively.

It is easy to implement the algorithm described in (16). Using an available fast Fourier transform (FFT) subroutine, a simple program was written to test the algorithm for the above example in Matlab. Firstly, some tests are implemented to verify the effectiveness of the regularization method under *a priori* parameter choice rule at several interior points. In addition, we would like to compare the *a posteriori* parameter choice rule (34) with the *a priori* parameter choice rule (22). The *a priori* bound E in (22) can be calculated easily in this example. The Newton's bisection is used to solve (34), where we choose $\tau = 1.1$.

Figures 1 and 2 provide the comparisons between the exact solutions and regularization solutions with error bound $\delta = 0.01$ and $\delta = 0.001$ at interior point $x = 0.1, 0.5, 0.9$ using *a priori* parameter choice rule (22), respectively. Figure 3 gives the comparison between the exact solutions and regularization solutions with error bound $\delta = 0.001$ at boundary point $x = 0$ using *a priori* parameter choice rule (32). And intuitively, it seems that there are better numerical effects for closer distance from $x = 1$. We list the error in Tables 1 and 2 to verify our result. Figure 4 demonstrates the comparisons between *a priori* and *a posteriori* parameter choice rules and Table 3 tells us that it has better effect using *a posteriori* parameter choice rule than *a priori* parameter choice rule. We also see that the difference of the numerical results between *a priori* choice rule and *a posteriori* choice rule is slight, which agrees with our theoretical results.

Acknowledgments

The authors are indebted to the referees of this paper for their most helpful comments and suggestions, which helped to improve the presentation greatly. This work was supported by the Fundamental Research Funds for the Central Universities ZYGX2011J104 and SWJTU11BR078 and the NSF of China (Nos. 11126102, 11226040, and 11126187).

References

[1] V. V. Kulish and J. L. Lage, "Fractional-diffusion solution for transient local temperature and heat flux," *Transactions of ASME*, vol. 122, pp. 372–376, 2000.

[2] K. B. Oldham and J. Spanier, *The Fractional Calculus: Theory and Application of Differential and Integration to Arbitrary Order*, Academic Press, 1974.

[3] K. B. Oldham and J. Spanier, "A general solution of the diffusion equation for semiinfinite geometries," *Journal of Mathematical Analysis and Applications*, vol. 39, pp. 655–669, 1972.

[4] K. B. Oldham and J. Spanier, "The replacement of Fick's law by a formulation involving semidifferentiation," *Journal of Electroanalytical Chemistry*, vol. 26, pp. 331–341, 1970.

[5] R. Gorenflo, F. Mainardi, D. Moretti, and P. Paradisi, "Time fractional diffusion: a discrete random walk approach," *Nonlinear Dynamics*, vol. 29, no. 1–4, pp. 129–143, 2002.

[6] V. V. Anh and N. N. Leonenko, "Non-Gaussian scenarios for the heat equation with singular initial conditions," *Stochastic Processes and their Applications*, vol. 84, no. 1, pp. 91–114, 1999.

[7] I. Podlubny, *Fractional Differential Equations*, Academic Press, 1999.

[8] M. M. El-Borai, "The fundamental solutions for fractional evolution equations of parabolic type," *Boletín de la Asociación Matemática Venezolana*, vol. 6, no. 1, pp. 29–43, 2004.

[9] F. Mainardi, Y. Luchko, and G. Pagnini, "The fundamental solution of the space-time fractional diffusion equation," *Fractional Calculus & Applied Analysis*, vol. 4, no. 2, pp. 153–192, 2001.

[10] F. Liu, V. Anhb, and I. Turnerb, "Numerical solution of the space fractional Fokker-Planck equation," *Journal of Computational and Applied Mathematics*, vol. 166, pp. 209–219, 2004.

[11] M. M. Meerschaert and C. Tadjeran, "Finite difference approximations for fractional advection-dispersion flow equations," *Journal of Computational and Applied Mathematics*, vol. 172, no. 1, pp. 65–77, 2004.

[12] S. B. Yuste, "Weighted average finite difference methods for fractional diffusion equations," *Journal of Computational Physics*, vol. 216, no. 1, pp. 264–274, 2006.

[13] G. J. Fix and J. P. Roop, "Least squares finite-element solution of a fractional order two-point boundary value problem," *Computers & Mathematics with Applications*, vol. 48, no. 7-8, pp. 1017–1033, 2004.

[14] J. P. Roop, "Computational aspects of FEM approximation of fractional advection dispersion equations on bounded domains in R^2," *Journal of Computational and Applied Mathematics*, vol. 193, no. 1, pp. 243–268, 2006.

[15] Z. M. Odibat and S. Momani, "Approximate solutions for boundary value problems of time-fractional wave equation," *Applied Mathematics and Computation*, vol. 181, no. 1, pp. 767–774, 2006.

[16] S. S. Ray and R. K. Bera, "Analytical solution of a fractional diffusion equation by Adomian decomposition method," *Applied Mathematics and Computation*, vol. 174, no. 1, pp. 329–336, 2006.

[17] J. J. Liu and M. Yamamoto, "A backward problem for the time-fractional diffusion equation," *Applicable Analysis*, vol. 89, no. 11, pp. 1769–1788, 2010.

[18] W. McLean, "Regularity of solutions to a time-fractional diffusion equation," *The ANZIAM Journal*, vol. 52, no. 2, pp. 123–138, 2010.

[19] J. Prüss, *Evolutionary Integral Equations and Applications*, vol. 87, Birkhäuser, Basel, Switzerland, 1993.

[20] A. A. Kilbas, H. M. Srivastava, and J. J. Trujillo, *Theory and Applications of Fractional Differential Equations*, Elsevier, 2006.

[21] S. G. Samko, A. A. Kilbas, and O. I. Marichev, *Fractional Integrals and Drivatives*, Gordon and Breach Science, Yverdon, Switzerland, 1993.

[22] D. A. Murio, "Stable numerical solution of a fractional-diffusion inverse heat conduction problem," *Computers & Mathematics with Applications*, vol. 53, no. 10, pp. 1492–1501, 2007.

[23] H. W. Engl, M. Hanke, and A. Neubauer, *Regularization of Inverse Problems*, Kluwer Academic, 1996.

[24] L. Yang, Z.-C. Deng, J.-N. Yu, and G.-W. Luo, "Two regularization strategies for an evolutional type inverse heat source problem," *Journal of Physics A*, vol. 42, no. 36, Article ID 365203, 16 pages, 2009.

[25] L. Yang, Z.-C. Deng, J.-N. Yu, and G.-W. Luo, "Optimization method for the inverse problem of reconstructing the source term in a parabolic equation," *Mathematics and Computers in Simulation*, vol. 80, no. 2, pp. 314–326, 2009.

[26] L. Yang, M. Dehghan, J.-N. Yu, and G.-W. Luo, "Inverse problem of time-dependent heat sources numerical reconstruction," *Mathematics and Computers in Simulation*, vol. 81, no. 8, pp. 1656–1672, 2011.

[27] L. Yang, J. N. Yu, G. W. Luo, and Z. C. Deng, "Reconstruction of a space and time dependent heat source from finite measurement data," *International Journal of Heat and Mass Transfer*, vol. 55, pp. 6573–6581, 2012.

[28] L. Yang, J. N. Yu, G. W. Luo, and Z. C. Deng, "Numerical identification of source terms for a two dimensional heat conduction problem in polar coordinate system," *Applied Mathematical Modelling*, vol. 37, pp. 939–957, 2013.

[29] F. F. Dou and Y. C. Hon, "Kernel-based approximation for Cauchy problem of the time-fractional diffusion equation," *Engineering Analysis with Boundary Elements*, vol. 36, no. 9, pp. 1344–1352, 2012.

[30] Y. C. Hon and T. Wei, "A fundamental solution method for inverse heat conduction problem," *Engineering Analysis With Boundary Elements*, vol. 28, no. 5, pp. 489–495, 2004.

[31] J. Cheng, J. Nakagawa, M. Yamamoto, and T. Yamazaki, "Uniqueness in an inverse problem for a one-dimensional fractional diffusion equation," *Inverse Problems*, vol. 25, no. 11, pp. 1-16, 2009.

[32] B. Jin and W. Rundell, "An inverse problem for a one-dimensional time-fractional diffusion problem," *Inverse Problems*, vol. 28, no. 7, Article ID 075010, 19 pages, 2012.

[33] D. N. Hào, "A mollification method for ill-posed problems," *Numerische Mathematik*, vol. 68, no. 4, pp. 469–506, 1994.

[34] U. Tautenhahn, "Optimality for ill-posed problems under general source conditions," *Numerical Functional Analysis and Optimization*, vol. 19, no. 3-4, pp. 377–398, 1998.

Cortex Effect on Vacuum Drying Process of Porous Medium

Zhijun Zhang,[1] Shiwei Zhang,[1] Tianyi Su,[1] and Shuangshuang Zhao[2]

[1] *School of Mechanical Engineering and Automation, Northeastern University, Shenyang 110004, China*
[2] *Shenyang Aircraft Design and Research Institute, Shenyang 110035, China*

Correspondence should be addressed to Zhijun Zhang; zhjzhang@mail.neu.edu.cn

Academic Editor: Jun Liu

Corns, fruits, and vegetables are usually used as porous medium in drying process. But in fact, it must be considered as the cortex effect on mass transfer because the mass transfer of cortex is very difficult than inner medium. Based on the theory of heat and mass transfer, a coupled model for the porous medium vacuum drying process with cortex effect is constructed. The model is implemented and solved using COMSOL software. The water evaporation rate is determined using a nonequilibrium method with the rate constant parameter K_r that has been studied. The effects of different vapor pressures (1000, 5000, and 9000 Pa), initial moisture contents (0.3, 0.4, and 0.5 water saturation), drying temperatures (323, 333, and 343 K), and intrinsic permeability for cortex part (10^{-13}, 10^{-14}, 10^{-15} m^2) on vacuum drying process were studied. The results facilitate a better understanding of the porous medium vacuum drying process that nearer to the reality.

1. Introduction

Scientists and engineers in China are currently studying vacuum drying equipment that could be used in corn drying [1–3]. However, the corn vacuum drying theory remains unclear. Hypothesized that corn is a porous medium, the vacuum drying of corn is a complicated heat and mass transfer process that has been the subject of intensive research [4–7]. All vacuum drying models have to address the water phase change during numerical solving. In one method, the vapor pressure is equal to its equilibrium value [8–11]. Another method is nonequilibrium method [12–16]. As the porous medium, the heat and mass transfer in vacuum drying process has been studied by nonequilibrium method [17]. The water evaporation rate is determined using a nonequilibrium method with the rate constant parameter K_r. K_r values of 1, 10, 1000, and 10000 are simulated. The effects of different vapor pressures (1000, 5000, and 9000 Pa), initial moisture contents (0.4, 0.5, and 0.6 water saturation), drying temperatures (323, 333, and 343 K), and intrinsic permeability (10^{-13}, 10^{-14}, 10^{-15} m^2) are studied. It was observed that the temperature increased quickly at the start of drying and then lowered gradually. As the drying process continued, the temperature increased slowly. In the absence of free water, temperature increased rapidly. As the drying process concluded, the temperature remained unchanged. The water evaporation rate could not be obtained during the porous medium vacuum drying process. The rate constant parameter is essential to the nonequilibrium method. When $K_r \geq 1000$, the simulation of the drying process was not evidently affected. Vapor pressure and heat transfer affected the transfer of mass. A similar effect was found in the initial moisture and the heat temperature. Intrinsic permeability had a greater effect on the drying process. In the study above, it is as the uniform porous medium. But in fact, it must be considered as the cortex effect on mass transfer because the mass transfer of cortex is more difficult than inner medium. The study of about cortex effect is very few.

In this paper, heat and mass transfer of porous medium with cortex in the vacuum drying process is implemented by using a nonequilibrium method. The effects of vapor pressure, initial moisture, heat temperature, and cortex intrinsic permeability on the drying process were then examined.

2. Physical Model

A physical one-dimensional (1D) model that explains the drying process is shown in Figure 1. The heat and mass

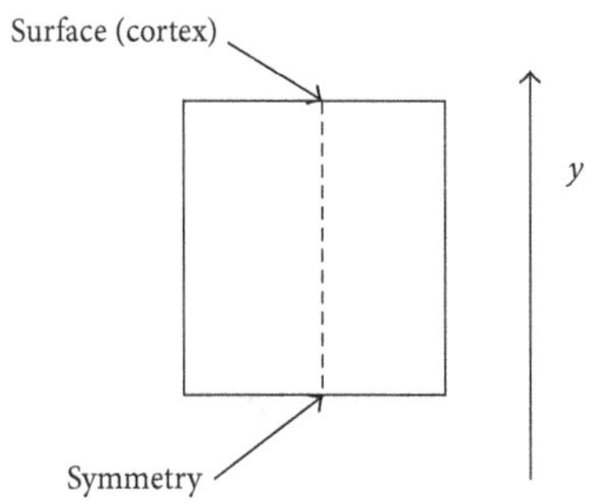

FIGURE 1: 1D model of porous medium with cortex.

transfer is considered only in the y direction. Because consider the cortex effect. The heat and mass transfer is through the surface of medium that is the cortex position. The total height of the porous medium is 1.05 cm, and the cortex height is 0.05 cm.

3. Mathematical Model

The porous medium consists of a continuous rigid solid phase, an incompressible liquid phase (free water), and a continuous gas phase that is assumed to be a perfect mixture of vapor and dry air, considered as ideal gases. For a mathematical description of the transport phenomenon in a porous medium, we adopt a continuum approach, wherein macroscopic partial differential equations are achieved through the volume averaging of the microscopic conservation laws. The value of any physical quantity at a point in space is given by its average value on the averaging volume centered at this point.

The moisture movement of the inner porous medium is liquid water and vapor movement; that is, the liquid water could become vapor, and the vapor and liquid water are moved by the pressure gradient. The heat and mass transfer theory could be found in everywhere [8].

The compressibility effects of the liquid phase are negligible, and the phase is homogeneous:

$$\overline{\rho}_l = \text{constant}. \tag{1}$$

The solid phase is rigid and homogeneous:

$$\overline{\rho}_s = \text{constant}. \tag{2}$$

The gaseous phase is considered an ideal gas. This phase ensures that

$$\begin{aligned} \overline{\rho}_a &= \frac{m_a P_a}{R\overline{T}}, \\ \overline{\rho}_v &= \frac{m_v \overline{P}_v}{R\overline{T}}, \\ \overline{P}_g &= \overline{P}_a + \overline{P}_v, \\ \overline{\rho}_g &= \overline{\rho}_a + \overline{\rho}_v. \end{aligned} \tag{3}$$

The assumption of the local thermal equilibrium between the solid, gas, and liquid phases involves

$$\overline{T}_s = \overline{T}_g = \overline{T}_l = \overline{T}. \tag{4}$$

Mass conservation equations are written for each component in each phase. Given that the solid phase is rigid, the following is given:

$$\frac{\partial \overline{\rho}_s}{\partial t} = 0. \tag{5}$$

The averaged mass conservation of the dry air yields

$$\frac{\partial \left(\varepsilon \cdot S_g \overline{\rho}_a\right)}{\partial t} + \nabla \cdot \left(\overline{\rho}_a \overline{V}_a\right) = 0. \tag{6}$$

For vapor,

$$\frac{\partial \left(\varepsilon \cdot S_g \overline{\rho}_v\right)}{\partial t} + \nabla \cdot \left(\overline{\rho}_v \overline{V}_v\right) = \dot{I}. \tag{7}$$

For free water,

$$\frac{\partial \left(\varepsilon \cdot S_w \overline{\rho}_l\right)}{\partial t} + \nabla \cdot \left(\overline{\rho}_l \overline{V}_l\right) = -\dot{I}. \tag{8}$$

For water, the general equation of mass conservation is obtained from the sum of the conservation equations of vapor (v) and free water (l). The general equation is written as follows:

$$\frac{\partial W}{\partial t} + \nabla \cdot \left\{\frac{1}{\overline{\rho}_s}\left(\overline{\rho}_l \overline{V}_l + \overline{\rho}_v \overline{V}_v\right)\right\} = 0, \tag{9}$$

$$W = \frac{\varepsilon \cdot S_w \overline{\rho}_l + \varepsilon \cdot S_g \overline{\rho}_v}{(1-\varepsilon)\,\overline{\rho}_s}. \tag{10}$$

For the Darcy flow of vapor,

$$\overline{\rho}_v \overline{V}_v = \overline{\rho}_v \overline{V}_g - \overline{\rho}_g D_{\text{eff}} \cdot \nabla \overline{\omega}. \tag{11}$$

For the Darcy flow of air,

$$\overline{\rho}_a \overline{V}_a = \overline{\rho}_a \overline{V}_g - \overline{\rho}_g D_{\text{eff}} \cdot \nabla \overline{\omega}, \tag{12}$$

where the gas and free water velocity is given by

$$\begin{aligned} \overline{V}_g &= \frac{k \cdot k_{rg}}{\mu_g} \cdot \left(\nabla \overline{P}_g - \rho_g \vec{g}\right), \\ \overline{V}_l &= \frac{k \cdot k_{rl}}{\mu_l} \cdot \left(\nabla \overline{P}_l - \rho_l \vec{g}\right). \end{aligned} \tag{13}$$

The effective diffusion coefficient [8] is given by

$$D_{\text{eff}} = D\underline{\underline{B}}. \tag{14}$$

The vapor fraction in mixed gas is given by

$$\overline{\omega} = \frac{\rho_v}{\rho_g}. \tag{15}$$

Table 1: Parameters used in the simulation process.

Parameter	Symbol	Value	Unit
Rate constant parameter	K_r	1000	s^{-1}
Intrinsic permeability of inner medium	k_{in}	10^{-13}	m^2
Intrinsic permeability of cortex	k_{cor}	$10^{-13}, 10^{-14}, 10^{-15}$	m^2
Initial water saturation	S_{l0}	0.5, 0.4, 0.3	
Vapor pressure of vacuum drying chamber	P_{vb}	1000, 5000, 9000	Pa
Heat temperature	T_h	323, 333, 343	K
Porosity	ε	0.615	
Solid density	ρ_s	476	$kg\,m^{-3}$
Air pressure of vacuum drying chamber	P_{ab}	0.001	Pa

The pressure moving the free water is given by

$$\overline{P}_l = \overline{P}_g - \overline{P}_c. \quad (16)$$

For capillary pressure,

$$\overline{P}_c = 56.75 \times 10^3 (1 - S_l) \exp\left(\frac{1.062}{S_l}\right). \quad (17)$$

The saturation of free water and gas is

$$S_g + S_l = 1. \quad (18)$$

Free water relative permeability is given by

$$k_{rl} = \begin{cases} \left(\dfrac{S_l - S_{cr}}{1 - S_{cr}}\right)^3 & S_w > S_{cr} \\ 0 & S_w \le S_{cr}. \end{cases} \quad (19)$$

Gas relative permeability is given by

$$k_{rg} = S_g. \quad (20)$$

The water phase change rate is expressed as

$$\dot{I} = K_r \frac{m_v (a_w P_{sat} - P_v) S_g \varepsilon}{RT}. \quad (21)$$

Water saturation vapor pressure is given by

$$P_{sat} = \frac{101325}{760} \times 10^{(8.07131-(1730.63/(233.426+(T-273))))}. \quad (22)$$

By considering the hypothesis of the local thermal equilibrium, the energy conservation is reduced to a unique equation:

$$\frac{\partial \overline{\rho h}}{\partial t} + \nabla \cdot \left\{ \left(\overline{\rho}_a \overline{V}_a \overline{h}_a + \overline{\rho}_v \overline{V}_v \overline{h}_v + \overline{\rho}_l \overline{V}_l \overline{h}_l - \lambda_e \cdot \nabla \overline{T} - \Delta H \cdot \dot{I} \right) \right\} = 0,$$

$$\lambda_e = (1-\varepsilon)\lambda_s + \varepsilon \left(S_l + S_g \left(\omega \lambda_v + (1-\omega) \lambda_a \right) \right), \quad (23)$$

$$\overline{\rho h} = \overline{\rho}_s \overline{h}_s + \varepsilon \cdot S_g \overline{\rho}_a \overline{h}_a + \varepsilon \cdot S_g \overline{\rho}_v \overline{h}_v + \varepsilon \cdot S_l \overline{\rho}_l \overline{h}_l.$$

4. Boundary Condition and Parameters

The air pressure on the external surface of the porous medium is fixed, and the boundary condition for air is given by

$$P_a = P_{av}. \quad (24)$$

The boundary condition for vapor at the surface of the porous medium is given by

$$P_v = P_{vb}. \quad (25)$$

To simulate the vapor pressure of the vacuum drying chamber effect on the drying process, four different vapor pressure boundary values are used.

The boundary condition for free water at the top of the porous medium is

$$n \cdot (-D \nabla S_w) = 0. \quad (26)$$

The boundary condition at the surface of the porous medium is

$$T = T_h. \quad (27)$$

Three different T_h values are used in the simulation.

The initial moisture of the porous medium is represented by the liquid water saturation; different initial water saturation values are used. To compare the effects, drying base moisture content (d.b.) was also used, as shown in (9). The water phase change rate is used as 1000 that has been studied before [17]. Intrinsic permeability of inner medium is 10^{-13} m^2. Intrinsic permeability of cortex is 10^{-14}, 10^{-15} m^2. In order to compare the results, the intrinsic permeability of cortex 10^{-13}, that is no cortex effect is simulated. In order to easily converge in, the parameter changer is used a smooth method. The modeling parameters are shown in Table 1.

5. Numerical Solution

COMSOL Multiphysics 3.5a was used to solve the set of equations. COMSOL is an advanced software used for modeling and simulating any physical process described by partial derivative equations. The set of equations introduced above was solved using the relative initial and boundary conditions of each. COMSOL offers three possibilities for writing the equations: (1) using a template (the Fick law and the Fourier law), (2) using the coefficient form (for mildly nonlinear problems), and (3) using the general form (for most nonlinear problems). Differential equations in the coefficient form were written using an unsymmetric-pattern multifrontal method. We used a direct solver for sparse matrices (UMFPACK), which involves significantly more complicated algorithms than solvers used for dense matrices. The main complication is the need to handle the fill-in in factors L and U efficiently.

A two-dimensional (2D) grid was used to solve the equations using COMSOL Multiphysics 3.5a. Given the symmetry condition setting at the left and the right sides, The 2D is applied to the the 1D model shown in Figure 1. The mesh consists of 2×200 elements (2D), and time stepping is 1 (from 0 s to 100 s of solution), 5 (from 100 s to 200 s of solution), 20 (from 200 s to 1000 s of solution), 30 (from 1000 s to 2000 s of solution), 40 (from 2000 s to 4000 s of solution), 50 (from 4000 s to 20000 s of solution), and 100 (from 20000 s to 50000 s of solution). Several grid sensitivity tests were conducted to determine the sufficiency of the mesh scheme and to ensure that the results are grid independent. The maximum element size was established as $1e^{-4}$. A backward differentiation formula was used to solve time-dependent variables. Relative tolerance was set to $1e^{-3}$, whereas absolute tolerance was set to $1e^{-4}$. The simulations were performed using a Tongfang PC with Intel Core 2 Duo processor with 3.0 GHz processing speed, and 4096 MB of RAM running Windows 7.

6. Results and Discussion

6.1. Effect of Boundary Condition. In the study before [17], the heat and mass transfer direction is the same. That is, the heat boundary is on the bottom of the medium. The heat transfer is from the bottom to top. The mass transfer boundary is on the top of bottom. The mass transfer is from the bottom to top by the pressure driving. But in this study, the cortex is considered. And the heat transfer and mass transfer are on the reversed direction as Figure 1. The heat and mass transfer with dryer is on the surface of the medium. The both condition was simulated to be compared as Figure 2. The simulation parameters are $S_{l0} = 0.5$, $K_r = 1000$, $T_h = 323$ K, $P_{vb} = 1000$ Pa, $k_{\text{in}} = 10^{-13}$ m^2, and no cortex. The moisture curve is just little different. The drying rate is little larger at the drying initial stage for same direction heat and mass transfer, and then it is little lower. The other condition is just reversed. The drying rate is little lower at the initial stage for reversed direction heat and mass transfer, and then it is little larger. The drying time for both is 6 hours.

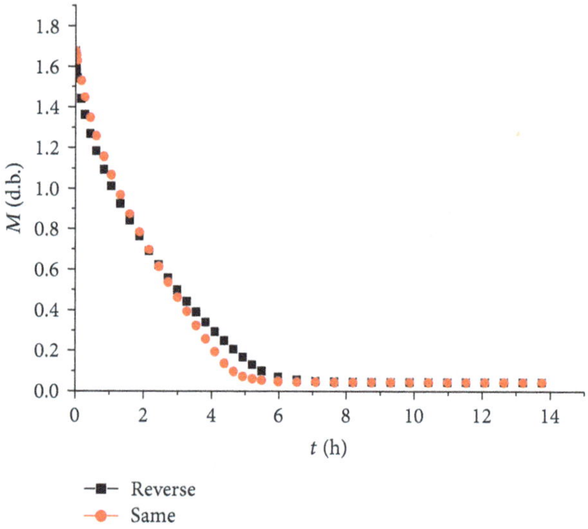

FIGURE 2: The effect of boundary condition, heat, and mass transfer is same and is reverse direction.

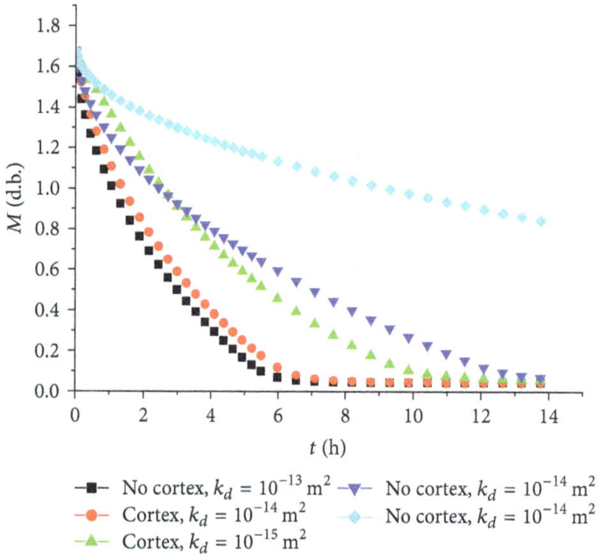

FIGURE 3: Moisture curves of no cortex ($k_{\text{in}} = 10^{-13}$, $k_{\text{cor}} = 10^{-13}$, 10^{-14}, and 10^{-15} m^2), ($k_{\text{cor}} = 10^{-14}$, 10^{-15}, $k_{\text{in}} = 10^{-13}$ m^2).

6.2. Effect of Cortex Resistance. Intrinsic permeability of porous medium is an inherent property and cannot be changed, and measuring it is difficult. Intrinsic permeability has a greater effect because the transfer of free water and vapor is affected by (13). Usually, the cortex intrinsic permeability is less than inner medium. Figure 3 is the moisture curves of no cortex (k_{in} and $k_{\text{cor}} = 10^{-13}$, 10^{-14}, and 10^{-15} m^2, resp.), with cortex ($k_{\text{in}} = 10^{-13}$ and $k_{\text{cor}} = 10^{-14}$ and 10^{-15} m^2 resp.). The other simulation parameters are $S_{l0} = 0.5$, $K_r = 1000$, $T_h = 323$ K, and $P_{vb} = 1000$ Pa. The drying time became evidently longer as the intrinsic permeability was reduced because the moisture movement velocity was lowered at the same pressure gradient. The different is not so obviously as the $k_{\text{cor}} = 10^{-14}$ m^2, that is lower 10-fold than inner medium. But when $k_{\text{cor}} = 10^{-15}$ m^2, the drying time is increased to 2-fold nearly. In order to compare, the no cortex, $k_{\text{cor}} = 10^{-14}$ m^2, and 10^{-15} m^2 was shown. Because the numerical calculation is not convergence, the cortex $k_{\text{cor}} = 10^{-16}$ m^2, $k_{\text{in}} = 10^{-13}$ m^2 result not gotten. The reason is that the material parameter changer is too big.

6.3. Effect of Vapor Pressure in Vacuum Drying Chamber. The pressure of a vacuum drying chamber, especially vapor

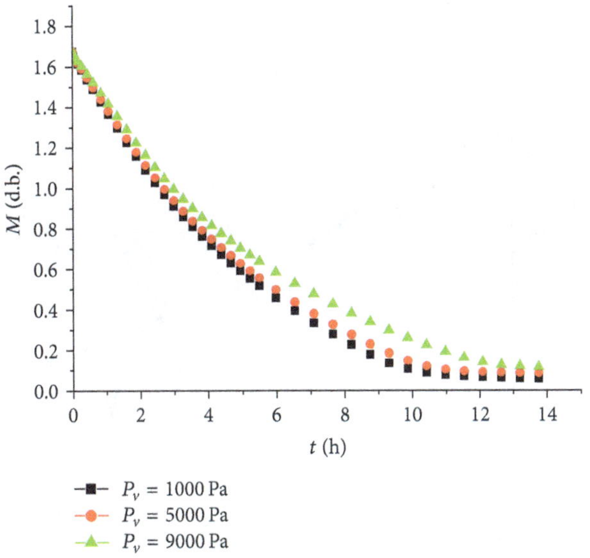

FIGURE 4: Moisture curves of different P_{vb} = 1000, 5000, 9000 Pa, $k_{\mathrm{cor}} = 10^{-15}\,\mathrm{m}^2$, $k_{\mathrm{in}} = 10^{-13}\,\mathrm{m}^2$.

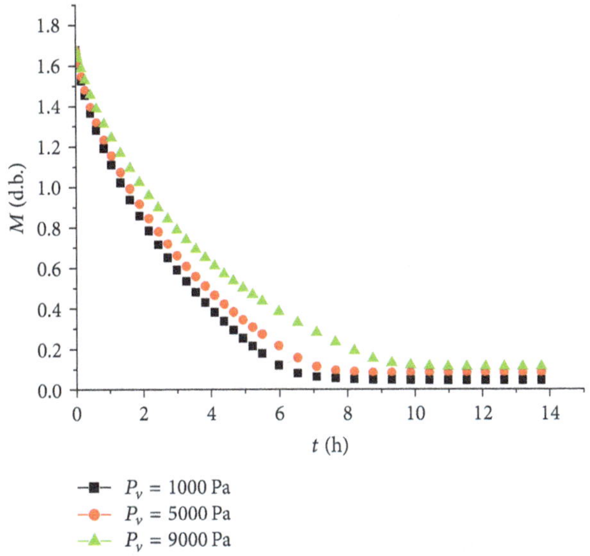

FIGURE 5: Moisture curves of different P_{vb} = 1000, 5000, 9000 Pa, $k_{\mathrm{cor}} = 10^{-14}\,\mathrm{m}^2$, $k_{\mathrm{in}} = 10^{-13}\,\mathrm{m}^2$.

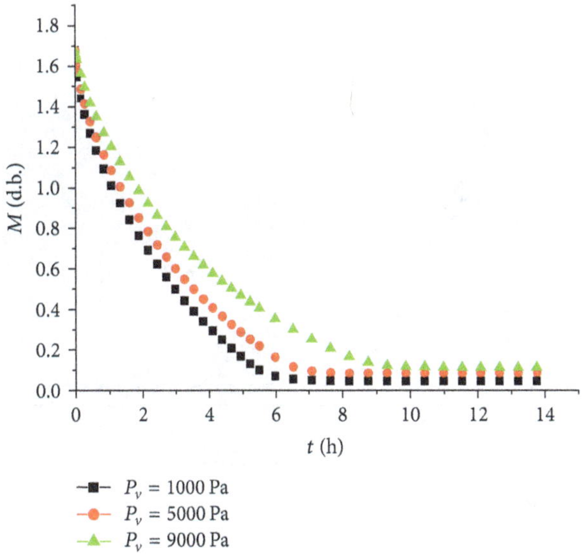

FIGURE 6: Moisture curves of different P_{vb} = 1000, 5000, 9000 Pa, no cortex, $k_{\mathrm{in}} = 10^{-13}\,\mathrm{m}^2$.

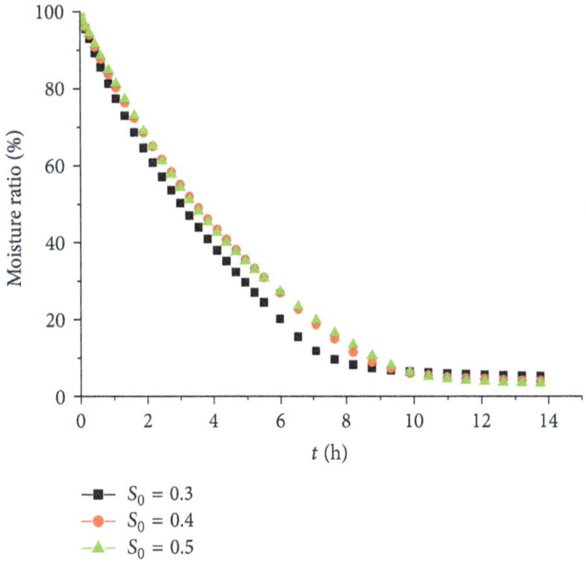

FIGURE 7: Moisture curves at different S_{l0} = 0.5, 0.4, 0.3, $k_{\mathrm{cor}} = 10^{-15}\,\mathrm{m}^2$, $k_{\mathrm{in}} = 10^{-13}\,\mathrm{m}^2$.

pressure, plays an important role in the vacuum drying process and is also linked to the drying cost. The moisture curves of P_{vb} = 1000, 5000, and 9000 Pa are shown in Figures 4, 5, and 6 with different cortex intrinsic permeability. The other simulation parameters are S_{l0} = 0.5, K_r = 1000, and T_h = 323 K. The vapor pressure has a greater effect on the drying process; a lower vapor pressure results in greater pressure degradation. The movements of free water and vapor as well as the free water evaporation rate are quicker, as given by (13) and (21), respectively. But the pressure effect is lower by the cortex intrinsic permeability reduced.

6.4. Effect of Initial Moisture Content. The effect of initial moisture content on the moisture curve is shown in Figures 7, 8, and 9 for S_{l0} = 0.5, 0.4, 0.3 with different cortex intrinsic permeability. To compare the results, moisture is represented by the moisture ratio M/M_0. The other parameters are K_r = 1000, P_{vb} = 1000 Pa, and T_h = 323 K. The drying time is about 10 hours, 6.5 hours, and 6 hours for S_{l0} = 0.5, 0.4, 0.3, respectively. In the same intrinsic permeability, the drying time is almost the same. The initial moisture has a less effect when initial moisture is 0.4 and 0.3, especially cortex intrinsic permeability $k_{\mathrm{cor}} = 10^{-15}\,\mathrm{m}^2$. The drying rate is little larger for S_{l0} = 0.5, and the drying rate is almost the same for S_{l0} = 0.4, 0.3.

6.5. Effect of Heat Temperature. The effect of heat temperature on moisture is shown in Figures 10, 11, and 12 for the heat

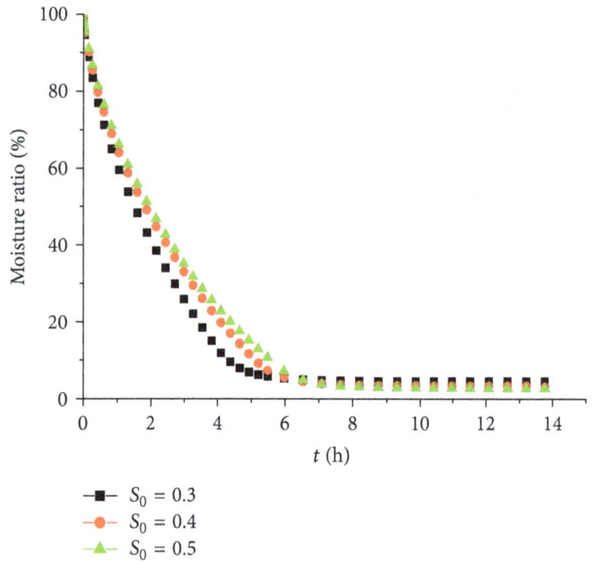

FIGURE 8: Moisture curves at different S_{l0} = 0.5, 0.4, 0.3, k_{cor} = 10^{-14} m^2, k_{in} = 10^{-13} m^2.

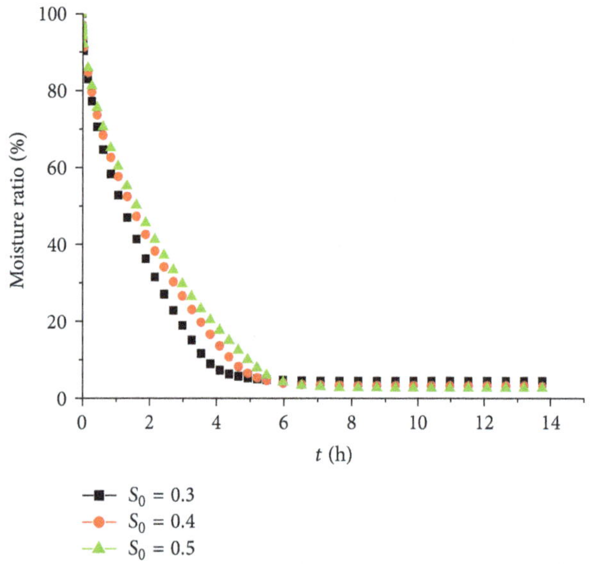

FIGURE 9: Moisture curves at different S_{l0} = 0.5, 0.4, 0.3, no cortex, k_{in} = 10^{-13} m^2.

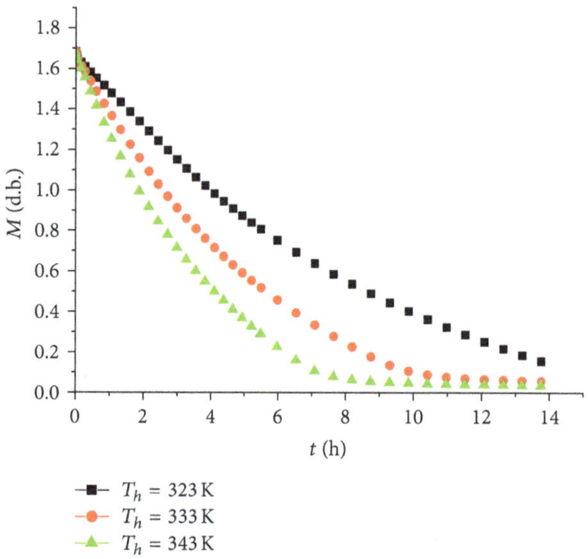

FIGURE 10: Moisture curves at different T_h = 323, 333, 343 K, k_{cor} = 10^{-15} m^2, k_{in} = 10^{-13} m^2.

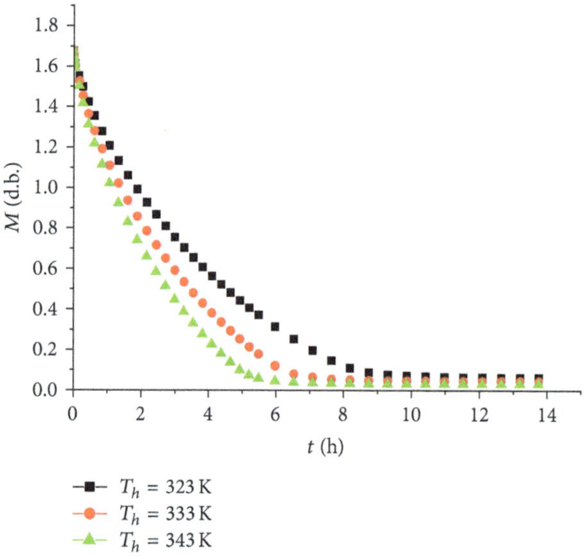

FIGURE 11: Moisture curves at different T_h = 323, 333, 343 K, k_{cor} = 10^{-14} m^2, k_{in} = 10^{-13} m^2.

temperature T_h = 323, 333, and 343 K. The other parameters are S_{l0} = 0.5, K_r = 1000, and P_v = 1000 Pa. The effect of heat temperature is obviously in all simulation. The drying time is more than 14 hours for T_h = 323 K, k_{cor} = 10^{-15} m^2, and k_{in} = 10^{-13} m^2. And it is about 10 hours and 8 hours for T_h = 333, 343 K, k_{cor} = 10^{-15} m^2, and k_{in} = 10^{-13} m^2. The drying time is more than 9 hours for T_h = 333 K, k_{cor} = 10^{-15} m^2, and k_{in} = 10^{-13} m^2. It is about 7 hours and 6 hours for T_h = 333, 343 K, k_{cor} = 10^{-15} m^2, and k_{in} = 10^{-13} m^2. The drying time is more than 7 hours for T_h = 333 K, k_{cor} = 10^{-15} m^2, and k_{in} = 10^{-13} m^2. It is about 6 hours and 5 hours for T_h = 333, 343 K, k_{cor} = 10^{-15} m^2, and k_{in} = 10^{-13} m^2.

7. Conclusion

A coupled model of porous medium vacuum drying with cortex effect based on the theory of heat and mass transfer was implemented in this paper. The drying rate is little higher at the drying initial stage for the same direction of heat and mass transfer, and then it is little lower. The drying time became evidently longer as the intrinsic permeability was reduced because the moisture movement velocity was lowered at the same pressure gradient. The difference is not so obvious as the k_{cor} = 10^{-14} m^2, that is lower 10-fold than inner medium. But when k_{cor} = 10^{-15} m^2, the drying time is increased to 200% nearly. The vapor pressure has a greater effect on

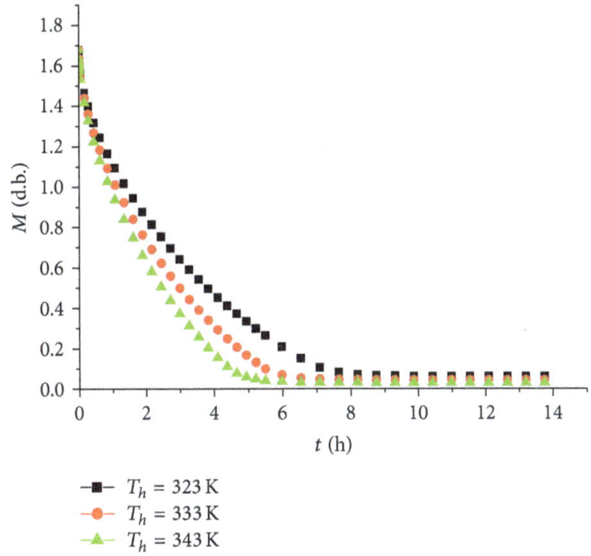

FIGURE 12: Moisture curves at different T_h = 323, 333, 343 K, no cortex, $k_{in} = 10^{-13}$ m^2.

the drying process; a lower vapor pressure results in greater pressure degradation. But the pressure effect is lower by the reduction of the cortex intrinsic permeability. For initial moisture effect, in the same intrinsic permeability, the drying time is almost the same. The initial moisture is a less effect when initial moisture is 0.4 and 0.3, especially cortex intrinsic permeability $k_{cor} = 10^{-15}$ m^2. The drying rate is little larger for S_{l0} = 0.5, and the drying rate is almost the same for S_{l0} = 0.4, 0.3. The effect of heat temperature is obviously in all simulation.

Nomenclature

B: Diagonal tensor
D: Diffusivity (m^2 s^{-1})
D_{eff}: Diffusion tensor (m^2 s^{-1})
g: Gravity vector (m s^{-2})
h: Intrinsic averaged enthalpy (Jkg^{-1})
I: Water phase rate (kg s^{-1} m^{-3})
k: Intrinsic permeability (m^2)
k_r: Relative permeability
m: Mass (kg)
n: Outer unit normal to the product
P: Pressure (Pa)
P_c: Capillary pressure (Pa)
R: Universal Gas constant (J kmol^{-1} K^{-1})
S: Saturation
t: Time (s)
T: Temperature (K)
W: Moisture content (in dry basis).

Greek Letters

ΔH: Latent of phase change (Jkg^{-1})
λ_{ef}: Effective thermal conductivity tensor (Wm^{-1}k^{-1})
μ: Viscosity (kg m^{-1}s^{-1})
ρ: Density (kg m^{-3})
ω: Vapor fraction.

Subscripts

a: Dry air
g: Gas
l: Liquid
s: Solid
v: Vapor
sat: Vapor saturation
in: Inner medium
cor: Cortex.

Mathematical Operators

Δ: Gradient operator
$\nabla\cdot$: Divergence operator.

Acknowledgment

This research was supported by the National Natural Science Foundation of China (Grant nos. 31000665 and nos. 51176027).

References

[1] C. H. Xu, Z. J. Zhang, S. W. Zhang, and X. He, "Probe into the structure of tower continuous vacuum dryer," in *Proceedings of the 5th Asia-Pacific Drying Conference (ADC '07)*, pp. 1261–1267, August 2007.

[2] Z. J. Zhang, C. H. Xu, S. W. Zhang, and X. He, "The study of corn low temperature continuous tower type vacuum dryer," in *Proceedings of the 5th Asia-Pacific Drying Conference (ADC '07)*, pp. 330–337, August 2007.

[3] Z. Zhang, C. Xu, S. Zhang, and L. Zhao, "Computer simulation of flow field in tower continuous vacuum dryer," in *Proceedings of the International Conference on Computer Science and Information Technology (ICCSIT '08)*, pp. 534–538, September 2008.

[4] I. Yasuaki and A. P. S. Selvadurai, *Transport Phenomena in Porous Media, Aspects of Micro/Macro Behaviour*, Springer, 2012.

[5] A. K. Haghi, "Transport phenomena in porous media: a review," *Theoretical Foundations of Chemical Engineering*, vol. 40, no. 1, pp. 14–26, 2006.

[6] S. J. Kowalski, *Drying of Porous Materials*, Springer, 2007.

[7] B. Jacob and Y. Bachmat, *Introduction to Modeling of Transport Phenomena in Porous Media*, Springer, 1990.

[8] A. Erriguible, P. Bernada, F. Couture, and M. A. Roques, "Simulation of vacuum drying by coupling models," *Chemical Engineering and Processing*, vol. 46, no. 12, pp. 1274–1285, 2007.

[9] A. Erriguible, P. Bernada, F. Couture, and M. A. Roques, "Modeling of heat and mass transfer at the boundary between a porous medium and its surroundings," *Drying Technology*, vol. 23, no. 3, pp. 455–472, 2005.

[10] K. Murugesan, H. N. Suresh, K. N. Seetharamu, P. A. Aswatha Narayana, and T. Sundararajan, "A theoretical model of brick drying as a conjugate problem," *International Journal of Heat and Mass Transfer*, vol. 44, no. 21, pp. 4075–4086, 2001.

[11] P. Perré and I. W. Turner, "A dual-scale model for describing drier and porous medium interactions," *AIChE Journal*, vol. 52, no. 9, pp. 3109–3117, 2006.

[12] S. S. Torres, W. Jomaa, J.-R. Puiggali, and S. Avramidis, "Multiphysics modeling of vacuum drying of wood," *Applied Mathematical Modelling*, vol. 35, no. 10, pp. 5006–5016, 2011.

[13] S. S. Torres, J. R. Ramírez, and L. Méndez-Lagunas, "Modeling plain vacuum drying by considering a dynamic capillary pressure," *Chemical and Biochemical Engineering Quarterly*, vol. 25, no. 3, pp. 327–334, 2011.

[14] A. Warning, A. Dhall, D. Mitrea, and A. K. Datta, "Porous media based model for deep-fat vacuum frying potato chips," *Journal of Food Engineering*, vol. 110, no. 3, pp. 428–440, 2012.

[15] A. Halder, A. Dhall, and A. K. Datta, "An improved, easily implementable, porous media based model for deep-fat frying. Part I: model development and input parameters," *Food and Bioproducts Processing*, vol. 85, no. 3 C, pp. 209–219, 2007.

[16] A. Halder, A. Dhall, and A. K. Datta, "An improved, easily implementable, porous media based model for deep-fat frying. Part II: results, validation and sensitivity analysis," *Food and Bioproducts Processing*, vol. 85, no. 3, pp. 220–230, 2007.

[17] Z. Zhang and N. Kong, "Nonequilibrium thermal dynamic modeling of porous medium vacuum drying process," *Mathematical Problems in Engineering*, vol. 2012, Article ID 347598, 22 pages, 2012.

10

A Noise-Insensitive Semi-Active Air Suspension for Heavy-Duty Vehicles with an Integrated Fuzzy-Wheelbase Preview Control

Zhengchao Xie, Pak Kin Wong, Jing Zhao, Tao Xu, Ka In Wong, and Hang Cheong Wong

Department of Electromechanical Engineering, University of Macau, Taipa, Macau, China

Correspondence should be addressed to Jing Zhao; zhaojing003@gmail.com

Academic Editor: Qingsong Xu

Semi-active air suspension is increasingly used on heavy-duty vehicles due to its capabilities of consuming less power and low cost and providing better ride quality. In this study, a new low cost but effective approach, fuzzy-wheelbase preview controller with wavelet denoising filter (FPW), is developed for semi-active air suspension system. A semi-active suspension system with a rolling lobe air spring is firstly modeled and a novel front axle vertical acceleration-based road prediction model is constructed. By adopting a sensor on the front axle, the road prediction model can predict more reliable road information for the rear wheel. After filtering useless signal noise, the proposed FPW can generate a noise-insensitive control damping force. Simulation results show that the ride quality, the road holding, the handling capability, the road friendliness, and the comprehensive performance of the semi-active air suspension with FPW outperform those with the traditional active suspension with PID-wheelbase preview controller (APP). It can also be seen that, with the addition of the wavelet filter, the impact of sensor noise on the suspension performance can be minimized.

1. Introduction

Suspension is one of the main components of a vehicle. Its role is to provide ride comfort for passengers, to offer road holding and competent handling capabilities, and to give support to the vehicle static weight [1]. Comparing to active suspension and passive suspension, semi-active suspension is currently the most popular type of vehicle suspension as it has the best compromise between the performance (handling, ride comfort, safety, etc.) and cost (sensors, weight, components, electronics, power consumption, etc.) [2, 3]. A semi-active suspension system generally consists of springs and some variable dampers that are controlled by various algorithms or controllers such that the damping coefficient can be adjusted instantly. In recent, air springs have been introduced in heavy-duty vehicles because they can effectively attenuate the effects of the disturbance from the road input and easily adjust the ride height on board. When air springs are used instead of conventional steel springs, the semi-active suspension system is also called a semi-active air suspension system, which is the case of this study.

In the current literature, many control algorithms for semi-active suspension have been introduced. The most common approach for both academic and industrial applications is the sky-hook (SH) damping control, which was first introduced by Crosby et al. [4]. Many researches on SH control [5–8] have been carried out as it provides an easy way to obtain good suspension performance. The SH control has also been extended to heavy-duty vehicles [9] and some enhanced SH control strategies [10–13] can still be found in recent years for more effective suspension improvement. Another ordinary example of suspension control algorithms is the ground-hook control strategy [14, 15], which can reduce the suspension deflection and increase the damping simultaneously. Other than these two methods, several nonlinear control techniques were also introduced recently, such as model predictive control [16–18] and the human simulated intelligent control [19]. However, the actuator of a semi-active suspension is actually a dissipation device which cannot be easily described by sets of differential equations, while an accurate and precise mathematical model is always required for those control techniques. Moreover, some parameters in the model are difficult to identify in real practice. Therefore, some artificial intelligence approaches have been applied to the semi-active suspension systems too as these approaches rely less on mathematical models. For instance, Hashiyama et al. used genetic algorithm to generate a fuzzy-based semi-active suspension controller [20]. Yoshimura et al. [21] proposed a semi-active controller using fuzzy reasoning. A neural-network-based fuzzy controller was also proposed by Eslaminasab et al. [22] to improve the suspension performance of heavy-duty vehicles. These studies showed that artificial intelligence approaches are very favorable for semi-active suspension control. As a result, this study also attempts to apply fuzzy logic-based controller to control the semi-active air suspension system.

Nevertheless, the existing fuzzy controller for semi-active suspension systems can only improve one performance, the road holding or handling capabilities. However, these two factors should be considered together. In order to deal with the ride quality, a strategy known as the preview control is adopted in this study. The principle of preview control is to predict the road condition for the rear wheel based on the road information obtained from the front of the vehicle so that the control action can be performed without too much delay. This idea was firstly introduced by Bender [23] and was then proved by several studies [24–29] to be a very promising approach with the use of hydraulic active suspensions. In general, two types of preview methods are available, namely, the front preview and the wheelbase preview [26]. For the first one, a road measurement sensor is placed in the front of the vehicle so that the road information can be obtained a few meters ahead of the two front wheels, whereas for the latter one, the road profile for the rear wheels is the road information obtained based on the displacement of the front wheels with a time delay [27]. In the semi-active suspension systems, wheelbase preview strategy is viewed as a low-cost solution because it only controls the rear suspension so as to save the number of variable dampers and sensors and control hardware for the front suspension.

Since more expensive sensors and accessories are required for the front preview, the wheelbase preview method is chosen for this research. The wheelbase preview control may not be better than the front preview, but the wheelbase preview control can at least provide an alternative and low-cost solution to the chassis engineer to choose. Nevertheless, in the existing wheelbase preview control strategy, the preview information is obtained by the vertical acceleration of the sprung mass in the front wheel. However, the vertical acceleration of the axle (or the unsprung mass) is actually more reliable than the vertical acceleration of the sprung mass in reflecting the road roughness as the latter one may be affected by the high frequency vibration of the spring and the damper [30]. Hence, the recent literature proved that sensors placed on the front axle should be more useful and convincing for the measurement of road roughness. With the application of the improved road measuring method, the road prediction process differs from the traditional one. Therefore, a new road prediction model that can handle the relationship between the vertical acceleration of front axle and the road input of rear wheel is demonstrated in this study as well.

Besides, during the road measuring process of traditional wheelbase preview method (and so as the front preview), all the state variables are assumed to be measured without any noise, yet the noise disturbance from the sensors during measurement cannot usually be avoided The noise disturbance, in fact, greatly influences the performance of the suspension system. Therefore, in order to improve the robustness of the system against noise, a wavelet denoising filter, or the so-called wavelet transform [31], is employed to filter out the useless measurement noise in the proposed semi-active air suspension. The wavelet transform is very suitable for nonstationary signal. Recent studies already applied this technique to isolate the automotive engine noise for signal analysis [32] and proved it to be superior to traditional denoising method such as Fourier analysis [33]. Hence, with the use of wavelet transform, fuzzy controller, and the new measurement idea in the wheelbase preview control, this paper proposes a new semi-active suspension with fuzzy-preview control and wavelet filter (SFPW).

The rest of this paper is organized as follows. Section 2 presents the general idea of the proposed controller. For the construction of a preview model, a half-vehicle model equipped with a semi-active air suspension system is then created in Section 3. As the springs used are air springs, a nonlinear rolling lobe air spring model is designed in this section too. Then for simulation purpose, a road input model is required. It is described in detail along with the novel road prediction model in Section 4. In Section 5, the proposed SPFW for semi-active air suspension system is described in detail. The controller can control not only the tire deflection, but also the sprung mass acceleration so that the semi-active damping force can be adjusted constantly for better ride comfort and road holding capacity. In fact, no existing work has integrated a fuzzy controller to semi-active air suspension with wheelbase preview control, and none of the previous studies has included the external signal correction between the fuzzy and preview controllers before the generation of the semi-active damping force. Moreover, the proposed SFPW is simulated under the environment of heavy-duty vehicles with semi-active suspensions. Therefore, this is an original research to deal with such issues in this area. In order to show effectiveness of the wavelet filter and wheelbase preview controller, a fuzzy-wheelbase preview controller without wavelet filter (SFP) and a sole fuzzy controller (SF) are employed, respectively, in the same semi-active suspension as comparisons. Also, to verify the superior performance of the proposed SFPW method with previous research, a traditional active suspension with PID-wheelbase preview controller (APP) [24] is introduced as the third comparison type. The corresponding results and analysis are discussed in Section 6, and a conclusion is provided in Section 7.

2. Proposed SPFW System

The purpose of the proposed SFPW is to control some variable dampers in the suspension system properly so as to provide comfort for passengers, eliminate the road damage, and protect the freight. In general, the rear suspensions

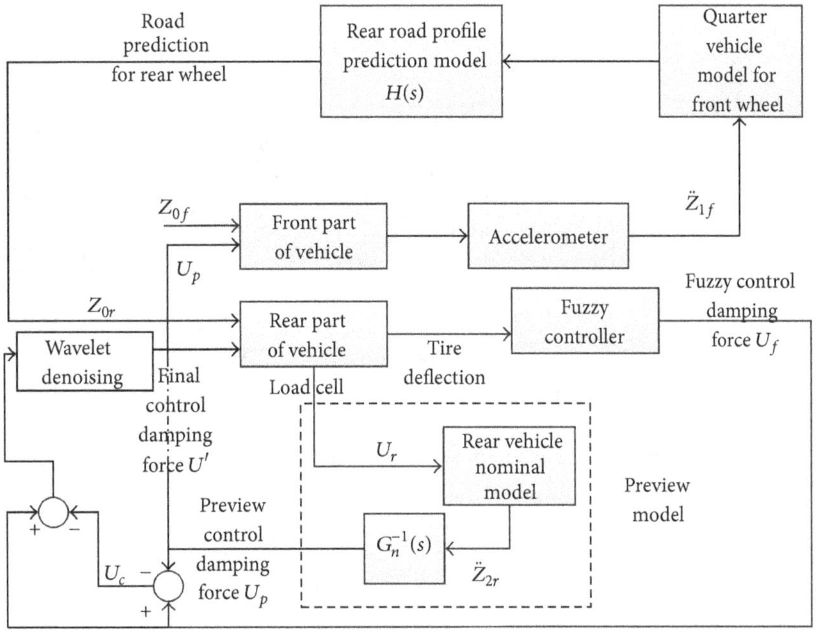

FIGURE 1: Schematic diagram of fuzzy-wheelbase preview control method.

undertake the main load of the heavy-duty vehicle, so the tire deflection and sprung mass acceleration are found in the rear wheels seriously. Therefore, reducing the tire deflection and optimizing the sprung mass acceleration are more desired than only concerning the sprung mass acceleration for the rear suspensions. Consequently, the proposed SFPW mainly consists of three parts—the preview model, the fuzzy controller, and the wavelet denosing filter. The preview model aims to optimize the sprung mass acceleration (i.e., the ride comfort) for front and rear suspensions, and the fuzzy controller deals with the tire deflection of the rear suspension, whereas the wavelet denoising filter suppresses the useless noise for producing a noise-insensitive control damping force. The schematic diagram of the proposed SFPW is depicted in Figure 1, in which the road input Z_{0f} is firstly provided to the front wheel of the vehicle.

With an accelerometer equipped on the front unsprung mass, the vertical acceleration can be detected, and then the road prediction model (Section 3.2) can predict the road information Z_{0r} for the rear wheel. With the predicted road profile, the damping force of the rear suspension U_r can be measured from the load cell, which is used as the input of the preview nominal model, then the sprung mass acceleration of rear suspension $\ddot{Z}_{2r}$ can be obtained, and finally the preview control damping force U_p for the rear suspension can then be determined using $G_n^{-1}(s)$. The preview control damping force is then forwarded as two equivalent output signals. One is sent to the front variable damper of the suspension system as its control damping force, and the other one is sent as an external correction signal for the fuzzy controller output.

Apart from the preview control damping force, the tire deflection and its derivative from the rear part of the vehicle are entered to the fuzzy controller. The controller then, based on the fuzzy rules, provides a fuzzy control damping force U_f. After that, this force is combined with the aforementioned external correction signal of the preview control damping force U_p, and an initial correction signal U_c is obtained. In consideration of the accuracy and stability of the final control damping force, the initial correction signal is then combined with the fuzzy control damping force U_f again, and the output of the above two signals is sent to the wavelet denoising filter. After filtering the useless noise, the final control damping force U' is then delivered to the rear variable damper to adjust the damping force of the rear suspension.

3. Half-Vehicle Modeling for Preview Control

3.1. Half Vehicle Model with Semi-Active Air Suspension System. There are currently two types of heavy-duty vehicle suspension models available in the literature. One is unconnected type and the other is connected type [34]. As the application of the air spring is usually found in the unconnected type, this research mainly focuses on the unconnected suspension. Since it is quite difficult to implement this proposed system on a real vehicle to test, computer simulation is employed to demonstrate the effectiveness of the proposed system. To simplify the simulation, a half-vehicle model is considered in this research. The free body diagram of a half vehicle with unconnected suspension is shown in Figure 2. The dot on the vehicle mass is the static equilibrium position. In the semi-active suspension system, the control damping force in each wheel is provided by the corresponding variable damper.

Considering the dot as the origin of the displacement of the mass center and the angular displacement of the vehicle

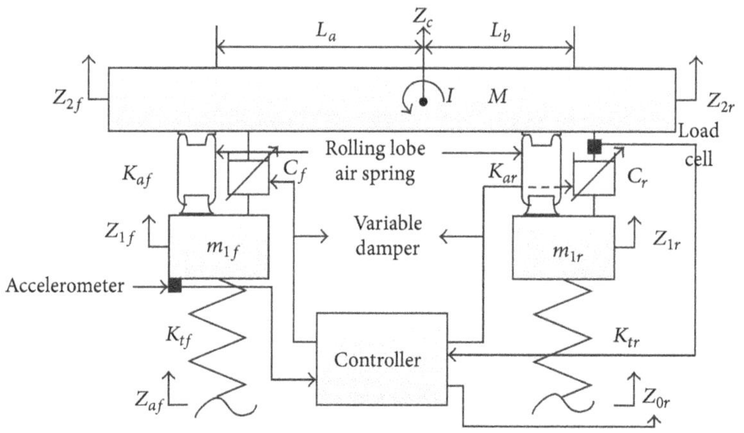

FIGURE 2: Half-vehicle model with semi-active air suspension.

body, the equations of motion for the front and rear unsprung masses are, respectively.

$$\begin{aligned} m_{1f}\ddot{Z}_{1f} &= K_{tf}\left(Z_{0f}-Z_{1f}\right) \\ &\quad - K_{af}\cdot\left(Z_{1f}-Z_{2f}\right) - C_f\left(\dot{Z}_{1f}-\dot{Z}_{2f}\right), \\ m_{1r}\ddot{Z}_{1r} &= K_{tr}\left(Z_{0r}-Z_{1r}\right) \\ &\quad - K_{ar}\cdot\left(Z_{1r}-Z_{2r}\right) - C_r\left(\dot{Z}_{1r}-\dot{Z}_{2r}\right) - U_p, \end{aligned} \tag{1}$$

where U_p is the wheelbase preview control damping force in rear wheel vehicle.

Moreover, the equations of linear motion and angular motion of the whole vehicle body are.

$$\begin{aligned} M\ddot{Z}_c = &- K_{af}\left(Z_{2f}-Z_{1f}\right) - K_{ar}\left(Z_{2r}-Z_{1r}\right) \\ &- C_f\left(\dot{Z}_{2f}-\dot{Z}_{1f}\right) - C_r\left(\dot{Z}_{2r}-\dot{Z}_{1r}\right) + U_p, \end{aligned} \tag{2}$$

$$\begin{aligned} I\ddot{\theta} = &La\left[K_{af}\left(Z_{2f}-Z_{1f}\right) + C_f\left(\dot{Z}_{2f}-\dot{Z}_{1f}\right)\right] \\ &- Lb\left[K_{ar}\left(Z_{2r}-Z_{1r}\right) + C_r\left(\dot{Z}_{2r}-\dot{Z}_{1r}\right) - U_p\right]. \end{aligned} \tag{3}$$

When the pitch angle θ is assumed to be very small, the displacement equations of front and rear sprung mass displacement are

$$\begin{aligned} Z_{2f} &= Z_c - L_a\theta, \\ Z_{2r} &= Z_c + L_b\theta. \end{aligned} \tag{4}$$

θ and Z_c in (4) can be obtained as.

$$\begin{aligned} \theta &= \frac{Z_{2r}-Z_{2f}}{L_a+L_b}, \\ Z_c &= \frac{L_aZ_{2r}+L_bZ_{2f}}{L_a+L_b}. \end{aligned} \tag{5}$$

Then, by combining (2) and (5), the accelerations of the front and rear sprung mass can be reformulated as follows:

$$\begin{aligned} \ddot{Z}_{2f} = &\left(\frac{1}{M}+\frac{L_a^2}{I}\right)\left[K_{af}\left(Z_{1f}-Z_{2f}\right)+C_f\left(\dot{Z}_{1f}-\dot{Z}_{2f}\right)\right] \\ &+\left(\frac{1}{M}-\frac{L_a\cdot L_b}{I}\right)\left[K_{ar}\left(Z_{1r}-Z_{2r}\right)+C_r\left(\dot{Z}_2-\dot{Z}_r\right)\right], \\ \ddot{Z}_{2r} = &\left(\frac{1}{M}-\frac{L_a\cdot L_b}{I}\right)\left[K_{af}\left(Z_{1f}-Z_{2f}\right)+C_f\left(\dot{Z}_{1f}-\dot{Z}_{2f}\right)\right] \\ &+\left(\frac{1}{M}+\frac{L_b^2}{I}\right) \\ &\times\left[K_{ar}\left(Z_{1r}-Z_{2r}\right)+C_r\left(\dot{Z}_{1r}-\dot{Z}_{2r}\right)+U_p\right]. \end{aligned} \tag{6}$$

By using (1) and (6), the semi-active air suspension for half-vehicle model can be constructed.

3.2. Air Spring Model. The type of air spring used in this study is the rolling lobe air spring, which is typically nonlinear. Yet most of the existing researches for semi-active suspension systems only use linear mechanical springs for their system models. Thus, a nonlinear spring model is used as one of original works in the modeling of the semi-active suspension system. Referring to Fox's description [35], the modeling of the rolling lobe air spring should be based on the effective volume, area, and other structural parameters. The variation of the interior air pressure and the cavity volume of air spring caused by the change of load should also be considered during the modeling process. Therefore, the rate of change of the effective area, as the most important characteristic, is derived first, and then the stiffness of the air spring system is derived accordingly. The whole air spring modeling procedure is conducted based on the laws of thermodynamics, and with the assumptions that the thermodynamic parameters do not vary with the position inside the air spring.

The effective area of air spring is simply a linear variable with the change of the height of air spring. For any given moment, the effective area of air spring can be defined as

$$A = A_0 + \beta x, \tag{7}$$

where A_0 is the initial effective area of air spring, β is the change of the effective area with respect to x, and$x$ is the instantaneous height variation of air spring (i.e., $x = Z_{1f} - Z_{2f}$ or $x = Z_{1r} - Z_{2r}$).

Apart from the effective area, the effective volume also changes at the same time. The effective volume of air spring can be defined as:

$$V = V_0 - \alpha x, \tag{8}$$

where V_0 is the initial effective volume of air spring and α is the change of the effective volume with respect to x.

When the air spring is moving slowly, it is generally assumed that the interior gas of air spring is isothermal condition, so the interior pressure of the spring obeys the Boyle's law:

$$(p + p_a) V = (p_a + p_0) V_0, \tag{9}$$

where p is the air pressure of final state, p_a is the standard atmospheric pressure, p_0 is the air pressure of initial state, V is the final effective volume of air spring, and V_0 is the initial effective volume of air spring.

If the compression or expansion stroke is rapid enough, it can be regarded as an adiabatic process. Then, the air state can be defined as

$$(p + p_a) V^\gamma = (p_a + p_0) V_0^\gamma, \tag{10}$$

where γ is specific heat ratio.

However, in practice, the interior gas in air spring is neither isothermal process nor adiabatic process. Rather, it is a multivariable process between the isothermal and adiabatic processes. If the air is under the isothermal process, the stiffness of the air spring is small. If the air is under adiabatic process, the stiffness of the air spring is large. Therefore, the polytropic coefficient γ is somewhere between the isothermal and adiabatic states. Since the current process tends to be an adiabatic process, a value around 1.3–1.4 is taken. Based on (10), the final interior air pressure of air spring in this study is defined as

$$p = (p_0 + p_a) \left(\frac{V_0}{V}\right)^\gamma - p_a. \tag{11}$$

The elastic force of air spring can be calculated as

$$F = pA. \tag{12}$$

By substituting (11) into (12), the equation becomes

$$F = \left[(p_0 + p_a) \left(\frac{V_0}{V}\right)^\gamma - p_a\right] A. \tag{13}$$

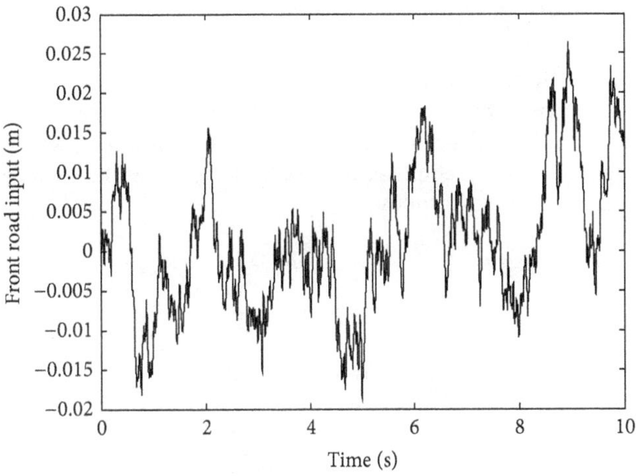

Figure 3: Road input of front wheel.

The effective area and the effective volume are functions of displacement, so the stiffness of air spring can be obtained with the derivative of displacement as follows:

$$k = \frac{dF}{dx} = -A\gamma (p + p_a) \frac{1}{V} \frac{dV}{dx} + p\frac{dA}{dx}. \tag{14}$$

The stiffness of air spring at standard height (when the displacement $x = 0$) is

$$k = \frac{dF}{dx} = -A\gamma\alpha \left(\frac{m_2 g}{A_0} + P_a\right) \frac{1}{V} + \frac{m_2 g}{A_0} \frac{dA}{dx}\bigg|_{x=0}. \tag{15}$$

4. Road Input and Road Prediction Models

4.1. Road Input Model. The function of the road input model is to provide the road information for the front wheel. To simulate the road roughness, a time domain displacement model is used in this study as the road input model, with the assumption that the input signal is band-limited white noise. The random road roughness (white noise) is used to represent the disturbance of road input. The standard mathematical model of a random excitation of road surface for single wheel with a constant vehicle speed v is expressed as

$$\dot{Z}_{0f} + \mu v Z_{0f} = \omega (t), \tag{16}$$

where μ is the road spatial frequency and $\omega(t)$ is the white noise. In this study, the road spatial frequency and vehicle speed v are defined as 0.1303 and 60 km/h, respectively. The corresponding time domain road profile is shown in Figure 3.

4.2. Road Prediction Model for Rear Wheel. The function of the road prediction model is to predict the road information for the rear wheel based on the road information sensed by the front wheel. In most of the previous studies, the road information for the rear wheel is assumed to have the same magnitude as that of the front wheel, but with different phases. Obviously, the vibration of the vehicle body is ignored because disturbance signals will be inevitably introduced during the prediction process. Therefore, to deal with this problem, the front wheel and the rear wheel of the vehicle should be considered separately, so a quarter vehicle model should be used as the base to create the road prediction model. Then, considering half of the free body diagram in Figure 2, the equations of motion for a front quarter vehicle can be obtained as follows:

$$\begin{aligned} m_{2f}\ddot{Z}_{2f} + C_f\left(\dot{Z}_{2f} - \dot{Z}_{1f}\right) + K_{af}\left(Z_{2f} - Z_{1f}\right) &= 0, \\ m_{1f}\ddot{Z}_{1f} + C_f\left(\dot{Z}_{1f} - \dot{Z}_{2f}\right) + K_{af}\left(Z_{1f} - Z_{2f}\right) & \\ + K_{tf}\left(Z_{0f} - Z_{1f}\right) &= 0. \end{aligned} \tag{17}$$

As mentioned in Section 1, the vertical acceleration of the unsprung mass is used instead of the vertical acceleration of the sprung mass in the road prediction procedure. Hence, the relationship between the road input and the vertical acceleration of the unsprung mass should be determined. The front unsprung mass acceleration is defined as

$$X(s) = \ddot{Z}_{1f}, \tag{18}$$

where $X(s)$ means the vertical acceleration of front axle.

By applying Laplace transform to (17) and (18) and combining them together, the relationship can first be obtained as the following transfer function:

$$H(s) = \frac{Z_{0r}(s)}{X(s)} = \frac{b_3 s^3 + b_2 s^2 + b_1 s + b_0}{a_4 s^4 + a_3 s^3 + a_2 s^2}, \tag{19}$$

where

$$\begin{aligned} a_2 &= K_{af}K_{tf}, \\ a_3 &= C_f K_{af}, \\ a_4 &= m_{2f}K_{tf}, \\ b_0 &= K_{af}K_{tf}, \\ b_1 &= C_f K_{tf} + m_{1f}K_{af}, \\ b_2 &= m_{2f}K_{af} + m_{2f}K_{tf} + C_f m_{1f}, \\ b_3 &= m_{1f}m_{2f} + C_f m_{2f}. \end{aligned} \tag{20}$$

With the constructed half-vehicle model in Section 3 and (20), the road information for the rear wheel, Z_{0r}, can be predicted for any given $X(s)$.

5. Design of Wavelet Fuzzy-Wheelbase Preview Controller

5.1. Wavelet Denoising Filter. There are three types of wavelet transform, which are continuous wavelet transform (CWT), wavelet packet transform (WPT), and discrete wavelet transform (DWT). The CWT is mainly used for signal analysis, the WPT is usually used for signal decomposition while the DWT is often used for data preconditioning and processing. The role of the wavelet denoising filter in this study is to suppress the signal disturbance before it is being entered into the rear half-vehicle suspension, so the DWT is used in this study and it can be represented as

$$\mathbf{w} = \mathbf{W}f, \tag{21}$$

where $\mathbf{w}$ is a vector containing wavelet transform coefficients, $\mathbf{W}$ is the matrix of the wavelet filter coefficients, and f is the discrete signal. Each $\mathbf{W}$ is a set of basic vectors, and these vectors can be derived from a mother wavelet by translation and dilation; the mother wavelet ψ is

$$\psi_{a,b}(t) = \frac{1}{\sqrt{|a|}}\psi\left(\frac{t-b}{a}\right), \tag{22}$$

where a is a scaling variable and b is a translation variable.

The main steps of wavelet denoising are signal decomposition, signal thresholding (used for elimination of small coefficients), and signal reconstruction. Regarding the decomposition and reconstruction of the signal, a dyadic dilations and translation mother wavelet method proposed by Mallat [36] is used:

$$\psi_{(s,l)} = 2^{-s/2}\psi\left(2^{-s}t - l\right), \tag{23}$$

where s is the scaling variable and l is the location index. Then, by constraining the ψ, s, and l, an orthogonal decomposition can be achieved. In this study, the composition level is 3 and the threshold method is Haar. Finally, the signal construction in the case of orthogonal W can be described with the following equation and the filtered signal f can be obtained:

$$f = \mathbf{W}^t\mathbf{w}. \tag{24}$$

5.2. Fuzzy Controller. In classical semi-active vehicle suspension system, the state feedback of disturbance signals (SFDS), which are important for the fuzzy controller, cannot be measured [37]. While the fuzzy controller in this study is combined with the preview model, by which the SFDS of the vehicle can be effectively introduced, the absence of the SFDS is effectively avoided.

The design of the fuzzy controller is as follows: the tire deflection "E" and its derivation "EC" are the two input variables, while the command signal for the variable damper "U" is corresponding output of the controller. The value of U can be viewed as the damping force in this study, The shape of membership function for all the three variables is the Gaussian function, in which six linguistic variables are defined, including Negative Big "NB", Negative Medium "NM", Negative Small "NS," Zero "Z," Positive Small "PS", Positive Medium "PM", and Positive Big "PB." The type of fuzzy inference system is Mamdani, and the fuzzy rules are given in Table 1.

5.3. Preview Model. In the preview control model, firstly, a nominal quarter vehicle model is used to obtain the response of vehicle state, and the transfer function is

$$G_n(s) = \frac{Z(s)}{U_r(s)}, \tag{25}$$

Table 1: Rules of the fuzzy controller.

U		EC						
		NB	NM	NS	Z	PS	PM	PB
	NB	PB	PB	PM	PM	PS	Z	Z
	NM	PB	PM	PM	PM	PS	Z	Z
	NS	PS	PS	PS	PS	PS	NS	NS
E	Z	PS	PS	Z	Z	NS	NM	NM
	PS	PS	Z	NS	NS	NS	NS	NM
	PM	PS	Z	NS	NM	NM	NM	NB
	PB	Z	NS	NS	NM	NM	NB	NB

where $Z(s)$ is the output in s-domain (i.e., the sprung mass acceleration $\ddot{Z}_{2r}$) and $U_r(s)$ is the real input in s-domain from the load cell (i.e., the damping force for rear suspension). In other words, the damping force is first provided to the nominal quarter vehicle model and the corresponding sprung mass acceleration of rear quarter vehicle can be predicted. Then, the sprung mass acceleration computed from the nominal model is used to calculate the preview control damping force U_p with the following transfer function:

$$\begin{aligned} G_n^{-1}(s) &= \frac{U_p(s)}{\ddot{Z}_{2r}(s)} \\ &= \left(m_{2r}m_{1r}s^4 + (m_{2r}+m_{1r})C_r s^3\right. \\ &\quad + (m_{1r}K_{ar} + m_{2r}K_{ar} + m_{2r}K_{tr})s^2 \\ &\quad \left.+C_r K_{tr} s + K_{tr}K_{ar}\right) \times \left(m_{1r}s^4 + K_{tr}s^2\right)^{-1}. \end{aligned} \quad (26)$$

Table 2: Input parameters of half-vehicle model.

Parameter	Unit	Value
C_f	N·m/s	15000
C_r	N·m/s	15000
I	kg·m^2	73603
K_{tf}	kN/m	592000
K_{tr}	kN/m	3000000
L_a	m	1.113
L_b	m	3.2
M	Kg	4174
m_{1f}	kg	145
m_{1r}	kg	312
m_{2f}	kg	917
m_{2r}	kg	3257
A_{0f^*}	m^2	0.0381
A_{0r^*}	m^2	0.0706
α_{f^*}	/	0.059
α_{r^*}	/	0.0827
β_{f^*}	/	0.0186
β_{r^*}	/	0.0207
V_{0f^*}	m^3	0.0078
V_{0r^*}	m^3	0.0233
p_{0f^*}	Mp$_a$	0.382
p_{0r^*}	Mp$_a$	0.3
p_a	Mp$_a$	0.1
γ	/	1.33

* (i) f stands for the front 1/4 vehicle model.
(ii) r stands for the rear 1/4 vehicle model.

6. Results and Analysis

In order to show the performance of the proposed SFPW on semi-active air suspension, simulations under heavy-duty vehicle environment were carried out based on the models created in the previous sections and parameters in Table 2. The simulations were also conducted using SFP, SF, and traditional APP, so that comparison can be made and the effectiveness of the proposed SFPW can be realized. In addition, the modeling of the fully active suspension system, which is used for the APP in comparison, has almost the same modeling process with the semi-active one, except the controller and generation of the active force. Moreover, five characteristics of the suspension are used to represent the performance of the suspension. They are the sprung mass acceleration, the tire deflection, the suspension deflection, the pitch, and the tire load.

Figures 4 and 5 show the sprung mass acceleration and the tire deflection, respectively. Normally, the smaller these values, the better the ride comfort and road holding performance. Therefore, it can be seen that the proposed SFPW could provide the best ride comfort and ride holding performance among all the other methods, regardless of the front or rear suspensions.

Figure 6 demonstrates the suspension deflection of the front and rear quarter vehicles. This time, the performance of APP is superior to the other three control methods. Although SPFW is the worst for the suspension deflection of the front quarter vehicle, it exhibits a better suspension deflection than SF for the rear quarter vehicle, meaning that the SPFW is still comparable in this category.

Figure 7 illustrates the pitch and pitch angle acceleration of the half-vehicle model. A good suspension system should be able to minimize the pitch motion and pitch angle acceleration. As it can be learnt from Figure 7, the APP has a better performance in terms of pitch angle while the SFPW behaves noticeably the best in the pitch angle acceleration in all the control methods.

Figure 8 shows the tire load performance (i.e., road friendliness) of the front and rear wheels. Road friendliness refers to the extent of damage exerted on the road by the vehicle, this performance is particularly important for heavy-duty vehicles, and this damage is mainly decided by the dynamic tire load [38]. From Figure 8(a), SFPW performs almost the same as SF for the tire load of front suspension, but from Figure 8(b), it can be learnt that SFPW outperforms the others for the rear suspension. Therefore, the road damage can be minimized with the SFPW. Besides, it can be observed from Figures 4 to 8 that SFPW has smoother results than SFP, indicating that the addition of wavelet filter can significantly reduce the noise in the sensor signals.

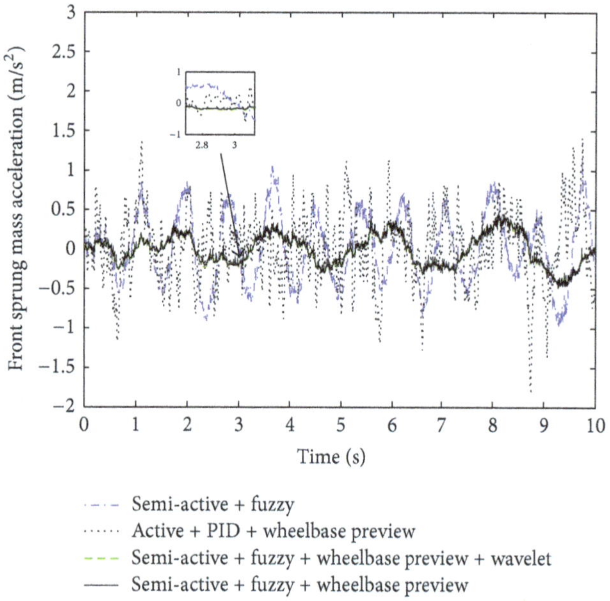

(a) Front sprung mass acceleration

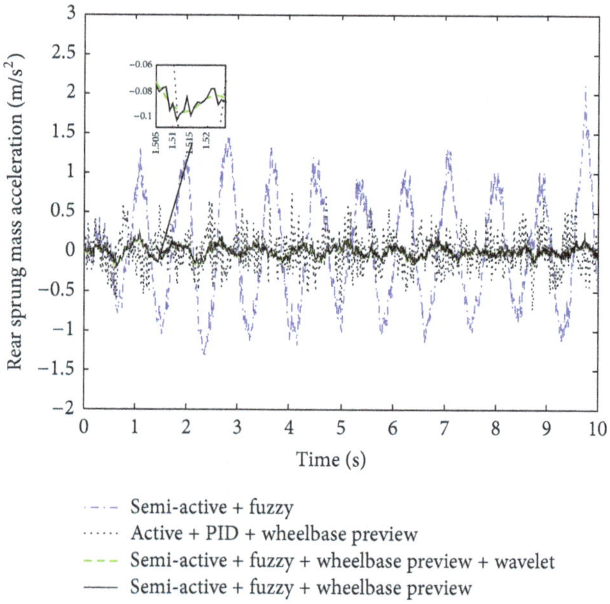

(b) Rear sprung mass acceleration

FIGURE 4: Comparison of sprung mass acceleration.

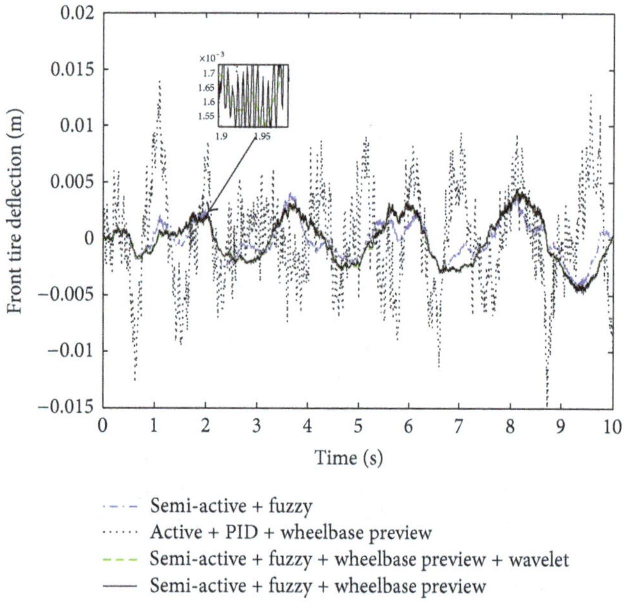

(a) Front tire deflection

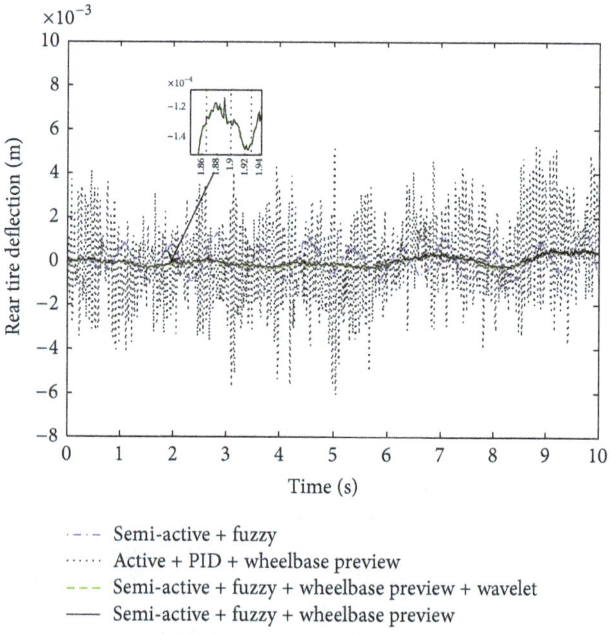

(b) Rear tire deflection

FIGURE 5: Comparison of tire deflection.

Other than comparing the five performance indexes individually, the overall performance evaluation of the controllers is also desired. However, many performance indexes are somewhat in conflict with each other and the direct comparison is difficult [39]. Hence, an overall fitness is used as the evaluation method, which is similar to the fitness function proposed by [40]. The basic idea is to sum up various normalized performance indexes with different user-defined weights. In general, the suspension on a vehicle typically has the following basic tasks, which are providing ride comfort, road holding, competent handling, and the support of the vehicle static weight. However, the handling capacity cannot be reasonably judged in a half-vehicle model, and the ride quality can be quantified by the sprung mass acceleration, the road holding capacity can be represented by the tire deflection, and the support of static weight can be specified

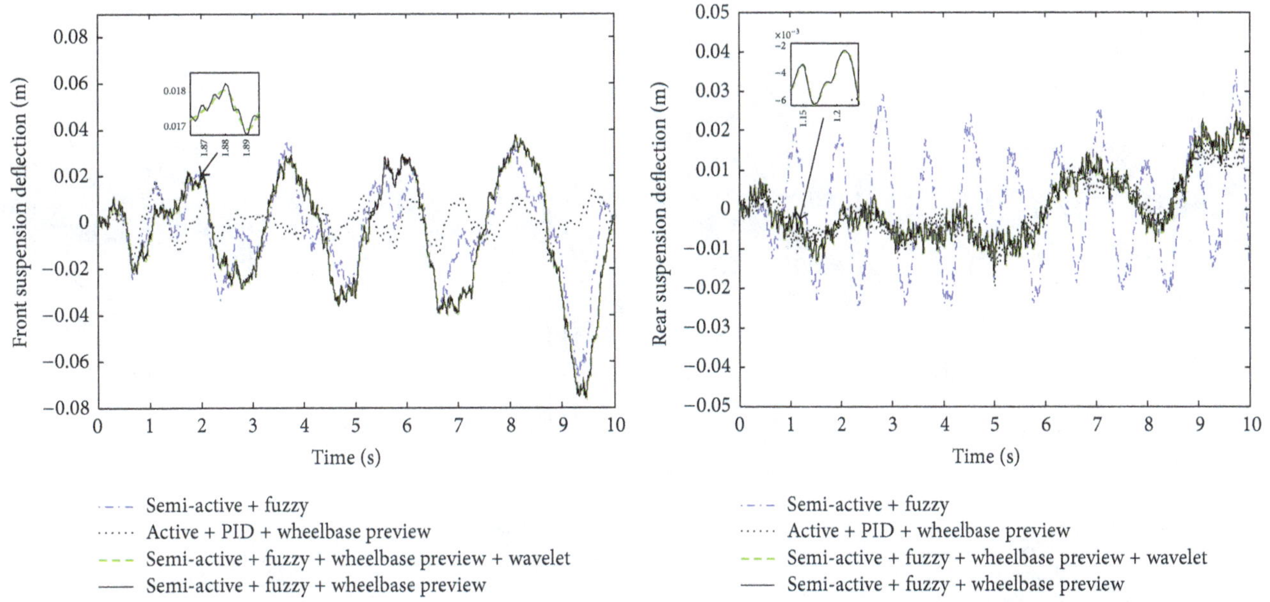

(a) Front suspension deflection

(b) Rear suspension deflection

FIGURE 6: Comparison of suspension deflection.

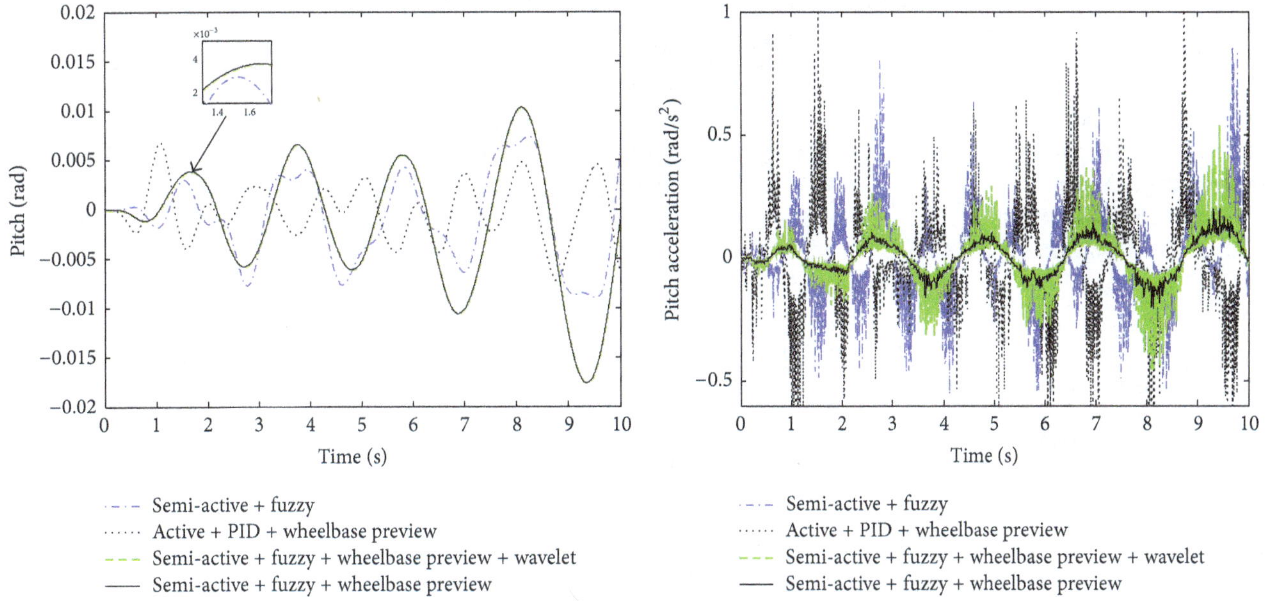

(a) Pitch

(b) Pitch acceleration

FIGURE 7: Comparison of pitch and pitch acceleration.

by the suspension deflection [1]. Thus, in this study, only three objectives were considered instead of five, which are the sprung mass acceleration, the tire deflection, and the suspension deflection.

Regarding the root mean square (RMS) rear suspension fitness (η_r), RMS tire deflection (η_{rt}), RMS sprung mass acceleration (η_{ra}), and RMS suspension deflection (η_{rs}) are ranked in the order of importance, and the corresponding weights w_{rt}, w_{ra}, and w_{rs} were set as 0.4, 0.35, and 0.25, respectively. For the RMS front suspension fitness (η_f), the weights were in the order of RMS sprung mass acceleration (η_{fa}), tire deflection (η_{ft}), and suspension deflection (η_{fs}), and the corresponding weights w_{fa}, w_{ft}, and w_{fs} are 0.45, 0.3, and 0.25, respectively. The weights of the front w_f and rear w_r

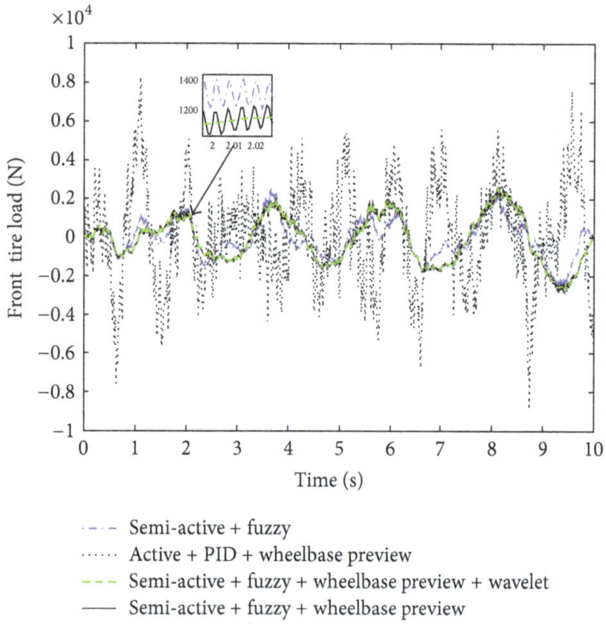

(a) Front tire load

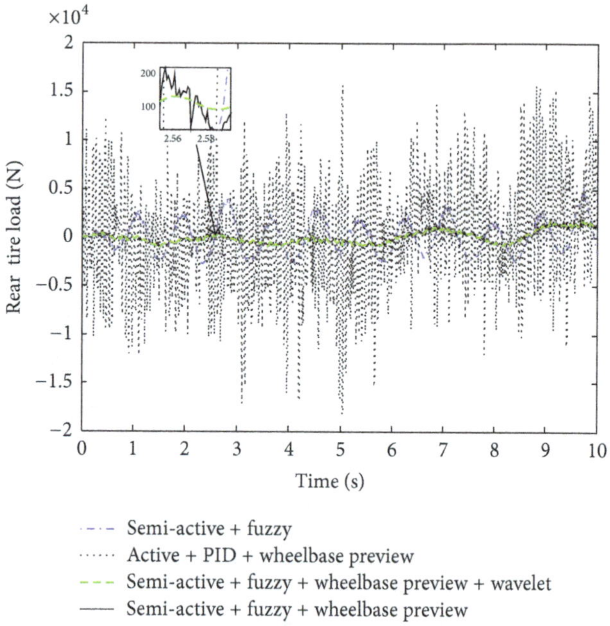

(b) Rear tire load

FIGURE 8: Comparison of tire load.

TABLE 3: Control performance with different control strategies.

Performance	SF*	APP*	SFP*	SFPW*	SFPW versus APP
RMS front sprung mass acceleration	0.4484	0.4477	0.1943	**0.1939**	56.69%
RMS rear sprung mass acceleration	0.7012	0.2217	0.0665	**0.0607**	72.62%
RMS front tire deflection	**0.0016**	0.0044	0.0020	0.0020	54.55%
RMS rear tire deflection	0.0005583	0.0018	0.0002192	**0.0002190**	87.83%
RMS front suspension deflection	0.0194	**0.0064**	0.0258	0.0250	−290.62%
RMS rear suspension deflection	0.0136	**0.0074**	0.0088	0.0085	−14.86%
η_f	5151.5871	**5122.4560**	5321.2273	5272.9123	−2.94%
η_r	5125.8603	4398.9401	3882.3721	**3748.8683**	14.78%
η_c	5136.1510	4688.3465	4457.9142	**4358.4859**	7.04%

*(i) SF: semi-active air suspension with fuzzy control.
(ii) APP: active suspension with PID-wheelbase preview control.
(iii) SFP: semi-active suspension with fuzzy-wheelbase preview control.
(iv) SFPW: semi-active suspension with fuzzy-wheelbase preview control and wavelet filter.

suspension are set as 0.4 and 0.6, respectively. The equations of the overall fitness function of the suspension performance η_c are shown as follows:

$$\eta_c = w_f \eta_f + w_r \eta_r,$$
$$\eta_f = w_{fa} N(\eta_{fa}) + w_{ft} N(\eta_{ft}) + w_{fs}(\eta_{fs}), \quad (27)$$
$$\eta_r = w_{rt}(\eta_{rt}) + w_{ra}(\eta_{ra}) + w_{rs}(\eta_{rs}),$$

where the $N(\bullet)$ is the normalization function which transforms the objective component value to the range $[0,1]$ so as to ensure each component has the same contribution to the fitness function. The general form of the normalization function is given in

$$v_t^* = \frac{(v_t - v_{\min})}{(v_{\max} - v_{\min})}, \quad (28)$$

where v_t^* is the normalized performance index and $v_{\max}$ and $v_{\min}$ are the upper limit and lower limit of the performance index before normalization.

With the above method, the RMS values of the individual performance index and the overall fitness from the simulation results are given in Table 3. It is shown that both the sprung mass acceleration and rear tire deflection of the half vehicle controlled by SFPW are improved remarkablly in contrast with the other three methods. Compared with the APP method, the SFPW method has the following improvements.

(1) The RMS values of front and rear sprung mass accelerations are reduced by 56.69% and 72.62%, respectively.

(2) The RMS values of the front and rear tire deflections are reduced by 54.55% and 87.73%, respectively.

(3) An apparent situation is that the RMS values of the front and rear suspension deflections by SFPW method are increased, respectively, by 290.62% and 14.86%, which are obviously worse than that of the APP method.

(4) The fitness value of the front suspension is decreased by 2.94%, while the fitness of the rear suspension is increased by 14.78%, and the comprehensive performance is improved by 7.04%.

7. Conclusion

In this paper, a new fuzzy-wheelbase preview controller integrated with wavelet filter has been developed for a semi-active air suspension. The wavelet filter is originally used for suppressing the noise of measured signals. Evaluation results show that the controller with wavelet filter can output smooth signals under sudden change of the frequency, which can make the system stable. Moreover, this paper also originally applies the idea of placing the acceleration sensors on the front axle to reasonably reflect the actual road roughness. A fitness function is constructed to formulate the weights of comprehensive performance indexes. The simulation results show that the integrated fuzzy-wheelbase preview controller of semi-active suspension can effectively improve the ride quality and road holding capacity. It is also noted that even though the rear suspension carries the main load in the heavy-duty vehicle, the road friendliness can still be improved significantly. Overall speaking, the comprehensive performance of the vehicle is greatly improved by the proposed controller.

Nomenclature

L_a: Longitudinal distance from front wheel center to center of gravity of vehicle
L_b: Longitudinal distance from rear wheel center to center of gravity of vehicle
C_f: Damping ratio of front suspension
C_r: Damping ratio of rear suspension
g: Acceleration of gravity
I: Total moment of inertia
K_{af}: Stiffness of front suspension
K_{ar}: Stiffness of rear suspension
K_{tf}: Stiffness of front tire
K_{tr}: Stiffness of rear tire
M: Total sprung mass
m_{1f}: Front unsprung mass
m_{1r}: Rear unsprung mass
m_2: Sprung mass of quarter vehicle
m_{2f}: Front sprung mass
m_{2r}: Rear sprung mass
v: Speed of vehicle
Z_{0f}: Road profile of front wheel
Z_{0r}: Road profile of rear wheel
Z_{1f}: Vertical displacement of front unsprung mass
Z_{1r}: Vertical displacement of rear unsprung mass
Z_{2f}: Vertical displacement of front sprung mass
Z_{2r}: Vertical displacement of rear sprung mass
Z_c: Vertical displacement of total sprung mass
θ: Pitch angle.

Acknowledgment

This work has been supported by the following multi-year research funds from University of Macau: MYRG042(Y2-L1)-FST12-XZC, MYRG077(Y1-L2)-FST13-WPK, and MYRG081 (Y2-L2)-FST12-WPK.

References

[1] R. Rajamani, *Vehicle Dynamics and Control*, 2nd edition, 2012.

[2] S. M. Savaresi, V. C. Poussot, and C. Spelta, *Semiactive Suspension Control Design for Vehicles*, Elsevier, 2010.

[3] L. BalaMurugan and J. Jancirani, "An investigation on semi-active suspension damper and control strategies for vehicle ride comfort and road holding," *Proceedings of the Institution of Mechanical Engineers I*, vol. 226, no. 8, pp. 1119–1129, 2012.

[4] M. J. Crosby, R. A. Harwood, and D. Karnopp, "Vibration control using semi-active force generators," *Lord Library of Technical Articles*, vol. 7004, pp. 619–626, 1973.

[5] X. B. Song, M. Ahmadian, and S. Southward, "Analysis and strategy for superharmonics with semiactive suspension control systems," *Journal of Dynamic Systems, Measurement and Control*, vol. 129, no. 6, pp. 795–803, 2007.

[6] K. S. Hong, H. C. Sohn, and J. K. Hedrick, "Modified skyhook control of semi-active suspensions: a new model, gain scheduling, and hardware-in-the-loop tuning," *Journal of Dynamic Systems, Measurement and Control*, vol. 124, no. 1, pp. 158–167, 2002.

[7] D. Sammier, O. Sename, and L. Dugard, "Skyhook and H∞ control of semi-active suspensions: some practical aspects," *Vehicle System Dynamics*, vol. 39, no. 4, pp. 279–308, 2003.

[8] H. C. Sohn, K. S. Hong, and J. K. Hedrick, "Semi-active control of the Macpherson suspension system: hardware-in-the-loop simulations," in *Proceedings of IEEE International Conference on Control Applications*, pp. 982–987, 2000.

[9] M. Ieluzzi, P. Turco, and M. Montiglio, "Development of a heavy truck semi-active suspension control," *Control Engineering Practice*, vol. 14, no. 3, pp. 305–312, 2006.

[10] S. M. Savaresi, E. Silani, and S. Bittanti, "Acceleration-Driven-Damper (ADD): an optimal control algorithm for comfort-oriented semiactive suspensions," *Journal of Dynamic Systems, Measurement and Control*, vol. 127, no. 2, pp. 218–229, 2005.

[11] S. M. Savaresi and C. Spelta, "A single-sensor control strategy for semi-active suspensions," *IEEE Transactions on Control Systems Technology*, vol. 17, no. 1, pp. 143–152, 2009.

[12] H. Bolandhemmat, C. M. Clark, and F. Golnaraghi, "Development of a systematic and practical methodology for the design of vehicles semi-active suspension control system," *Vehicle System Dynamics*, vol. 48, no. 5, pp. 567–585, 2010.

[13] R. Morselli and R. Zanasi, "Control of port Hamiltonian systems by dissipative devices and its application to improve the semi-active suspension behaviour," *Mechatronics*, vol. 18, no. 7, pp. 364–369, 2008.

[14] M. Valasek and W. Kortum, "Road-friendly trucks," *Structural Dynamics*, vol. 1-2, pp. 855–860, 1999.

[15] J. Koo, M. Ahmadian, M. Setareh, and T. M. Murray, "In search of suitable control methods for semi-active tuned vibration absorbers," *Journal of Vibration and Control*, vol. 10, no. 2, pp. 163–174, 2004.

[16] N. Giorgetti, A. Bemporad, H. E. Tseng, and D. Hrovat, "Hybrid model predictive control application towards optimal semi-active suspension," *International Journal of Control*, vol. 79, no. 5, pp. 521–533, 2006.

[17] M. Canale, M. Milanese, and C. Novara, "Semi-active suspension control using "fast" model-predictive techniques," *IEEE Transactions on Control Systems Technology*, vol. 14, no. 6, pp. 1034–1046, 2006.

[18] X. M. Dong, M. Yu, C. R. Liao, and W. M. Chen, "Comparative research on semi-active control strategies for magneto-rheological suspension," *Nonlinear Dynamics*, vol. 59, no. 3, pp. 433–453, 2010.

[19] M. Yu, X. M. Dong, S. B. Choi, and C. R. Liao, "Human simulated intelligent control of vehicle suspension system with MR dampers," *Journal of Sound and Vibration*, vol. 319, no. 3–5, pp. 753–767, 2009.

[20] T. Hashiyama, T. Furuhashi, and Y. Uchikawa, "Study on finding fuzzy rules for semi-active suspension controllers with genetic algorithm," in *Proceedings of IEEE International Conference on Evolutionary Computation*, vol. 1-2, pp. 279–282, December 1995.

[21] T. Yoshimura, K. Nakaminami, and J. Hino, "A semi-active suspension with dynamic absorbers of ground vehicles using fuzzy reasoning," *International Journal of Vehicle Design*, vol. 18, no. 1, pp. 19–34, 1997.

[22] N. Eslaminasab, M. Biglarbegian, W. W. Melek, and M. F. Golnaraghi, "A neural network based fuzzy control approach to improve ride comfort and road handling of heavy vehicles using semi-active dampers," *International Journal of Heavy Vehicle Systems*, vol. 14, no. 2, pp. 135–157, 2007.

[23] E. K. Bender, "Optimum linear preview control with application to vehicle suspension," *Journal of Basic Engineering*, vol. 100, pp. 213–221, 1968.

[24] Z. C. Xie, P. K. Wong, X. Z. Huang, and H. C. Wong, "Design of an active vehicle suspension based on an enhanced PID control with wheelbase preview and tuning using genetic algorithm," *Journal of the Chinese Society of Mechanical Engineers*, vol. 33, pp. 103–112, 2012.

[25] P. K. Wong, S. J. Huang, T. Xu, H. C. Wong, and Z. C. Xie, "Design of a new suspension system controlled by fuzzy-PID with wheelbase preview," *Advanced Mechanical Engineering II*, vol. 192, pp. 106–110, 2012.

[26] P. K. Wong, Z. C. Xie, H. C. Wong, and X. Z. Huang, "Design of a fuzzy preview active suspension system for automobiles," in *Proceedings of the International Conference on System Science and Engineering (ICSSE '11)*, pp. 525–529, June 2011.

[27] M. M. ElMadany, "Control and evaluation of slow-active suspensions with preview for a full car," *Mathematical Problems in Engineering*, vol. 2012, Article ID 375080, 19 pages, 2012.

[28] A. Hac, "Optimal linear preview control of active vehicle suspension," *Vehicle System Dynamics*, vol. 21, no. 3, pp. 167–195, 1992.

[29] M. M. Elmadany, Z. Abduljabbar, and M. Foda, "Optimal preview control of active suspensions with integral constraint," *Journal of Vibration and Control*, vol. 9, no. 12, pp. 1377–1400, 2003.

[30] F. Richter, C. Sourkounis, and N. Bingchang, "Prediction of stochastically fluctuating values using the example of road surface roughness," in *Procedings of the 17th Mediterranean Conference on Control and Automation*, pp. 1009–1013, 2009.

[31] A. Grossmann and J. Morlet, "Decomposition of hardy functions into square integrable wavelets of constant shape," *SIAM Journal on Mathematical Analysis*, vol. 15, pp. 723–736, 1984.

[32] C. M. Vong and P. K. Wong, "Engine ignition signal diagnosis with Wavelet Packet Transform and Multi-class Least Squares Support Vector Machines," *Expert Systems with Applications*, vol. 38, no. 7, pp. 8563–8570, 2011.

[33] B. S. Wu and C. Z. Cai, "Wavelet denoising and its implementation in LabVIEW," in *Proceedings of the 2nd International Congress on Image and Signal Processing (CISP '09)*, vol. 1–9, pp. 400–403, October 2009.

[34] D. P. Cao, S. Rakheja, and C. Y. Su, "Heavy vehicle pitch dynamics and suspension tuning. Part I: unconnected suspension," *Vehicle System Dynamics*, vol. 46, no. 10, pp. 931–953, 2008.

[35] M. N. Fox, R. L. Roebuck, and D. Cebon, "Modelling rolling-lobe air springs," *International Journal of Heavy Vehicle Systems*, vol. 14, no. 3, pp. 254–270, 2007.

[36] S. G. Mallat, "Theory for multiresolution signal decomposition: the wavelet representation," *IEEE Transactions on Pattern Analysis and Machine Intelligence*, vol. 11, no. 7, pp. 674–693, 1989.

[37] D. Krokavec and A. Filasova, "Optimal fuzzy control for a class of nonlinear systems," *Mathematical Problems in Engineering*, vol. 2012, Article ID 481942, 29 pages, 2012.

[38] D. J. Cole and D. Cebon, "Truck suspension design to minimize road damage," *Proceedings of the Institution of Mechanical Engineers D*, vol. 210, no. 2, pp. 95–107, 1996.

[39] L. Chai and T. Sun, "The design of LQG controller for active suspension based on analytic hierarchy process," *Mathematical Problems in Engineering*, vol. 2010, Article ID 701951, 19 pages, 2010.

[40] P. K. Wong, L. M. Tam, and L. Ke, "Automotive engine power performance tuning under numerical and nominal data," *Control Engineering Practice*, vol. 20, no. 3, pp. 300–314, 2012.

11

An Extended Non-Lane-Based Optimal Velocity Model with Dynamic Collaboration

Zhipeng Li and Run Zhang

The Key Laboratory of Embedded System and Service Computing of Ministry of Education, Tongji University, Shanghai 201804, China

Correspondence should be addressed to Zhipeng Li; lizhipeng@tongji.edu.cn

Academic Editor: Valentina E. Balas

Incorporating the effects of the lane width in traffic, in this paper, we propose a dynamical model based on the strategy of three-vehicle cooperation driving. We obtain the smoother acceleration distribution in the new model through considering the dynamic collaboration with the nearest preceding vehicle and the nearest following vehicle. It is proved that the stability of the new model is greatly improved compared to the early non-lane-based car following model by using the linear stability theory. We find that when the parameter of lateral separation distance is identified, the amplitude of traffic congestion decreases with increasing the strength of dynamic collaboration in the simulation experiments. In addition, we apply the new extended model to simulate the motions of cars starting from a traffic signal and the dissipating of the traffic congestion; it is found that our new model can predict realistic delay time and kinematic wave speed and obtained a faster dissipation speed of traffic congestion than the traffic flow model without considering the dynamic collaboration.

1. Introduction

Transportation has become one of the most concern issues in our daily life because the rapid development of modern society relies on the mobility of staff severely. In order to study the problems existing in the traffic, many mathematical models have been done in describing the dynamics of discrete groups of cars by an increasing number of investigators with different backgrounds and points of views. From the research perspective there are essentially two different types of approaches to studying traffic dynamics, namely, macroscopic and microscopic models. As one of the latter models, the car following model in traffic flow has been theoretically studied for nearly 50 years, from the earliest car following model proposed by Pipes in 1953 [1]. Its basic idea is that the current vehicle accelerates when it is slower than its leading vehicle, and decelerates as it is faster than its leading vehicle. This is the easiest and the most basic car following model. Later, in 1995 Bando et al. [2] proposed the optimal velocity model (OVM), which can successfully describe the dynamical formation of traffic congestion. Thereafter, there is a spring tide about the study of the car following model in traffic flow, based on optimal velocity model (OVM). For example, Helbing [3] and Treiber et al. [4] presented the generalized force model (GFM), in which high acceleration and unrealistic deceleration in simulating real traffic do not appear. In 2009, Gaididei et al. [5] conducted a detailed analysis of GFM and found out that although the GFM model has better agreement with the field data than original OVM, it cannot predict correctly the delay time of car motion and the kinematic wave speed at jam density; then they suggested a full velocity difference model (FVDM).

Based on the fact that real-time reporting systems of car information are now becoming widely available, in order to aid emergency dispatch assistance and traffic control management as being an important part of intelligent transportation system (ITS), the study has been focused on using the information of many other cars to suppress the appearance of traffic jams efficiently in traffic flow modeling. By taking the next nearest neighbor interaction into traffic flow modeling, a lattice model of optimized traffic flow with the consideration of optimal current was proposed by Xue [6]. Besides, taking the effect of information about the two leading vehicles to the following vehicle, Peng et al. [7] advised a two-car following model (TCFM) and Ge et al. [8] proposed a two-velocity difference model (TVDM). The difference in the two model is that the headway is one of the crucial parameters in the traffic flow between the second nearest neighbor vehicle and the following vehicle. Moreover, considering the effect of multi-velocity difference ahead, an extended multi-look-ahead driving model is presented by Mo et al. [9]. Considering the multi-velocity difference ahead, it is indicated that the stability is improved greatly and the energy consumption of the traffic flow is reduced obviously. However, all of their work focuses on the information multivehicles in front [10–12]. And it has been proved that the stability of traffic flow will be improved by taking preceding vehicles into consideration. The effect of the following vehicle is not taken into account until Gaididei et al. [5] and Nakayama et al. [13] raised an extended OV model, looking at the following vehicle, which can improve the stability of traffic flow as well [14, 15].

Apart from considering the factor of other vehicle's motion information, the effects of lane width have been studied to participate in the traffic flow modeling. In 2010 Hadiuzzaman and Rahman [16] pointed out the importance of the lane width on vehicle driving behaviors and Jin et al. [17] have been sparing no effort to focus on the effects of the lane width on car following model and conclude that lane width effects have great influence on the stability of traffic flow and the capacity of traffic flow. This is quite natural and very necessary in construction of traffic flow model because in actual traffic fields the influence on the real traffic states is uninterrupted; traffic flow modeling without considering this factor will not truly describe the real traffic flow evolution.

However, to our knowledge, the existing researches using the information of many other cars to suppress the appearance of traffic jams efficiently in traffic flow modeling mostly concentrate on the speeds and the sites information of other vehicles. On the other hand, traffic flow modeling is to help building the vehicular dynamical equation, which determines the power output of each vehicle running on road. So the introduction of the other vehicle's motion information into microscopic modeling of each vehicle will finally convert to the power fine-tuning of a single vehicle in traffic [18, 19]. In fact, the process of slowing down of the traffic congestion of traffic flow is the result of power fine-tuning of individual vehicle in road network, which gives us the hint that direct adjusting of power output of each vehicle based on the other vehicle's power assist may be helpful to result in the stabilization of the traffic flow. So in this paper we build a new traffic flow model by considering the dynamic collaboration with the nearest preceding vehicle and the nearest following vehicle. In order to describe the real traffic characteristics, the extended model also takes the effects of the lane width on vehicle driving behavior into consideration. We will examine whether the new consideration can improve the stability of the traffic flow, and we will also check whether our new model can predict realistic delay time and kinematic wave speed and obtain a faster dissipation speed of traffic congestion than the traffic flow model without considering the dynamic collaboration.

This paper is organized as follows. Firstly, considering the dynamic collaboration, we propose the model based on the non-lane-based car following model. Secondly, by means of the liner stability theory, the model's stability condition is derived. Thirdly, numerical simulations are carried out about the traffic jams, and flux-density fundamental diagram is obtained. In addition, the research about delay times and dissipation speed of traffic congestion is discussed later. Finally, the obtained results are summarized.

2. Model

The optimal velocity model, which is one of the most classical car following models, is based on the assumption that the acceleration of the nth vehicle at time t is determined by the difference between the actual velocity, $v_n(t)$, and an optimal velocity, $V(\Delta x_n(t))$, which depends on the headway $\Delta x_n(t)$ to the car in the front:

$$\frac{dv_j(t)}{dt} = a\left[V\left(\Delta x_j(t)\right) - v_j(t)\right], \tag{1}$$

where a is the sensitivity coefficient representing the driver's reaction time, applied to the difference between the optimal velocity and the current velocity. $\Delta x_j(t) = x_{j-1}(t) - x_j(t)$ is the headway between vehicle $j-1$ and vehicle j at time t.

All the vehicles in the original OVM assumptions are supposed in a straight line exactly in a single lane with no passing. But in real traffic fields the vehicles are usually not in a straight line and appear to canine in most time in real traffic situations as shown in Figure 1. Hence the effects of the lane width are taken into account in the non-lane-based car following model to improve the linear stability just as considered in [11]. So we get the new acceleration

$$\frac{dv_j(t)}{dt} = a\left[V\left(\Delta x_{j,j-1}(t), \Delta x_{j,j-2}(t)\right)\right], \tag{2}$$

where $\Delta x_{j,j-1}(t) = x_{j-1}(t) - x_j(t)$ indicates the distance headway between the vehicle $j+1$ and the vehicle j at the time t; $\Delta x_{j,j-2}(t) = x_{j-2}(t) - x_j(t)$ is headway between the vehicle $j+2$ and the vehicle j at the time t. The function $V[\Delta x_{j,j-1}(t), \Delta x_{j,j-2}(t)]$ can be represented as

$$\begin{aligned} &V\left[\Delta x_{j,j-1}(t), \Delta x_{j,j-2}(t)\right] \\ &\quad = V\left[\left(1-p_j\right)\Delta x_{j,j-1} + p_j \Delta_{j,j-2}\right], \end{aligned} \tag{3}$$

where p_j is the effect parameter of lateral separation distance, which is obtained by the equation $p_j = \mathrm{LS}_j/\mathrm{LS}_{\max}$, and LS_j is the lateral separation distance between vehicle j and vehicle $j-1$. When the leading vehicle was not on the same lane, it would not have impact on the following vehicle at all by the lateral separation distance increasing to the maximal lateral separation distance $\mathrm{LS}_{\max}$. Generally, $\mathrm{LS}_{\max}$ can be set as the typical lane width (3.6 m).

Similarly, we can get the acceleration of vehicle $j-1$ and vehicle $j+1$, respectively, as follows:

$$\begin{aligned} \frac{dv_{j-1}(t)}{dt} &= a\left[V\left(\Delta x_{j-1,j-2}(t), \Delta x_{j-1,j-3}(t)\right)\right], \\ \frac{dv_{j+1}(t)}{dt} &= a\left[V\left(\Delta x_{j+1,j}(t), \Delta x_{j+1,j-1}(t)\right)\right]. \end{aligned} \tag{4}$$

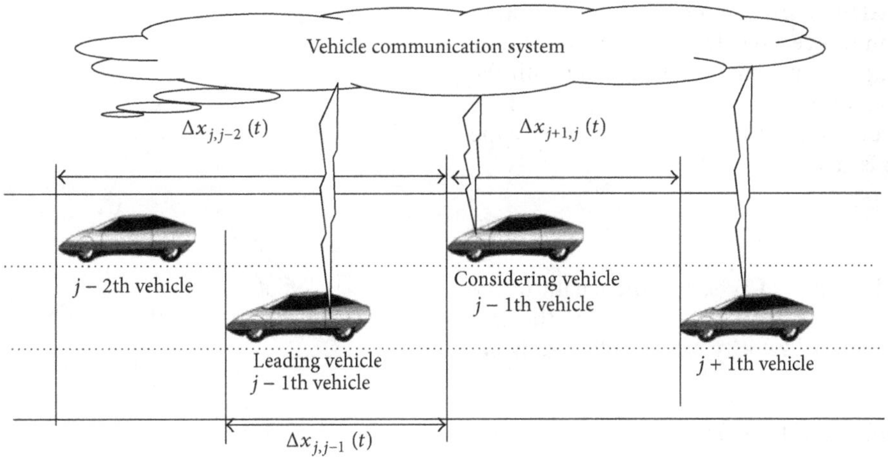

FIGURE 1: Schematic diagram for running vehicles.

Based on the discussion in Section 1, a new car following model is proposed here by extending the non-lane-based car following model to consider dynamic collaboration of the nearest preceding vehicle and the nearest following vehicle (shows in Figure 1). The acceleration of the considering vehicle and its nearest leading vehicle and following vehicle are obtained by the non-lane-based car following model. Besides, the movement of considering vehicle is influenced by its nearest leading vehicle and following vehicle greatly. Depending on both the acceleration of the nearest leading vehicle and that of the following vehicle, the acceleration of the considering vehicle is given as

$$\begin{aligned}\frac{d^2x_j(t)}{dt^2} &= k_l\left[V\left(\Delta x_{j-1,j-2}(t), \Delta x_{j-1,j-3}(t)\right) - x'_{j-1}(t)\right]\\ &\quad + k_c\left[V\left(\Delta x_{j,j-1}(t), \Delta x_{j,j-2}(t)\right) - x'_j(t)\right]\\ &\quad + k_f\left[V\left(\Delta x_{j+1,j}(t), \Delta x_{j+1,j-1}(t)\right) - x'_{j+1}(t)\right],\end{aligned} \tag{5}$$

where the coefficients k_l, k_c, and k_f, are the sensitivity of vehicle $j+1$, j, $j-1$ separately. Generally, k_c must be positive and (6) should be satisfied:

$$k_c - \left(|k_l| + |kf|\right) > 0. \tag{6}$$

Dynamic collaboration is adopted in our extended model, and its own acceleration must hold the dominant component, for it is obtained by the headway between its nearest leading and following vehicles, which have the greatest influence on the considering vehicle. So the weighting coefficients must hold (6).

From Figure 1, we find that every vehicle communicates with the vehicle communication system. Besides, each vehicle uploads its position and velocity to vehicle communication system and the acceleration is computed by (3) with the headway between its nearest leading and following vehicles. Then the acceleration is uploaded and the acceleration of the considering vehicle is adjusted by coupling the dynamics of its nearest leading and following vehicles. What is more, the weighting coefficients can be adjusted. Therefore, (5) is obtained. This is called dynamic collaboration.

Two important elements are considered in our proposed model (5). On one hand, the lateral space among the vehicles affects its moving states. On the other hand, dynamic collaboration has taken the information of the nearest leading vehicle and the nearest following vehicle into account with different weighting coefficients k_l, k_c, k_f.

There is a wide variety of optimal velocities to be selected in the model. Helbing and Tilch carried out a calibration of OVM with respect to the empirical data and suggested an optimal velocity function (OVF) as

$$V\left(\Delta x_j(t)\right) = V_1 + V_2 \tanh\left[C_1\left(\Delta x_j(t) - l_c\right) - C_2\right], \tag{7}$$

where $l_c = 5\,\text{m}$ is the length of vehicle. The other parameter values are determined as follows:

$$\begin{aligned}V_1 &= 6.75\,\text{m/s}, \qquad V_2 = 7.91\,\text{m/s},\\ C_1 &= 0.13, \qquad C_2 = 1.57.\end{aligned} \tag{8}$$

From our new extended model (5), we can observe that not only the lane width effects in car following model are incorporated to describe the real traffic fields, but also the dynamic collaborations of the nearest leading vehicle and the nearest following vehicle are used to smooth the acceleration. With the smoother acceleration, the stability of the traffic flow is expected to be further improved and the traffic jams at the high density will be efficiently suppressed by the new consideration.

3. Linear Stability Analysis

Next we explore whether the proposed model can further stabilize traffic flow with the new consideration of dynamic collaborations of the nearest preceding vehicle and the nearest following vehicle. We apply the linear stability method to the extended model described by (5). Besides, the linear stability method is adopted to estimate the other car following models, such as OVM and GHR. The same method and derivation process are applied in this section. The homogeneous traffic flow is defined by such a state that all vehicles run with the same headway h, the same optimal velocity $V(h, 2h)$, and the same lateral separation LS_j to ensure the same p_j. Hence the steady state solution is given as

$$x_j^{(0)}(t) = V(h, 2h)\,t + jh, \quad h = \frac{L}{N}, \tag{9}$$

where N is the total number of cars, L is the length of the road, and h is the steady headway. What is more, in the uniform traffic flow the optimal velocity can be obtained by the equation below

$$V(h, 2h) = V\left[\left(1 - p_j\right)h + p_j 2h\right], \tag{10}$$

and suppose $y_j(t)$ to be a small deviation from the steady state solution $x_j^0(t)$, so

$$x_j(t) = x_j^{(0)}(t) + y_j(t). \tag{11}$$

Substituting (11) and (9) into (5), it is turned into

$$\begin{aligned} y_j''(t) = {} & k_l\left[V'(h, 2h)\left(\left(1 - p_{j-1}\right)\Delta y_{j-1,j-2}(t) + p_{j-1}\Delta y_{j-1,j-3}(t)\right) - y_{j-1}'(t)\right] \\ & + k_c\left[V'(h, 2h)\left(\left(1 - p_j\right)\Delta y_{j,j-1}(t) + p_j\Delta y_{j,j-2}(t)\right) - y_j'(t)\right] \\ & + k_f\left[V'(h, 2h)\left(\left(1 - p_{j+1}\right)\Delta y_{j+1,j}(t) + p_{j+1}\Delta y_{j+1,j-1}(t)\right) - y_{j+1}'(t)\right], \end{aligned} \tag{12}$$

where $\Delta y_{j,j-1}(t) = y_{j-1}(t) - y_j(t)$ and $V'(h, 2h) = dV(\Delta x_j)/d\Delta x_j \mid \Delta x_j = (1 + p_j)h$.

The $y_j(t)$ can be supposed in the Fourier models, that is, $y_j(t) = A\exp(ikj + zt)$ as it is a small deviation. Substituting it into (12), we obtain

$$\begin{aligned} z^2 = {} & k_c\left[V'(h)\left(\exp(ik) - 1\right) - z\right] \\ & + k_l\left[V'(h)\left(\exp(2ik) - \exp(ik)\right) - z\exp(ik)\right] \\ & + k_f\left[V'(h)\left(1 - \exp(-ik)\right) - z\exp(-ik)\right]. \end{aligned} \tag{13}$$

Expanding $z = z_1(ik) + z_2(ik)^2 + \cdots$ and substituting it into (13), the first- and second-order terms of coefficients in the expression of z are given as

$$\begin{aligned} z_1 &= V'(h, 2h)\left(1 + p_j\right), \\ z_2 &= \frac{V'(h, 2h)\left(1 + 3p_j\right)}{2} - \frac{\left[V'(h, 2h)\right]^2\left(1 + p_j\right)^2}{k_l + k_c + k_f}. \end{aligned} \tag{14}$$

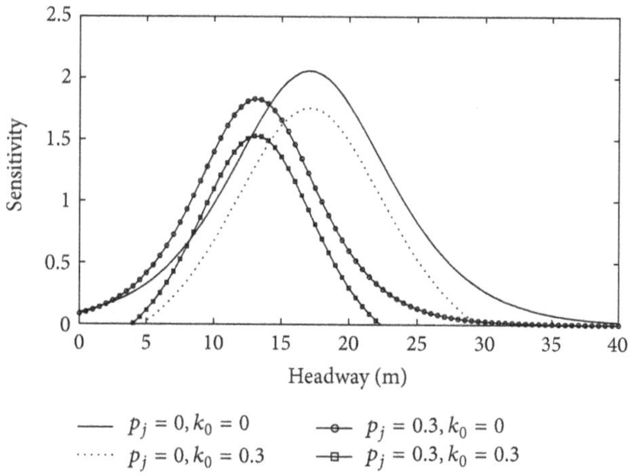

FIGURE 2: Neutral stability lines in the headway-sensitivity space with different (p_j, k_0).

When $z_2 > 0$, the traffic flow is stable; then the neutral stability is obtained by

$$a = k_c = \frac{2V'(h, 2h)\left(1 + p_j\right)^2}{1 + 3p_j} - k_l - k_f. \tag{15}$$

If $a < 2V'(h, 2h)(1 + p_j)^2/(1 + 3p_j) - k_l - k_f$, the traffic flow is unstable in the conditions, for small disturbances with long wavelengths.

To our knowledge, the proof method is widely used in the analysis of car following system and the derivation process is reasonable and concise. Finally, the method is applicative in our extended model.

Figure 2 shows that the neutral stability lines in the headway-sensitivity space with different p_j and k_0 ($k_0 = k_l + k_f$), which are the circumscription between stable region and unstable region of the traffic flow. In the stable region, the traffic flow is stable and no jams occur. In addition, $k_0 = k_l + k_f$ is set to study the influence of the proposed model by the dynamic collaboration of both the nearest leading vehicle and the following vehicle on the traffic flow. Furthermore, it is obtained that the parameters k_l and k_f have the same influence on the stability from (15). However, the traffic flow is unstable below the neutral stability lines and a traffic jam will appear. From Figure 2, we can also find that the stability regions are enlarged with the increasing of p_j, which had been proved by Jin et al. [17]. Besides, the more we consider the dynamic collaboration of the nearest leading vehicle and the nearest following vehicle, the critical points and curves are more lowered in our new model based on the non-lane-based car following model. This means that when the parameter p_j is fixed to a certain value, we also can enlarge the stable region with increasing the weight of dynamic collaboration by the augmentation of the parameter k_0. Apparently, the greater values of parameter k_0 can bring about the greater stability region. Then the simulation results will be listed in the next section.

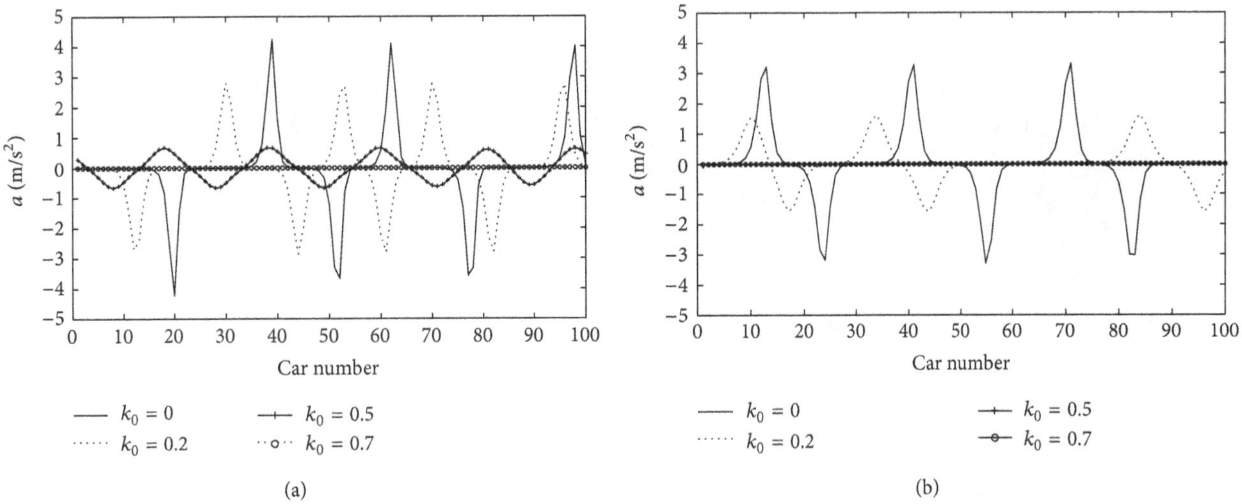

FIGURE 3: Acceleration profiles at sufficiently large time $t = 10000$ s for (a) $p_j = 0.0$ and (b) $p_j = 0.1$.

4. Results

To check the validity of our analysis above, we apply (5) to study numerical simulation for our model described by (5) under the periodic boundary condition. At the beginning, there are 100 vehicles moving with the same optimal velocity $V(h, 2h)$ on a ring road of 1700 m. According to empirical data, we set the value of the parameter k_c, $k_c = a = 1.5$. What is more, the initial simulation conditions and small perturbations are given as

$$\Delta x_1(0) = 18\,\text{m}, \qquad \Delta x_2(0) = 16\,\text{m},$$
$$\Delta x_j(0) = 17\,\text{m} \quad (j \neq 1, 2), \tag{16}$$
$$x'_j(0) = V(h, 2h).$$

Figure 3 shows the acceleration profiles at sufficiently large time $t = 10000$ s with the different parameters p_j and k_0. From this figure, we can observe that the accelerations of these vehicles are smoother with the increasing of the p_j with the same value of k_0, which shows that the lateral separation effects can strengthen the stability of the uniform traffic flow and the traffic jams are suppressed efficiently as proved by Jin [17]. Moreover, if we fixed the lateral separation effects with $p_j = 0.1$, we can obtain the typical spatiotemporal evolution of the headway after sufficiently long time $t = 10^4$ s for different value of the k_0, which is the weight of dynamic collaboration of the nearest leading vehicle and the nearest following vehicle as shown in Figure 4. We can see that the phase transit from free flow to jammed traffic is observed and the jams propagate backward as the kink-antikink density waves. In Figure 4(a), our new model will turn to the non-lane-based optimal velocity car following model as $k_l = k_f = 0$ is set. Figure 5 shows the headway profile obtained at sufficiently large time $t = 10^4$ s. With increasing k_0, and the weight of dynamic collaboration of the nearest leading vehicle and the nearest following vehicle, the amplitude of the density wave is weakened, initial small disturbances decay, and initial traffic flow with a nonuniform density profile evolves to a uniformly traffic flow which is shown in pattern (b) of Figure 3 if we set $p_j = 0.1$ and $k_0 = 0.13$. Compared to the model without dynamic collaboration, all the results show that the new model considering the dynamic collaboration of the nearest leading vehicle and the nearest following vehicle can improve the stability of the traffic flow.

Figure 6 shows the flux density diagram plots of the traffic flow for different value of k_0. Just as we see from the picture, applying the smooth acceleration obtained by dynamic collaboration, we can receive the glossy curve of flux density in the new model. The three states, divided by the neutral stability lines, are free flow (state I), synchronized flow (state II), and homogeneous traffic flow (state III). Besides, vehicles move freely with low density in state I, and the traffic flux mutates with the density among a certain reign. With the higher density, the traffic flow enters into homogeneous flow. However, the traffic flux mutates when the density is higher than the large critical value and is lower than the little critical value in other car following models. Clearly, there is no traffic flux mutation in our extended model, that is to say, our traffic flow is stable and the new model can suppress the mutation with increasing the k_0. Besides, the congestion dissipation is discussed in Section 5.

5. Delay Times and Dissipating Speed of the Traffic Congestion

The two parameters, delay times and dissipating speed, are important to measure car following model. However, lateral space and dynamic collaboration with its nearest leading and following vehicles are considered in our extended model, which are not hoped to worsen the two important parameters. In this section, we apply the new extended model to simulate the motions of cars starting from a traffic signal to find out the difference between our extended model and other classical models like FVD about the delay times and the kinematic wave speed. If velocities of two successive cars are given by

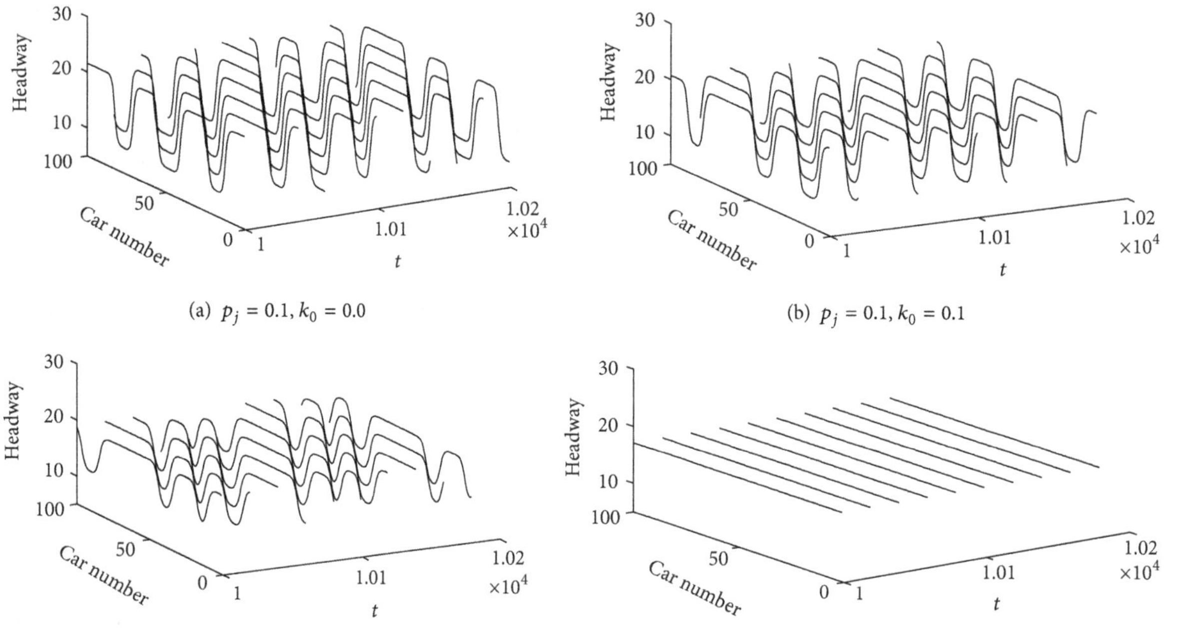

(a) $p_j = 0.1, k_0 = 0.0$ (b) $p_j = 0.1, k_0 = 0.1$

(c) $p_j = 0.1, k_0 = 0.125$ (d) $p_j = 0.1, k_0 = 0.13$

FIGURE 4: Space-time evolution of the headway after 1000000 time steps for different p_j and k_0.

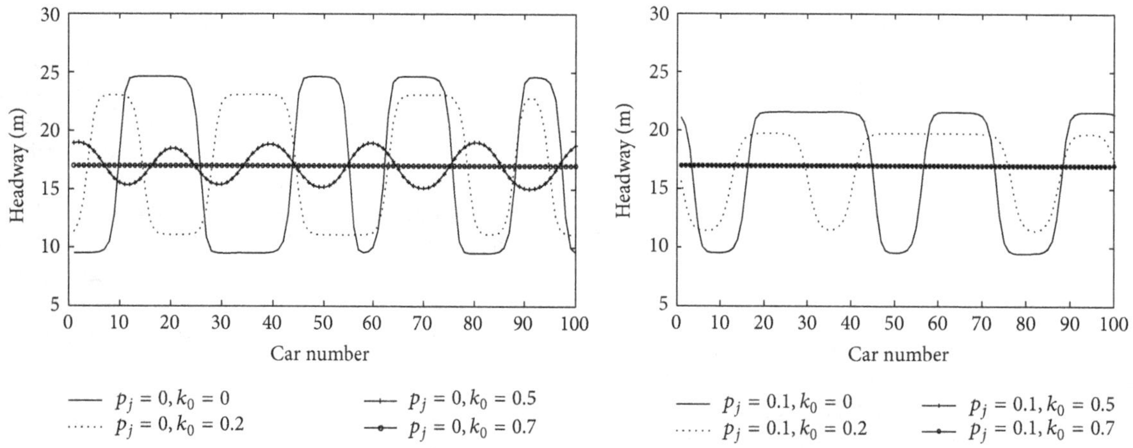

(a) (b)

FIGURE 5: A snapshot of headway of all cars at $t = 10,000$ s for (a) $p_j = 0.0$ and (b) $p_j = 0.1$.

$v(t)$ and $v(t-T)$, respectively, the delay time of car motion is T and the kinematic wave speed v_k is equal to the quotient of the headway 7.4 m divided by T. The delay time of car motion is about one second in actual observation pointed out by Bando. Besides, Del Castillo and Benitez put forward that the range of disturbance propagation speed of car motions is from 17 km/s to 23 km/s. We assume that the traffic signal is red and all cars are waiting with a headway of 7.4 m except the first car for the optimal velocity is zero with the headway obtained by (7), that there are 20 vehicles waiting at the traffic signal with the identical headway h, and that all cars start and pass through the bottleneck region one by one at time $t = 0$.

Figure 7 shows the results about delay times and disturbance propagation speed in the simulation of cars motion, from which the delay time is obtained with the two different parameters k_l and k_f in patterns (a) and (b), respectively. The new extended model degenerates to that without the effect of the dynamic collaboration of the nearest leading vehicle and the nearest following vehicle at parameters $k_l = k_f = 0.0$ in pattern (a); that is to say, the current vehicle moves freely without constraints of dynamic collaboration with the nearest forward and nearest backward vehicles. Compared to pattern (a), the delay time becomes shorter and disturbance propagation speed is still in the region which was pointed out

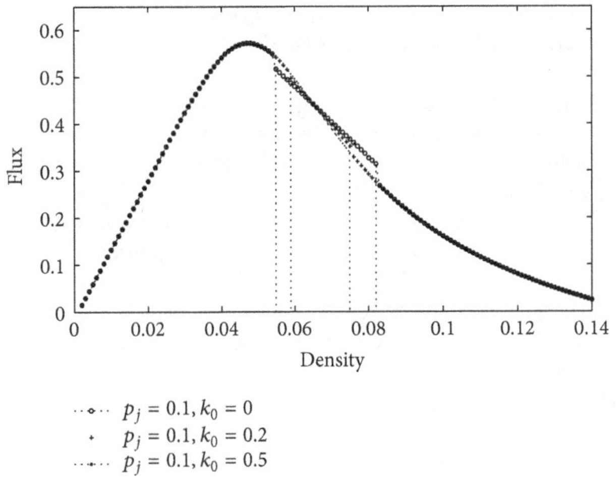

FIGURE 6: Flux-density profiles plot for different k_0.

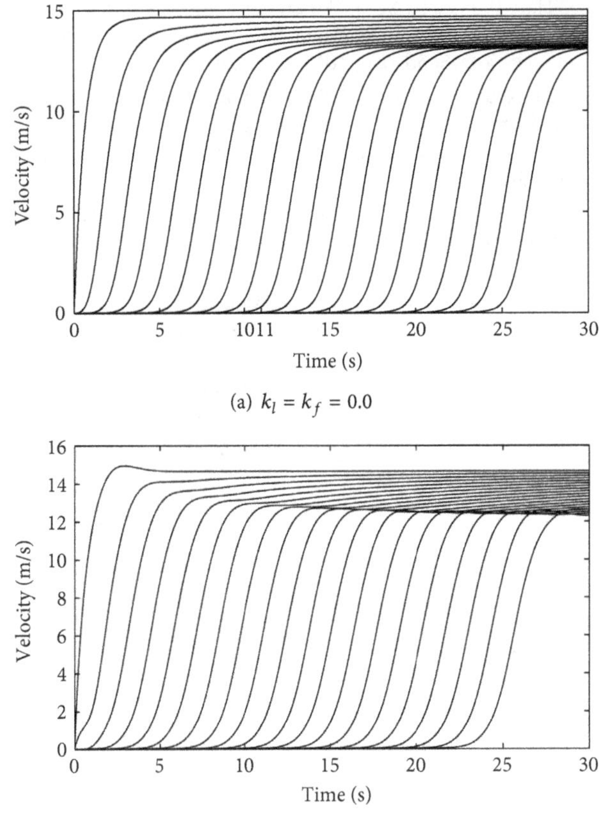

(a) $k_l = k_f = 0.0$

(b) $k_l = k_f = 0.3$

FIGURE 7: Motions of cars starting from a traffic signal.

by Del Castillo and Benitez with dynamic collaboration by the new extended model in pattern (b), in which we set the parameters $k_l = k_f = 0.3$.

The difference between the model without the dynamic collaboration and the new extended model with the dynamic collaboration in delay times and disturbance propagation speed is revealed in Table 1 intuitively. The longer delay time with the dynamic collaboration is not hoped to appear. For comparison, we use the same sensitivity coefficient as that in the new model. From Table 1, we can see that the both models are in accord with the theoretical value, and the delay time is improved a little even more. Therefore, the new extended model is successful in anticipating the delay times and disturbance propagation speed with the shorter delay times.

TABLE 1: Delay times and disturbance propagation speed of car motion from a traffic in different model.

Model	δt (s)	C_j (km/h)
Without the dynamic collaboration	1.38	19.30
With the dynamic collaboration	1.32	20.18

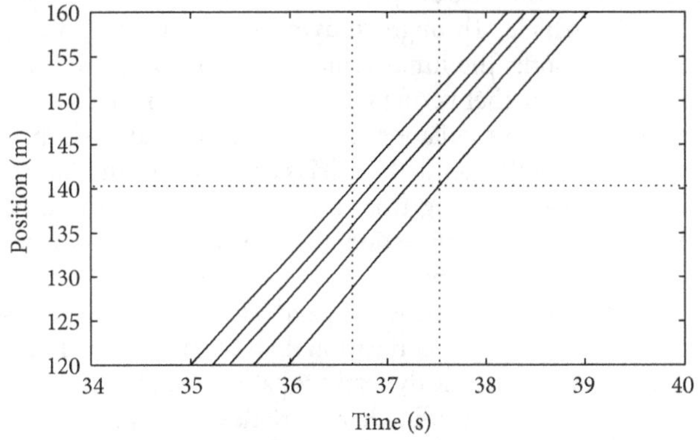

FIGURE 8: Congestion dissipation for different k_0.

What's more, our extended model is useful to dissipating the traffic congestion which is proved in the simulation from Figure 8 as the traffic congestion is becoming more and more serious, especially in Beijing, Shanghai, and other large cities with more vehicles and prosperous economy. In the simulation, we assume that a sufficient number of vehicles wait in the queue in a traffic bottleneck at the beginning and start to accelerate with the smooth acceleration in dynamic collaboration one by one at the time $t > 0$. Here, we take the point that the departure of vehicles from a traffic light and the dissipation from a traffic jam have similar dissipation process. We keep an account of the leaving time by the 20th vehicle and compare the total time of congestion dissipation for the new model with and without the dynamical collaboration.

From Figure 8, it is found that the traffic congestion dissipates faster with increasing the weight of dynamic collaboration. The line of the rightmost represents the parameters $k_l = k_f = 0.0$ with no consideration of dynamic collaboration and the leftmost line stands for the displacement evolution of the twentieth vehicle at the parameters $k_l = k_f = 0.5$. Apparently, the congestion dissipation is improved successfully by the consideration of dynamic collaboration.

6. Conclusion

Incorporating the effects of the lane width in traffic, in this paper, we propose a new model based on the strategy of three-vehicle cooperation driving. Through theoretical analysis and simulations, it has been verified that by introducing the dynamic collaboration of the nearest preceding vehicle and the nearest following vehicle and the effects of the lane width in traffic, the extended model can further improve the stability of traffic flow. Through analyzing the evolution of stop-and-go waves, it has been observed that with the same effect of the line width, traffic congestions can be suppressed efficiently by increasing the strength of dynamic collaboration of the nearest preceding vehicle and the nearest following vehicle. Through studying traffic states and jamming transitions, the improvement of stability of the traffic flow has been further verified along with the increasing intensity of the dynamic collaboration and effect of lane width. In addition, the new extended model is arranged to simulate the motions of cars starting from a traffic signal and the dissipating of the traffic congestion; it is concluded that our new model with the dynamic collaboration can predict realistic delay time and kinematic wave speed, and obtain a faster dissipation speed of traffic congestion than the traffic flow model without considering the dynamic collaboration. In future, the research about the traffic characteristics influenced by the none-lane-change model and the different models taken into the lane changing condition will be our direction.

Acknowledgments

This work is supported by the Natural Science Foundation of China under Grant no. 60904068, Natural Science Foundation of China under Grant no. 10902076, Natural Science Foundation of China under Grant no. 11072117, Natural Science Foundation of China under Grant no. 61004113, the Fundamental Research Funds for the Central Universities under Grant no. 0800219198 and Natural Science Foundation of Shanghai under Grant no. 12ZR1433900.

References

[1] L. A. Pipes, "An operational analysis of traffic dynamics," *Journal of Applied Physics*, vol. 24, no. 3, pp. 274–281, 1953.

[2] M. Bando, H. Hasebe, and A. Nakayama, "Dynamical model of traffic congestion and numerical-simulation," *Physical Review E*, vol. 51, no. 2, pp. 1035–1042, 1995.

[3] D. Helbing, "Generalized force model of traffic dynamics," *Physical Review E*, vol. 58, no. 1, pp. 133–138, 1998.

[4] M. Treiber, A. Hennecke, and D. Helbing, "Derivation, properties, and simulation of a gas-kinetic-based, nonlocal traffic model," *Physical Review E*, vol. 59, no. 1, pp. 239–253, 2003.

[5] Y. B. Gaididei, R. Berkemer, and J. G. Caputo, "A Analytical solutions of jam pattern formation on a ring for a class of optimal velocity traffic models," *New Journal of Physics*, vol. 11, 2009.

[6] Y. Xue, "A car-following model with stochastically considering the relative velocity in a traffic flow," *Acta Physica Sinica*, vol. 52, no. 11, pp. 2750–2756, 2003.

[7] G. H. Peng, D. H. Sun, and H. P. He, "Two-car following model of traffic flow and numerical simulation," *Acta Physica Sinica*, vol. 57, no. 12, pp. 7541–7546, 2008.

[8] H. X. Ge, R. J. Cheng, and Z. P. Li, "Two velocity difference model for a car following theory," *Physica A*, vol. 387, pp. 5239–5245, 2008.

[9] Y. L. Mo, H. D. He, and Y. Xue, "Effect of multi-velocity-difference in traffic flow," *Chinese Physics B*, vol. 17, no. 12, pp. 4446–4450, 2008.

[10] H. Ge, Y. Liu, and R. Cheng, "A modified coupled map car following model and its traffic congestion analysis," *Communications in Nonlinear Science and Numerical Simulation*, vol. 17, no. 11, pp. 4439–4445, 2012.

[11] R. Manzo, B. Piccoli, and L. Rarità, "Optimal distribution of traffic flows in emergency cases," *European Journal of Applied Mathematics*, vol. 23, no. 4, pp. 515–535, 2012.

[12] N. Sugiyama and T. Nagatani, "Multiple-vehicle collision induced by a sudden stop in traffic flow," *Physics Letters A*, vol. 376, no. 22, pp. 1806–1806, 1803.

[13] A. Nakayama, Y. Sugiyama, and K. Hasebe, "Soliton solutions of exactly solvable dissipative systems," *Computer Physics Communications*, vol. 142, no. 1–3, pp. 259–262, 2001.

[14] S. Jin, D. Wang, and C. Xu, "Staggered car-following induced by lateral separation effects in traffic flow," *Physics Letters A*, vol. 376, no. 3, pp. 153–157, 2012.

[15] J. Tian, B. Jia, and X. Li, "A new car following model: comprehensive optimal velocity model," *Communications in Theoretical Physics*, vol. 55, no. 6, pp. 1119–1126, 2011.

[16] M. Hadiuzzaman and M. M. Rahman, "Capacity analysis for fixed-time signalized intersection for non-lane based traffic condition," *Advances in Materials and Processing*, vol. 83, pp. 904–913, 2010.

[17] S. Jin, D. Wang, and P. Tao, "Non-lane-based full velocity difference car following mode," *Physica A*, vol. 389, pp. 4654–4662, 2010.

[18] M. Batista and E. Twrdy, "Optimal velocity functions for car-following models," *Journal of Zhenjiang University Science A*, vol. 11, no. 7, pp. 520–529, 2010.

[19] N. Jia and S. Ma, "Comparison between the optimal velocity model and the Nagel-Schreckenberg model," *Acta Physical Sinica*, vol. 59, no. 2, pp. 832–841, 2010.

12

New Developments in Mathematical Control and Information for Fuzzy Systems

Hamid Reza Karimi,[1] Mohammed Chadli,[2] and Peng Shi[3,4]

[1] *Department of Engineering, Faculty of Engineering and Science, University of Agder, 4898 Grimstad, Norway*
[2] *Laboratory of Modeling, Information and Systems, University of Picardie Jules Verne, 80039 Amiens, France*
[3] *College of Engineering and Science, Victoria University, Melbourne, VIC 8001, Australia*
[4] *School of Electrical and Electronic Engineering, The University of Adelaide, Adelaide, SA 5005, Australia*

Correspondence should be addressed to Hamid Reza Karimi; hamid.r.karimi@uia.no

1. Discussion and Up-To-Date Overview

Intelligence techniques, such as fuzzy logic approaches, have long been applied to dynamical systems with many important theoretical solutions and successful applications. The overall aim of this special issue is both to promote discussion among researchers actively working on applied fuzzy systems and to provide an up-to-date overview of the research directions and advance observer and controllers methods in the field. Of particular interest, the papers in this special issue are devoted to the development of mathematical methodologies analysis, control and observer problems of fuzzy systems, including nonlinear dynamics, fuzzy logic, and interdisciplinary topics with artificial intelligence. Potential topics include, but are not limited to (1) fuzzy controller design (robust control, adaptive control, and supervisory control), (2) fuzzy observer design (adaptive observers, sliding mode observers, and unknown inputs observers), (3) cascaded fuzzy systems and observers, (4) synchronization in fuzzy systems, (5) stochastic fuzzy systems, (6) robust fault detection and isolation, fault diagnosis, and fault-tolerant control of fuzzy systems, (7) applications of fuzzy controllers and observers to complex systems (including hardware and software development environments for fuzzy systems), (8) fuzzy logics, and (9) fuzzy modeling and optimization. For this special issue we solicit submissions from mathematicians, electrical/control/mechanical engineers, and computer scientists. After a rigorous peer review process, 31 papers have been selected. These papers have covered both the theoretical and practical aspects of fuzzy systems in the broad areas of dynamical systems, mathematics, statistics, operational research, and engineering.

During the past decades, the problem of stability analysis of fuzzy systems has received significant attentions. In the paper entitled "*Delay-dependent stability analysis of uncertain fuzzy systems with state and input delays under imperfect premise matching*" by Z. Zhang et al., the robust stability and stabilization problems are studied for general nonlinear fuzzy systems with time-varying state and input delays. For obtaining a less conservative delay-dependent robust stability criterion, the authors introduce a novel augmented Lyapunov function with an additional triple-integral term to the stability analysis for the systems. Moreover, for improving the design flexibility and reducing the implementation cost of the fuzzy controller, they propose a new design approach different from the traditional PDC design technique, which means that the fuzzy model and the fuzzy controllers share different membership functions. Some less conservative stability conditions are obtained, and a new design approach is also proposed. The proposed stability conditions are less conservative in the sense of getting larger allowable time-delay and obtaining smaller feedback control gains, and the design flexibility is enhanced by arbitrarily selecting simple fuzzy membership functions. In the paper entitled "*Stability analysis and stabilization of T-S fuzzy delta operator systems with time-varying delay via an input-output approach*" by Z. Zhong et al., the stability analysis and stabilization problems are investigated for fuzzy delta operator systems with time-varying delay. The delta operator method is introduced to transform T-S fuzzy continuous-time systems into discrete-time systems, and the input-output (IO) approach is employed to deal with the stability analysis and control design of T-S fuzzy delta operator systems with time-varying delay. The main contribution of the paper is that the stability analysis and stabilization problems for fuzzy delta operator systems with time-varying delay are investigated by the IO Approach. A model approximation method is employed to transform the original system into an equivalent interconnected system, which is comprised of a forward subsystem with constant time-delays and a feedback one with delayed uncertainties, such that less conservative results are ensured.

The problems of modeling and identification have long been the main stream of research topics, and much effort has been made for fuzzy systems. In order to approximate any nonlinear systems, not just affine nonlinear systems, generalized T-S fuzzy systems where the control variables and the state variables are all premise variables are introduced in the paper entitled "*A novel identification method for generalized T-S fuzzy systems*" by L. Huang et al.. In order to improving the identification effect, a new method combined colony algorithm and genetic algorithm is proposed. Firstly, fuzzy spaces and rules are determined by using ant colony cluster algorithm to get the best one, and colony cluster algorithm has general optimizing capability, which is better than general cluster algorithm. Secondly, the state-space model parameters are optimized by genetic algorithm based on the least square method, which provided the better consequence parameters. In the paper entitled "*Approximate analytic and numerical solutions to Lane-Emden equation via fuzzy modeling method*" by D. Wang et al., a novel algorithm, called variable weight fuzzy marginal linearization (VWFML) method, is proposed to obtain the approximate analytic and numerical solutions to Lane-Emden equations. The main ideas of the VWFML method are as follows. Firstly, the authors need to transfer a group of data into fuzzy rules. Then, they, respectively, use two partition methods to divide the input universes and utilize fuzzy marginal linearization method to obtain the corresponding fuzzy system. By transferring initial value technology, the authors can solve these two fuzzy systems. Finally, they take weighted sum of these two solutions and obtain the approximate solutions of the Lane-Emden equations. When the Lane-Emden equation is unknown and only data information can obtained, many traditional numerical approximation methods could not solve it. The main contribution of this paper is that the authors can only use some data information to obtain the approximate analytic and numerical solutions to Lane-Emden equation with high accuracy. And this method is easy to be implemented and extended for solving other nonlinear differential equations. In the paper entitled "*Bi-objective optimization method and application of mechanism design based on pigs' payoff game behavior*" by L. Wang et al., a new bi-objective optimization game method is proposed. Two design goals can be regarded as two game players, the design variables set can be regarded as strategy subsets named S_1 and S_2, and the constraints in multiobjective problems can be regarded as constraints in the game method. Through the specific technological means, the design variables can be divided into each game players strategy subsets (S_1 and S_2), and two payoff functions u are constructed based on pigs' payoff game behavior. For optimization problems with preferred target, the designers need to emphasize one design goal. For this problem, there exist traditional methods such as weighting method (by adjusting the weight of each goal), hierarchical sequence method (by adjusting the objective optimization order), and goal programming method. In this paper, one new bi-objective optimization game method is proposed based on pigs' payoff game behavior for solving optimization problems with preferred target. It takes bi-objective optimization of luffing mechanism of compensative shave block; for example, the results show that the method can effectively solve the bi-objective optimization problems with preferred target (designers need to take the preferred target as the small pig side and take another target as the big pig side), the efficiency and accuracy are well, and the solution is obtained only through fewer game rounds. Although the Markowitz mean-variance (MV) portfolio optimization theory has been widely used, which leads the investment theory to a new era, in the paper entitled "*Fuzzy investment portfolio selection models based on interval analysis approach*" by H. Guo et al., the authors found that this model has unintuitive problem in investment portfolio selection in economic decision. Thus this paper employs fuzzy set theory and extends it to a fuzzy investment portfolio selection model to solve this problem. Our model establishes intervals for expected returns and risk preference, which can take into account investors' different investment appetite and thus can find the optimal resolution for each interval. In the empirical part, we test this model in Chinese stocks investment and find that this model can fulfill different kinds of investors' objectives. Furthermore, investment risk can be decreased when we add investment limit to each stock in our model. The results indicate that fuzzy set theory is useful to avoid the problems of Markowitz mean-variance portfolio model and takes into account different expected return levels and risk preference levels. In the paper entitled "*On interval valued supra fuzzy syntopogenous structure*" by F. Sleim and H. Mustafa, the authors generalize the concept of supra fuzzy syntopogenous space by using the notion of interval valued set. Topology and its generalization proximity and syntopogenous are branches of mathematics which have many real life applications. They believe that the generalized topological structure suggested in this paper will be important base for modification of medical diagnosis, decision making, and knowledge discovery.

In the past decades, the issue of control design for fuzzy systems has received considerable research interests and has found successful applications in a variety of areas. An adaptive fuzzy sliding mode controller for a class of uncertain nonlinear systems is proposed in the paper entitled "*Fuzzy sliding mode controller design using Takagi-Sugeno modelled nonlinear systems*" by S. Bououden et al.. The unknown system dynamics and upper bounds of the minimum approximation errors are adaptively updated with stabilizing adaptive laws. The closed-loop system driven by the proposed controllers is shown to be stable with all the adaptation parameters being bounded. The performance and stability of the proposed control system is achieved analytically using the Lyapunov stability theory. In the paper entitled "*Fuzzy variable structure control for uncertain systems with disturbance*" by B. Wang et al., the authors focus on the fuzzy variable structure control for uncertain systems with disturbance. Specifically, the fuzzy control is introduced to estimate the control disturbance, and the switching control is included to compensate the approximation error and possess the characteristic of simpleness in design and effectiveness in attenuating the control chattering. In the paper entitled "*Robust H_∞ control for a class of uncertain switched fuzzy time-delay systems based on T-S models*" by Y. Cui et al., the authors have developed a robust H_∞ control with disturbance attenuation level γ approach for a class of uncertain switched fuzzy time-delay systems based on T-S models. Each and every subsystem of the switched systems is an uncertain fuzzy one to which the PDC (parallel distributed compensation) controller of every subfuzzy system is proposed with its main condition given in a more solvable form of convex combinations. The closed-loop stability of the proposed robust H_∞ control scheme is rigorously proved using Lyapunov theory. Finally, switching law of the state-dependent form achieving system quadratic stability of the switched fuzzy system is given. In the paper entitled "*direct adaptive fuzzy sliding mode control with variable universe fuzzy switching term for a class of MIMO nonlinear systems*" by G. Haigang et al., the authors developed a novel framework for the control of the MIMO nonlinear with model uncertainty and external disturbances. They also proposed a high-precision controller by the variable universe fuzzy control theory. In the paper entitled "*A Fuzzy approach to robust control of stochastic nonaffine nonlinear systems*" by T. Gang et al., the authors investigate the stabilization problem for a class of discrete-time stochastic nonaffine nonlinear systems (SNNS) based on generalized stochastic T-S fuzzy models. By using the function approximation capability of the stochastic T-S fuzzy models, the original SNNS can be exactly represented by a stochastic T-S fuzzy model with some norm bounded approximation errors as the uncertainty term on any compact set. In this way, the stabilization problem of the SNNS can be solved as a robust stabilization problem of the obtained uncertain stochastic T-S fuzzy models. By using a class of piecewise dynamic feedback fuzzy controllers and piecewise quadratic Lyapunov functions, robust semiglobally stabilization of SNNS can be formulated in terms of linear matrix inequalities. In the paper entitled "*adaptive fuzzy tracking control for uncertain nonlinear time-delay systems with unknown dead-zone input*" by C. Chiang, the problem of output tracking control is investigated for a class of uncertain nonlinear state time-delay systems containing unknown dead-zone input and unmatched uncertainties. In general systems, there exist some nonsmooth nonlinearities in the actuators, such as dead zone, saturation, and backlash. However, the dead-zone characteristics in actuators may severely limit the performance of the systems. Also, time-delay characteristic and the existence of uncertain elements usually confronted in engineering systems may degrade the control performance and make the systems unstable. The main features of the proposed robust adaptive fuzzy controller are summarized as follows. (i) An adaptive law is used to estimate the properties of the dead-zone model intuitively and mathematically, without constructing a dead-zone inverse. (ii) Fuzzy logic systems with some appropriate learning laws are applied to approximate the nonlinear gain function and the upper bounded functions of uncertainties. (iii) The unknown upper bound of the uncertainties caused by approximation (or fuzzy modeling) error is estimated by a simple adaptive law. (iv) By means of Lyapunov stability theorem, the proposed controller cannot only guarantee the robust stability of the whole closed-loop system, but also obtain the good tracking performance. In this paper entitled "*Fuzzy PD control of networked control systems based on CMAC neural network*" by L. Huang and J. Chen, the main problem is how to design the fuzzy PD controller and combine with the cerebellar model articulation controller to compose a integral controller which used in the networked control systems effectively. In order to solve this problem, the switching control system between fuzzy PD and PD control is proposed. PD control in the small deviation is applied to obtain higher static control accuracy, and fuzzy PD control in large deviations is applied to obtain faster dynamic response and smaller overshoot. The paper entitled "*Fuzzy control and connected region marking algorithm-based SEM nanomanipulation*" by D. Li et al. is motivated for manipulating the nanocomponent in SEM with telepresence as in macroscale. With the help of virtual reality and haptic technology, the SEM-based master-slave telenanomanipulation platform is established having the performance of security, reliable, and real time without force sensor. The CRM algorithm is introduced to process the 2D SEM image which provides effective position data of the objects for updating the virtual environment. The fuzzy control algorithm is adopted in the master-slave control to obtain relatively stable control variable to avoid damage of platform.

Over the past decades, the observer/filter problems of fuzzy systems have been investigated extensively, since they are very useful in signal processing and engineering applications. In the paper entiled "*Robust observer design for Takagi-Sugeno fuzzy systems with mixed neutral and discrete delays and unknown inputs*" by H. Karimi and M. Chadli, a robust observer design is proposed for Takagi-Sugeno fuzzy neutral models with unknown inputs. The model consists of a mixed neutral and discrete delay, and the disturbances are imposed on both state and output signals. Delay-dependent sufficient conditions for the design of an unknown input T-S observer with time delays are given in terms of linear matrix inequalities. In the paper entitled "*Observer-based robust adaptive fuzzy control for MIMO nonlinear uncertain systems with delayed output*" by C. Chiang, the problem of controller design is considered for a class of MIMO nonlinear uncertain output-delay systems whose states are not available. The common feature of most previous results is the assumption that the controlled systems are free of uncertainties, or the uncertainties are assumed to be a bounded external disturbance. Therefore, the motivation of this paper is to synthesize an observer-based robust adaptive fuzzy control scheme to deal with the tracking control problem for a class of MIMO nonlinear uncertain systems with delayed output in the presence of uncertainties including the structural uncertainty. In the paper entitled "*Unknown input observer design for fuzzy bilinear system: an LMI approach*" by D. Saoudi et al., a new method to design a Fuzzy Bilinear Observer (FBO) with unknown inputs is developed for a class of nonlinear systems. The nonlinear system is modeled as a fuzzy bilinear model (FBM). This kind of T-S fuzzy model is especially suitable for a nonlinear system with a bilinear term. The proposed fuzzy bilinear observer subject to unknown inputs is developed to ensure the asymptotic convergence of the error dynamic using the Lyapunov method. An unknown input fuzzy bilinear fault diagnosis observer design is also proposed. Specifying for a class of nonlinear systems described by Takagi-Sugeno (T-S) model, in the paper, entitled "*A reduced-order TS fuzzy observer scheme with application to actuator faults reconstruction*" by D. Krokavec and A. Filasova, is newly defined the reduced-order T-S fuzzy observer, and it is demonstrated that the matching requirement, under which it can be disposed to actuator faults estimation, can be reflected in the observer design stipulation. The stability conditions, relying on the feasibility of an associated system of linear matrix inequalities (LMI), are derived and guarantee that the observer scheme asymptotically estimates actuator faults. In terms of fuzzy systems, the article provides a suitable new methodology and expands the theoretical basis of TS fuzzy model applications in control system design. The paper entitled "*Terminal sliding mode control using adaptive fuzzy-neural observer*" by D. Xu et al. proposed a dynamic approximation algorithm which is used to simplify the nonaffine nonlinear systems as affine nonlinear systems with time-varying parameters. And the time-varying parameters can be obtained by a filter. Next, an adaptive fuzzy-neural observer is designed to estimate the signals of position, velocity, and unknown functions. Terminal sliding mode control is used to design based on the observer. Stability analysis and simulations can show that the method is effective. The paper entitled "*Filtering for discrete fuzzy stochastic time-delay systems with sensor saturation*" by J. You et al. investigates the H_∞ filtering problem for T-S fuzzy systems with time varying delay and sensor saturation. The communication channel between the plant and the filer is supposed to be imperfect, and random noise depending on the state and external disturbance is taken into account. The key method employed to handle with the time-varying delay is to develop the Scaled Small Gain (SSG) theorem to the stochastic systems. The main contribution is that this paper establishes a research approach that could handle with time varying delay and sensor saturation together, and both characteristics are always involved in many theoretical and practical problems. The paper entitled "*Robust stabilization for continuous Takagi-Sugeno fuzzy system based on observer design*" by Y. Manai and M. Benrejeb investigates the influence of new Parallel Distributed Controller (PDC) on the stabilization region of continuous Takagi-Sugeno (T-S) fuzzy models. Using a nonquadratic Lyapunov function, new sufficient stabilization criterion is established in terms of linear matrix inequality. The criterion examines the derivative membership function; an approach to determine state variables is given based on observer design.

The applications of various control schemes have received considerable research interests in the past decades. In the work entitled "*An iterative procedure for optimizing the performance of the fuzzy-neural job cycle time estimation approach in a wafer fabrication factory*" by T. Chen and Y. Wang, a classifying fuzzy-neural approach, based on the combination of principal component analysis (PCA), fuzzy c-means (FCM), and back propagation network (BPN), is proposed to estimate the cycle time of a job in a wafer fabrication factory. The paper entitled "*Switched two-level H_∞ and robust fuzzy learning control of an overhead crane*" by K. Hung et al. investigates the use of fuzzy techniques for modeling nonlinear plants as a combination of a nominal linear system and a T-S fuzzy blending of affine terms. This type of dynamic model significantly simplifies subsequent analysis and control designs, because assumptions on the plant dynamics can be significantly reduced. In the paper entitled "*A two-wheeled self-balancing robot with the fuzzy PD control method*" by J. Wu et al., the utility and effectiveness of soft computing approaches for a two-wheeled self-balancing robot with structured and unstructured uncertainties are presented. In this approach, precompensation of a hybrid fuzzy PD controller is proposed. The control scheme consists of a fuzzy logic-based precompensator followed by fuzzy PD control. Moreover, a fuzzy supervisory controller is used to supervise conventional proportional and derivative actions, such that the conventional gains are adapted online through fuzzy reasoning. Due to the influence of nonlinear friction, creep phenomenon occurs when the moving speed is low in the telescopic boom system of heavy-load transfer robot. To solve this problem and to improve control precision, B. You et al. in the paper entitled "*Low-speed control of heavy-load transfer robot with long telescopic boom based on stribeck friction model*" built a three-loop control nonlinear model of the system with the Stribeck friction disturbance model to simulate the motion of the telescopic boom in low speed. Fuzzy PID control was used to solve the problem of "flat-top" position tracking and "dead-zone" speed tracking. The creep phenomenon was eliminated, and the tracking accuracy and robustness of the system were also improved. The paper entitled "*Using metaheuristic and fuzzy system for optimization of material-pull in a push pull flow logistics network*" by A. Mehrsai et al. generally complies with a known problem in manufacturing environment, which is the coordination of heterogeneous material flows throughout supply chains and within every production plant. Alternative flows of material, following push and pull principles of control cause collection of inventory and WIP of materials as well as lags in delivery times. Model-based control of material flows regarding the interference of human in the entire process seems relatively very complex to practitioners in industries. In contrast, in this paper some material flow strategies are recommended to simplify and improve the flows throughout. Besides, some heuristics (i.e., genetic algorithm and simulated annealing) to solve the flow coordination are experimented here as well. Fuzzy logic as a suitable solution for solving human centered operations is reasonably explained and experimented in two applications, that is, control of pallets within an assembly system as well as normalization of multiobjective optimization problem. In the paper entitled "*A novel evaluation model for hybrid power system based on vague set and dempster-shafer evidence theory*" by D. Niu et al., due to advantages of vague set processing abundant uncertain and fuzzy information, it is chosen to determine basic decision matrix of evaluation model. Combining vague set with D-S evidence theory, a novel evaluation algorithm is established and applied into the comprehensive benefit evaluation of hybrid power system. In the paper entitled "*H_∞ fuzzy control of DC-DC converters with input constraint*," D. Saifia et al. study fuzzy control of DC-DC converters under actuator saturation. Because linear control design methods do not take into account the nonlinearity of the system, a T-S fuzzy models and a controller design approach are used. The input constraint is first transformed into a symmetric saturation which is represented by a polytopic model. Stabilization conditions for the H_∞ state feedback system of DC-DC converters under actuator saturation are established using the Lyapunov approach. In the paper entitled "*A compound fuzzy disturbance observer based on sliding modes and its application on flight simulator*," Y. Wu et al. present a compound fuzzy disturbance observer based on sliding modes. The proposed method improves the performance of the disturbance inhibition when there exists huge modeling mismatch. Traditional methods use high-gain control, which may cause resonance in controlling elastic electromechanical systems, to inhibit equivalent disturbance. The proposed method employs fuzzy tools to deal with the primary part of the disturbance, and the residual disturbance is compensated by sliding mode control. The proposed method improves the robustness and performance of the system while avoiding the disadvantages of traditional methods. Finally, the paper entitled "*Fuzzy formation control and collision avoidance for multiagent systems*" by Y. Chang et al. investigates the formation control of leader-follower multiagent systems, where the problem of collision avoidance is considered. Based on the graph-theoretic concepts and locally distributed information, a neural fuzzy formation controller is designed with the capability of online learning. The learning rules of controller parameters can be derived from the gradient descent method. To avoid collisions between neighboring agents, a fuzzy separation controller is proposed such that the local minimum problem can be solved.

Appendix

A. Accepted Papers According to Classified Topics

A.1. Papers on the Topic of Stability

(1) Delay-dependent Stability Analysis of Uncertain Fuzzy Systems with State and Input Delays under Imperfect Premise Matching.

(2) Stability Analysis and Stabilization of T-S Fuzzy Delta Operator Systems with Time-Varying Delay via an Input-Output Approach.

A.2. Papers on the Topic of Modeling/Identification

(3) A Novel Identification Method for Generalized T-S Fuzzy Systems.

(4) Approximate analytic and numerical solutions to Lane-Emden equation via fuzzy modeling method.

(5) Bi-objective Optimization Method and Application of Mechanism Design Based on Pigs' Payoff Game Behavior.

(6) Fuzzy investment portfolio selection models based on interval analysis approach.

(7) On interval valued supra fuzzy syntopogenous structure.

A.3. Papers on the Topic of Control for Fuzzy Systems

(8) Fuzzy Sliding Mode Controller Design Using Takagi-Sugeno modelled Nonlinear Systems.

(9) Fuzzy variable structure control for uncertain systems with disturbance.

(10) Robust H_∞ Control for A Class of Uncertain Switched Fuzzy Time-Delay Systems based on T-S Models.

(11) Direct Adaptive Fuzzy Sliding Mode Control with Variable Universe Fuzzy Switching Term for a class of MIMO Nonlinear Systems.

(12) A Fuzzy Approach to Robust Control of Stochastic Non-Affine Nonlinear Systems.

(13) Adaptive Fuzzy Tracking Control for Uncertain Nonlinear Time-Delay Systems with Unknown Dead-Zone Input.

(14) Fuzzy PD Control of Networked Control Systems Based on CMAC Neural Network.

(15) Fuzzy Control and Connected Region Marking Algorithm-based SEM Nanomanipulation.

A.4. Papers on the Topic of Observer/Filter Design for Complex Systems

(16) Robust Observer Design for Takagi-Sugeno Fuzzy Systems with Mixed Neutral and Discrete delays and Unknown Inputs.

(17) Observer-based Robust Adaptive Fuzzy Control for MIMO Nonlinear Uncertain Systems with Delayed Output.

(18) Unknown Input Observer Design for Fuzzy Bilinear System: an LMI Approach.

(19) A reduced-order TS fuzzy observer scheme with application to actuator faults reconstruction.

(20) Terminal Sliding Mode Control using Adaptive Fuzzy-Neural Observer.

(21) Filtering for Discrete Fuzzy Stochastic Time-Delay Systems with Sensor Saturation.

(22) Robust Stabilization for Continuous Takagi-Sugeno Fuzzy System based on Observer Design.

A.5. Papers on the Topic of Applications

(23) An Iterative Procedure for Optimizing the Performance of the Fuzzy-neural Job Cycle Time Estimation Approach in a Wafer Fabrication Factory.

(24) Switched Two-Level H_∞ and Robust Fuzzy Learning Control of an Overhead Crane.

(25) A Two-Wheeled Self-Balancing Robot with the Fuzzy PD Control Method.

(26) Low-Speed Control of Heavy-Load Transfer Robot with Long Telescopic Boom Based on Stribeck Friction Model.

(27) Using Meta-Heuristic and Fuzzy System for Optimization of Material-Pull In A Push-Pull Flow Logistics Network.

(28) A Novel Evaluation Model for Hybrid Power System Based on Vague Set and Dempster-Shafer Evidence Theory.

(29) H_∞ Fuzzy Control of DC-DC Converters with Input Constraint.

(30) A Compound Fuzzy Disturbance Observer based on Sliding Modes and Its Application on Flight Simulator.

(31) Fuzzy Formation Control and Collision Avoidance for Multi-Agent Systems.

Acknowledgments

The guest editors would like to thank all the authors of this special issue for contributing the high quality papers, and we hope the reader will share our joy and find this special issue very useful. We are also very grateful to the reviewers for their efforts in reviewing the papers on time.

Hamid Reza Karimi
Mohammed Chadli
Peng Shi

13

Event-Triggered Reliable Control in Networked Control Systems with Probabilistic Actuator Faults

Yuming Zhai,[1,2] Ruixia Yan,[3] Haifeng Liu,[4] and Jinliang Liu[5]

[1] *School of Economics and Management, Shanghai Institute of Technology, Shanghai 201418, China*
[2] *Antai College of Economics & Management, Shanghai Jiao Tong University, Shanghai 200052, China*
[3] *School of Management, Shanghai University of Engineering Science, Shanghai 201620, China*
[4] *College of Humanities, Donghua University, Shanghai 200051, China*
[5] *Department of Applied Mathematics, Nanjing University of Finance and Economics, Nanjing, Jiangsu 210046, China*

Correspondence should be addressed to Jinliang Liu; liujinliang2009@163.com

Academic Editor: Engang Tian

This paper introduces a novel event-triggered scheme into networked control systems which is used to determine when to transmit the newly sampled state information to the controller. Considering the effect of the network transmission delay and probabilistic actuator fault with different failure rates, a new actuator fault model is proposed under this event-triggered scheme. Then, criteria for the exponential mean square stability (EMSS) and criteria for codesigning both the feedback and the trigger parameters are derived by using Lyapunov functional method. These criteria are obtained in the form of linear matrix inequalities. A simulation example is employed to show that our event-triggered scheme can lead to a larger release period than some existing ones.

1. Introduction

Nowadays, more and more attention has been paid to the study of stability analysis and control design of networked control systems (NCSs). NCSs have a relatively new structure where the links from sensor to controller and from controller to actuator are not connected directly, but through a network [1, 2]. The application of networks into control systems can be advantageous in terms of simplicity scalability, and cost-effectiveness. However, the introduction of a communication network can also bring about many problems, such as network-induced delay and packet dropout. Therefore, the tasks in traditional systems, such as the control problems and signal estimation problem, should be reconsidered. In recent years, the stability analysis and control design for NCS have been invested, and lots of outstanding results have been obtained [3, 4]. In these works, most are based on the periodic triggered control method, which is called a time-trigged control. In this triggering scheme, the fixed sampling period is determined under worst conditions such as external disturbances, uncertainties, and time-delays. However, in practical systems, the worst cases seldom occur. Thus, this kind of triggering method may often lead to transmitting many unnecessary signals through the network, which in turn will increase the load of network transmission and waste the network bandwidth. Hence, it is an important problem to reduce communication requirements.

Recently, event-triggered scheme for control design, advocating the use of actuation only when some function of the system state exceeds a threshold, has received considerable attention, and many important results have been reported [5–9]. Event triggering method provides a useful way of determining when the sampling action is carried out, compared with periodic sampling methods, and it has the following advantages: (1) it only samples when necessary; (2) the burden of the network communication is reduced; (3) the computation cost of the controller and the occupation of the sensor and actuator are reduced.

More specifically, a networked estimator problem for event-triggered sampling systems with packet dropouts was solved in [7]. Refrence[9] studied event design in event-triggered feedback systems, and a novel event triggering scheme was presented to ensure exponential stability of the resulting sampled-data system. Refrence [8] studied how the event triggered as well as the self-triggered control systems could be reformed in the case of discrete-time systems. The methods for design or implication of controllers in the event-triggered form based on dissipation inequalities were proposed for both linear and nonlinear systems in [6]. The work in [5] examined event-triggered data transmission in distributed networked control systems with packet loss and transmission delays.

Moreover, in NCSs, the temporary measurements failure and probabilistic distortion are usually unavoidable for a variety of reasons, for example, networked delay, sensor/actuators aging, electromagnetic interference, and zero shift, which may lead to intolerable system performance [10]. Therefore, from a safety as well as performance point of view, it is required to design a reliable controller that can tolerate actuators failures as well as networked delay. Recently, the fault model has received a lot of interest, and lots of outstanding results have been obtained [11–13]. In [11], the authors considered the problem of delay-dependent adaptive reliable H_∞ controller design against actuator faults for linear time-varying delay systems. By using a linear matrix inequality technique and an adaptive method, they established a new delay-dependent reliable H_∞ controller, which guaranteed the stability and adaptive H_∞ performance of closed-loop systems in normal and faulty cases. Considering the different failure rates of each sensor or actuator, the authors studied the reliable controller for networked control systems in [12, 13], reliable controllers were designed, and sufficient conditions for the exponentially mean square stability of NCS were obtained. Up to now, to the best of authors' knowledge, there are no papers to deal with the reliable control for event-triggered networked control systems with probabilistic actuator faults, which still remains as a challenging problem.

In this paper, firstly, an new event-triggered scheme is introduced to networked control systems, which can reduce the burden of the network communication. Then, considering different failure rates of actuators and the measurements distortion of every actuator, a new probabilistic actuator fault model is proposed under the proposed event-triggered scheme. By using Lyapunov functional method, criteria for the exponential stability and criteria for codesigning both the feedback and the trigger parameters are derived in terms of linear matrix inequalities. A simulation example is given to show that the proposed event-triggered scheme is superior to some existing ones.

Notation. $\mathbb{R}^n$ and $\mathbb{R}^{n\times m}$ denote the n-dimensional Euclidean space and the set of $n \times m$ real matrices; the superscript "T" stands for matrix transposition; I is the identity matrix of appropriate dimension; $\|\cdot\|$ stands for the Euclidean vector norm or the induced matrix 2-norm as appropriate; the notation $X > 0$ (resp., $X \geq 0$) for $X \in \mathbb{R}^{n\times n}$ means that the matrix X is real symmetric positive definite (resp., positive semi definite). When x is a stochastic variable, $\mathbb{E}\{x\}$ stands for the expectation of x. For a matrix B and two symmetric matrices A and C, $\left[\begin{smallmatrix} A & * \\ B & C \end{smallmatrix}\right]$ denotes a symmetric matrix, where $*$ denotes the entries implied by symmetry.

2. System Description

In this paper, we consider the following system:

$$\dot{x}(t) = Ax(t) + Bu(t), \tag{1}$$

where $x(t) \in \mathbb{R}^n$ and $u(t) \in \mathbb{R}^m$ denote the state vector and control vector, respectively; A and B are parameter matrices with appropriate dimensions.

Throughout this paper, we assume that the system (1) is controlled though a network.

As is well known, periodic sampling mechanism has been widely used in practical systems; however, it may often lead to transmitting many unnecessary signals through the network, which in turn increases the load of network transmission and wastes the network bandwidth. Therefore, for the control of networked control systems shown in Figure 1, in order to save network resources such as network bandwidth, it is significant to introduce an event triggered mechanism which decides whether the newly sampled state should be sent out to the controller. As is shown in Figure 1, the state are sampled regularly by the sampler of the smart sensor with period h and feeds into an event generator that decides when to transmit the states to the controller via a network medium by a specified trigger condition, which will be given in sequel. The following function of network architecture in Figure 1 is expected.

(1) The states are sampled at time kh by sampler with a given period h. The next state is at time $(k+1)h$.

(2) As is shown in Figure 1, the event generator is constructed between the sensor and the controller which is used to determine whether the newly sampled state will be sent out to the controller or not by using the following judgement algorithm:

$$\begin{aligned} &[x((k+j)h) - x(kh)]^T \Omega [x((k+j)h) - x(kh)] \\ &\quad \leq \rho x^T((k+j)h)\,\Omega x((k+j)h). \end{aligned} \tag{2}$$

(3) Under the event-triggered scheme (2), the release times are assumed to be $t_0h, t_1h, t_2h, \ldots,$ where t_0 is the initial time. $s_ih = t_{i+1}h - t_ih$ denotes the release period of event generator in (2). Considering the effect of the transmission delay on the network system, we suppose that the time-varying delay in the network communication is τ_k and $\tau_k \in [0, \overline{\tau})$, where $\overline{\tau}$ is a positive real number. Therefore, the states $x(t_0h), x(t_1h), x(t_2h), \ldots$ will arrive at the controller side at the instants $t_0h + \tau_0, t_1h + \tau_1, t_2h + \tau_2, \ldots,$ respectively. Notice that the set of the release instants, that is, $\{t_0, t_1, t_2, \ldots\}$ is a subset of $\{0, 1, 2, \ldots\}$. The amount of $\{t_0, t_1, t_2, \ldots\}$, depends on not only the value of ρ, but also the variation of the state. When $\rho = 0$, $\{t_0, t_1, t_2, \ldots\} = \{0, 1, 2, \ldots\}$, it reduces to the case with periodic release times.

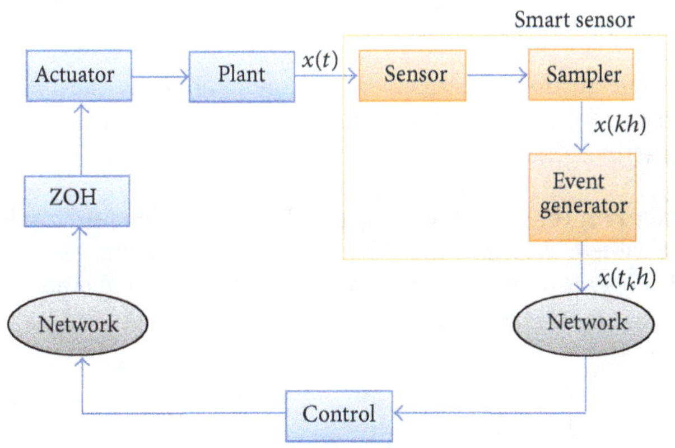

FIGURE 1: The structure of an event-triggered networked control system.

Based on the previous analysis, considering the effect of the transmission delay, the system model under the event generator with (2) can be described as

$$\dot{x}(t) = Ax(t) + Bu(t_k h), \quad t \in [t_k h + \tau_k, t_{k+1}h + \tau_{k+1}). \tag{3}$$

Assumption 1. The actuators in the closed-loop systems have different failure rates because of different working conditions. Furthermore, the measurements distortion of every actuator is also take into consideration.

Under Assumption 1, the control $u(t) = Kx(t)$ can be described as

$$u^F(t) = \Xi Kx(t) = \sum_{i=1}^{m} \xi_i C_i Kx(t), \tag{4}$$

where $\Xi = \operatorname{diag}\{\xi_1, \ldots, \xi_m\}$ and ξ_i $(i = 1, 2, \ldots, m)$ are m unrelated variables taking values on the interval $[0, \theta]$, where $\theta \geq 1$, the mathematical expectation and variance of ξ_i are μ_i and σ_i^2 $(i = 1, 2, \ldots, m)$, and $C_i = \operatorname{diag}\{\underbrace{0, \ldots, 0}_{i-1}, 1, \underbrace{0, \ldots, 0}_{m-i}\}$.

Define $\overline{\Xi} = \operatorname{diag}\{\mu_1, \ldots, \mu_m\} = \sum_{i=1}^{m} \mu_i C_i$, and obviously, $\mathbb{E}(\Xi) = \overline{\Xi}$, $\mathbb{E}(\Xi - \overline{\Xi}) = 0$, $\mathbb{E}(\xi_i - \mu_i)^2 = \sigma_i^2$.

Under (4), for $t \in [t_k h + \tau_k, t_{k+1}h + \tau_{k+1})$, (3) can be rewritten as

$$\dot{x}(t) = Ax(t) + B\Xi Kx(t_k h). \tag{5}$$

Remark 2. $u(t)$ is the control input without actuator failure, and $u^F(t)$ is the control input after actuator failures occur. $\xi_i = 0$ and $\xi_i = 1$ represent the meaning of completely failure or completely normal [14]. $0 < \xi_i < 1$ means partial failure [15, 16]. $\xi_i > 1$ represents the condition of data distortion [12, 13].

Remark 3. $\mu_i = 1$ does not represent that the ith actuator is always in good work condition, but it means that the expectation of ith actuator is 1.

For technical convenience, consider the following two cases.

Case 1. If $t_k h + h + \overline{\tau} \geq t_{k+1}h + \tau_{k+1}$, where $\overline{\tau} = \max \tau_k$, define a function $\tau(t)$ as

$$\tau(t) = t - t_k h, \quad t \in [t_k h + \tau_k, t_{k+1}h + \tau_{k+1}); \tag{6}$$

clearly,

$$\tau_k \leq \tau(t) \leq (t_{k+1} - t_k)h + \tau_{k+1} \leq h + \overline{\tau}. \tag{7}$$

Case 2. If $t_k h + h + \overline{\tau} < t_{k+1}h + \tau_{k+1}$, consider the following two intervals:

$$[t_k h + \tau_k, t_k h + h + \overline{\tau}), \qquad [t_k h + ih + \overline{\tau}, t_k h + ih + h + \overline{\tau}). \tag{8}$$

Since $\tau_k \leq \overline{\tau}$, it can be easily shown that there exists d_M such that

$$t_k h + d_M h + \overline{\tau} < t_{k+1}h + \tau_{k+1} \leq t_k h + d_M h + h + \overline{\tau}. \tag{9}$$

Moreover, $x(t_k h)$ and $t_k h + ih$ with $i = 1, 2, \ldots, d_M$ satisfy (2). Let

$$\begin{aligned} I_0 &= [t_k h + \tau_k, t_k h + h + \overline{\tau}), \\ I_i &= [t_k h + ih + \overline{\tau}, t_k h + ih + h + \overline{\tau}), \\ I_{d_M} &= [t_k h + d_M h + \overline{\tau}, t_{k+1}h + \tau_{k+1}), \end{aligned} \tag{10}$$

where $i = 1, 2, \ldots, d_M - 1$. One can see that

$$[t_k h + \tau_k, t_{k+1}h + \tau_{k+1}) = \bigcup_{i=0}^{i=d_M} I_i. \tag{11}$$

Define

$$\tau(t) = \begin{cases} t - t_k h, & t \in I_0, \\ t - t_k h - ih, & t \in I_i, \ i = 1, 2, \ldots, d_M - 1, \\ t - t_k h - d_M h, & t \in I_{d_M}. \end{cases} \tag{12}$$

then, we have

$$\begin{gathered} \tau_k \leq \tau(t) < h + \overline{\tau}, \quad t \in I_0, \\ \tau_k \leq \overline{\tau} \leq \tau(t) < h + \overline{\tau}, \quad t \in I_i, \ i = 1, 2, \ldots, d_M - 1, \\ \tau_k \leq \overline{\tau} \leq \tau(t) < h + \overline{\tau}, \quad t \in I_{d_M}, \end{gathered} \tag{13}$$

where the third row in (13) holds because $t_{k+1}h + \tau_{k+1} \leq t_k h + (d_M + 1)h + \overline{\tau}$. Obviously,

$$\begin{gathered} 0 \leq \tau_k \leq \tau(t) \leq h + \overline{\tau} \triangleq \tau_M, \\ t \in [t_k h + \tau_k, t_{k+1}h + \tau_{k+1}). \end{gathered} \tag{14}$$

In Case 1, for $t \in [t_k h + \tau_k, t_{k+1}h + \tau_{k+1})$, define $e_k(t) = 0$. In Case 2, define

$$e_k(t) = \begin{cases} 0, & t \in I_0, \\ x(t_k h) - x(t_k h + ih), & t \in I_i, \ i = 1, 2, \ldots, d_M - 1, \\ x(t_k h) - x(t_k h + d_M h), & t \in I_{d_M}. \end{cases} \tag{15}$$

From the definition of $e_k(t)$ and the triggering algorithm (2), it can be easily seen that, for $t \in [t_k h + \tau_k, t_{k+1}h + \tau_{k+1})$,

$$e_k^T(t)\,\Omega e_k(t) \le \rho x^T(t-\tau(t))\,\Omega x(t-\tau(t)). \quad (16)$$

Utilizing $\tau(t)$ and $e_k(t)$, (5) can be rewritten as

$$\begin{aligned}\dot{x}(t) &= Ax(t) + B\Xi K[x(t-\tau(t)) + e_k(t)] \\ &= Ax(t) + B\overline{\Xi}Kx(t-\tau(t)) + B\overline{\Xi}Ke_k(t) \\ &\quad + B(\Xi-\overline{\Xi})Kx(t-\tau(t)) + B(\Xi-\overline{\Xi})Ke_k(t),\end{aligned} \quad (17)$$

where $t \in [t_k h + \tau_k, t_{k+1}h + \tau_{k+1})$.

For the system (17), we supplement the initial condition of the state $x(t)$ on $[\tau_M, 0]$ as

$$x(t) = \phi(t), \quad t \in [\tau_M, 0], \quad (18)$$

where $\phi(t)$ is a continuous function on $[\tau_M, 0]$.

Remark 4. If $\tau_k = 0$, it means that no transmission delay exists or transmission delay can be ignored, and the maximum sampling period is τ_M. Note that $\tau_M = h + \overline{\tau}$. If $\tau_k > 0$, the selecting sampling period $h < \tau_M$, $\overline{\tau} = \tau_M - h$, is the allowable maximum transmission delay.

Remark 5. Notice that if $\rho = 0$, then $e_k(t) = 0$; in this situation, the model (17) reduces to a model for a networked control system with a time-triggered scheme. It means that, if $\rho \to 0^+$, the dynamic of the system under event-triggered scheme will approach to the one with a time-triggered scheme.

In the following, we will introduce the following definitions and lemmas, which are needed in the next section.

Definition 6. For a given function $V : C^b_{F_0}([-\tau_M, 0], \mathbb{R}^n) \times S$, its infinitesimal operator $\mathscr{L}$ is defined as

$$\mathscr{L}(V\eta(t)) = \lim_{\Delta\to 0^+}\frac{1}{\Delta}[\mathbb{E}(V(\eta_t+\Delta)\mid\eta_t) - V(\eta_t)]. \quad (19)$$

Definition 7. System (17) is said to be exponentially of mean square stability (EMSS) if there exist constants $\alpha > 0$ and $\epsilon > 0$ such that for $t \ge 0$,

$$\mathbb{E}\left\{\|x(t)\|^2\right\} \le \alpha e^{-\epsilon t}\mathbb{E}\left\{\sup_{-\tau_M\le s\le 0}\|\psi(s)\|^2\right\}. \quad (20)$$

Lemma 8 (see [17]). *For any vectors $x, y \in R^n$ and positive definite matrix $Q \in R^{n\times n}$, the following inequality holds:*

$$2x^T y \le x^T Qx + y^T Q^{-1} y. \quad (21)$$

Lemma 9 (see [18]). *Ξ_1, Ξ_2, and Ω are matrices with appropriate dimensions, $\tau(t)$ is a function of t, and $0 \le \tau(t) \le \tau_M$; then*

$$\tau(t)\Xi_1 + (\tau_M - \tau(t))\Xi_2 + \Omega < 0 \quad (22)$$

if and only if

$$\begin{aligned}\tau_M\Xi_1 + \Omega &< 0, \\ \tau_M\Xi_2 + \Omega &< 0.\end{aligned} \quad (23)$$

3. Main Results

In this section, the following theorem provides the EMSS criteria for system (17) with the controller (4) under the event generator (2).

Theorem 10. *For given μ_i, σ_i $(i = 1, \ldots, m)$, ρ, and matrix K, the system described by (17) is EMSS, if there exists matrices $P > 0$, $Q > 0$, $R > 0$, $\Omega > 0$, and N, M with which appropriate dimensions such that for $s = 1, 2$,*

$$\Sigma(s) = \begin{bmatrix}\Sigma_{11}+\Gamma+\Gamma^T & * & * & * \\ \Sigma_{21} & -R & * & * \\ \Sigma_{31} & 0 & \Sigma_{33} & * \\ \Sigma_{41}(s) & 0 & 0 & -R\end{bmatrix} < 0, \quad s = 1, 2, \quad (24)$$

where

$$\begin{gathered}\Sigma_{11} = \begin{bmatrix}PA + A^TP + Q & * & * & * \\ K^T\overline{\Xi}^TB^TP & \rho\Omega & * & * \\ 0 & 0 & -Q & * \\ K^T\overline{\Xi}^TB^TP & 0 & 0 & -\Omega\end{bmatrix}, \\ \Gamma = [N \quad -N+M \quad -M \quad 0], \\ \Sigma_{21} = \left[\sqrt{\tau_M}RA \quad \sqrt{\tau_M}RB\overline{\Xi}K \quad 0 \quad \sqrt{\tau_M}RB\overline{\Xi}K\right], \\ \Sigma_{31} = \begin{bmatrix}0 & \Pi & 0 & 0 \\ 0 & 0 & 0 & \Pi\end{bmatrix}, \\ \Pi = \begin{bmatrix}\sigma_1\sqrt{\tau_M}RBC_1K \\ \sigma_2\sqrt{\tau_M}RBC_2K \\ \vdots \\ \sigma_m\sqrt{\tau_M}RBC_mK\end{bmatrix}, \\ \Sigma_{33} = \operatorname{diag}\{\underbrace{-R,\ldots,-R}_{2m}\}, \\ \Sigma_{41}(1) = \sqrt{\tau_M}N^T, \qquad \Sigma_{41}(2) = \sqrt{\tau_M}M^T.\end{gathered} \quad (25)$$

Proof. Choose the following Lyapunov functional candidate as

$$V(x_t) = V_1(x_t) + V_2(x_t) + V_3(x_t), \quad (26)$$

where

$$\begin{aligned}V_1(x_t) &= x^T(t)Px(t), \\ V_2(x_t) &= \int_{t-\tau_M}^{t} x^T(s)Qx(s)\,ds, \\ V_3(x_t) &= \int_{t-\tau_M}^{t}\int_{s}^{t}\dot{x}^T(v)R\dot{x}(v)\,dv\,ds,\end{aligned} \quad (27)$$

in which P, Q, and R are symmetric positive definite matrices.

From the definition of Ξ, it can be concluded that $\mathbb{E}\{\Xi - \overline{\Xi}\} = \text{diag}\{0, \ldots, 0\}$, and using the infinitesimal operator (19) for $V(x_t)$ and taking expectation on it, we obtain

$$\begin{aligned}\mathbb{E}\{\mathscr{L}V_1(x_t)\} &= 2x^T(t)P\Upsilon(t),\\ \mathbb{E}\{\mathscr{L}V_2(x_t)\} &= x^T(t)Qx(t) - x^T(t-\tau(t))Qx(t-\tau(t)),\\ \mathbb{E}\{\mathscr{L}V_3(x_t)\} &= \mathbb{E}\left[\tau_M\dot{x}^T(t)R\dot{x}(t)\right] - \int_{t-\tau_M}^{t}\dot{x}^T(s)R\dot{x}(s)\,ds,\end{aligned} \tag{28}$$

where $\Upsilon(t) = Ax(t) + B\overline{\Xi}Kx(t-\tau(t)) + B\overline{\Xi}Ke_k(t)$.

Notice that

$$\begin{aligned}\mathbb{E}\left[\tau_M\dot{x}^T(t)R\dot{x}(t)\right] &= \tau_M\Upsilon^T(t)R\Upsilon(t)\\ &\quad+\sum_{i=1}^{m}\tau_M\sigma_i^2e_k^T(t)K^TC_i^TB^TRBC_iKe_k(t)\\ &\quad+\sum_{i=1}^{m}\tau_M\sigma_i^2x^T(t-\tau(t))K^TC_i^TB^TRBC_i\\ &\qquad\times Kx(t-\tau(t)).\end{aligned} \tag{29}$$

Combining (28) and (29), we obtain

$$\begin{aligned}\mathbb{E}\{\mathscr{L}V(x_t)\} &= 2x^T(t)P\Upsilon(t) + x^T(t)Qx(t)\\ &\quad- x^T(t-\tau(t))Qx(t-\tau(t))\\ &\quad+\tau_M\Upsilon^T(t)R\Upsilon(t)\\ &\quad+\sum_{i=1}^{m}\tau_M\sigma_i^2e_k^T(t)K^TC_i^TB^TRBC_iKe_k(t)\\ &\quad+\sum_{i=1}^{m}\tau_M\sigma_i^2x^T(t-\tau(t))K^TC_i^TB^TRBC_i\\ &\qquad\times Kx(t-\tau(t))\\ &\quad-\int_{t-\tau_M}^{t}\dot{x}^T(s)R\dot{x}(s)\,ds + \Gamma_1 + \Gamma_2,\end{aligned} \tag{30}$$

where Γ_1 and Γ_2 are introduced by employing free weight matrix method [19, 20]

$$\begin{aligned}\Gamma_1 &= 2\zeta^T(t)N\left[x(t) - x(t-\tau(t)) - \int_{t-\tau(t)}^{t}\dot{x}(s)\,ds\right] = 0,\\ \Gamma_2 &= 2\eta^T(t)M\left[x(t-\tau(t)) - x(t-\tau_M) - \int_{t-\tau_M}^{t-\tau(t)}\dot{x}(s)\,ds\right] = 0,\end{aligned} \tag{31}$$

where N and M are matrices with appropriate dimensions, and

$$\zeta^T(t) = \left[x^T(t)\,x^T(t-\tau(t))\,x^T(t-\tau_M)\,e_k^T(t)\right]. \tag{32}$$

By Lemma 8, we have

$$\begin{aligned}-2\zeta^T(t)N\int_{t-\tau(t)}^{t}\dot{x}(s)\,ds &\le \tau(t)\zeta^T(t)NR^{-1}N^T\zeta(t)\\ &\quad+\int_{t-\tau(t)}^{t}\dot{x}^T(s)R\dot{x}(s)\,ds,\\ -2\zeta^T(t)M\int_{t-\tau_M}^{t-\tau(t)}\dot{x}(s)\,ds &\le (\tau_M-\tau(t))\zeta^T(t)NR^{-1}M^T\zeta(t)\\ &\quad+\int_{t-\tau_M}^{t-\tau(t)}\dot{x}^T(s)R\dot{x}(s)\,ds.\end{aligned} \tag{33}$$

Combining (16), substitute (33) into (30), and we obtain that

$$\begin{aligned}\mathbb{E}\{\mathscr{L}V(x_t)\} &\le 2x^T(t)P\Upsilon(t) + x^T(t)Qx(t)\\ &\quad- x^T(t-\tau(t))Qx(t-\tau(t))\\ &\quad+\tau_M\Upsilon^T(t)R\Upsilon(t)\\ &\quad+\sum_{i=1}^{m}\tau_M\sigma_i^2e_k^T(t)K^TC_i^TB^TRBC_iKe_k(t)\\ &\quad+\sum_{i=1}^{m}\tau_M\sigma_i^2x^T(t-\tau(t))K^TC_i^TB^TRBC_i\\ &\qquad\times Kx(t-\tau(t))\\ &\quad+2\zeta^T(t)N\left[x(t)-x(t-\tau(t))\right]+2\zeta^T(t)M\\ &\quad\times\left[x(t-\tau(t))-x(t-\tau_M)\right]\\ &\quad-e_k^T(t)\Omega e_k(t)+\rho x^T(t-\tau(t))\Omega x(t-\tau(t))\\ &\quad+(\tau_M-\tau(t))\zeta^T(t)MR^{-1}M^T\zeta(t)\\ &\quad+\tau(t)\zeta^T(t)NR^{-1}N^T\zeta(t);\end{aligned} \tag{34}$$

that is,

$$\mathbb{E}\{\mathscr{L}V(x_t)\} \le \zeta^T(t)\Theta\zeta(t), \tag{35}$$

where $\Theta = \Sigma_{11} + \Gamma + \Gamma^T + \Sigma_{21}^TR^{-1}\Sigma_{21} - \Sigma_{31}^T\Sigma_{31}^{-1}\Sigma_{31} + (\tau_M - \tau(t))MR^{-1}M^T + \tau(t)NR^{-1}N^T$.

By using Schur complement and Lemma 9, we have $\Theta < 0$, if and only if the following holds:

$$\begin{bmatrix}\Sigma_{11}+\Gamma+\Gamma^T+\Sigma_{21}^TR^{-1}\Sigma_{21}-\Sigma_{31}^T\Sigma_{31}^{-1}\Sigma_{31} & *\\ \Sigma_{41}(s) & -R\end{bmatrix} < 0,\quad s = 1, 2. \tag{36}$$

By using Schur complement, we can obtain that (36) is equivalent to (24). Furthermore,

$$\mathbb{E}\{\mathscr{L}V(x_t)\} \le -\lambda\zeta^T(t)\zeta(t), \tag{37}$$

where $\lambda = \min\{\lambda_{\min}[\Theta]\}$. Define a new function as

$$W(x_t) = e^{\epsilon t} V(x_t). \tag{38}$$

Its infinitesimal operator $\mathscr{L}$ is given by

$$\mathscr{L}W(x_t) = \epsilon e^{\epsilon t} V(x_t) + e^{\epsilon t} \mathscr{L}V(x_t). \tag{39}$$

From (39), we can obtain that

$$\mathbb{E}W(x_t) - \mathbb{E}W(x_0) = \int_0^t \epsilon e^{\epsilon s} \mathbb{E}\{V(x_s)\} + \int_0^t e^{\epsilon s} \mathbb{E}\{\mathscr{L}V(x_s)\}. \tag{40}$$

Then using the similar method of [15], we can observe that there exists a positive number α such that for $t \geq 0$

$$\mathbb{E}\{V(x_t)\} \leq \alpha \sup_{-\tau_M \leq s \leq 0} e^{-\epsilon s} \mathbb{E}\{\|\psi(s)\|^2\}. \tag{41}$$

Since $V(x_t) \geq \lambda_{\min}(P) x^T(t) x(t)$, it can be shown from (41) that, for $t \geq 0$,

$$\mathbb{E}\{x^T(t) x(t)\} \leq \overline{\alpha} e^{-\epsilon t} \sup_{-\tau_M \leq s \leq 0} \mathbb{E}\{\|\psi(s)\|^2\}, \tag{42}$$

where $\overline{\alpha} = \alpha / \lambda_{\min}(P)$. The proof can be completed. □

In the following, based on Theorem 10, we will design the feedback gain K in (4) under the event-trigger (2).

Theorem 11. *For given μ_i, σ_i $(i = 1, \ldots, m)$, ρ and ε, system (17) with the feedback gain $K = YX^{-1}$ under the event trigger condition (2) is EMSS if there exist matrices $X > 0$, $\widetilde{Q} > 0$, $\widetilde{R} > 0$, $\widetilde{\Omega} > 0$, and $\widetilde{N}$, $\widetilde{M}$ with appropriate dimensions such that $s = 1, 2$*

$$\begin{bmatrix} \widetilde{\Sigma}_{11} + \widetilde{\Gamma} + \widetilde{\Gamma}^T & * & * & * \\ \widetilde{\Sigma}_{21} & -2\varepsilon X + \varepsilon^2 \widetilde{R} & * & * \\ \widetilde{\Sigma}_{31} & 0 & \widetilde{\Sigma}_{33} & * \\ \widetilde{\Sigma}_{41}(s) & 0 & 0 & -2\varepsilon X + \varepsilon^2 \widetilde{R} \end{bmatrix} < 0, \quad s = 1, 2, \tag{43}$$

where

$$\widetilde{\Sigma}_{11} = \begin{bmatrix} AX + XA^T + \widetilde{Q} & * & * & * \\ Y^T \overline{\Xi}^T B^T & \rho \widetilde{\Omega} & * & * \\ 0 & 0 & -\widetilde{Q} & * \\ Y^T \overline{\Xi}^T B^T & 0 & 0 & -\widetilde{\Omega} \end{bmatrix},$$

$$\widetilde{\Gamma} = \begin{bmatrix} \widetilde{N} & -\widetilde{N} + \widetilde{M} & -\widetilde{M} & 0 \end{bmatrix},$$

$$\widetilde{\Sigma}_{21} = \begin{bmatrix} \sqrt{\tau_M} AX & \sqrt{\tau_M} B\overline{\Xi} Y & 0 & \sqrt{\tau_M} B\overline{\Xi} Y \end{bmatrix},$$

$$\widetilde{\Sigma}_{31} = \begin{bmatrix} 0 & \widetilde{\Pi} & 0 & 0 \\ 0 & 0 & 0 & \widetilde{\Pi} \end{bmatrix}, \quad \widetilde{\Pi} = \begin{bmatrix} \sigma_1 \sqrt{\tau_M} BC_1 Y \\ \sigma_2 \sqrt{\tau_M} BC_2 Y \\ \vdots \\ \sigma_m \sqrt{\tau_M} BC_m Y \end{bmatrix},$$

$$\widetilde{\Sigma}_{33} = \operatorname{diag}\{\underbrace{-2\varepsilon X + \varepsilon^2 \widetilde{R}, \ldots, -2\varepsilon X + \varepsilon^2 \widetilde{R}}_{2m}\},$$

$$\widetilde{\Sigma}_{41}(1) = \sqrt{\tau_M} \widetilde{N}^T, \quad \widetilde{\Sigma}_{41}(2) = \sqrt{\tau_M} \widetilde{M}^T. \tag{44}$$

Proof. By using Schur complement, we can obtain that the following is equivalent to (24):

$$\begin{bmatrix} \Sigma_{11} + \Gamma + \Gamma^T & * & * & * \\ \Phi_{21} & -PR^{-1}P & * & * \\ \Phi_{31} & 0 & \Phi_{33} & * \\ \Sigma_{41}(s) & 0 & 0 & -R \end{bmatrix} < 0, \quad s = 1, 2, \tag{45}$$

where

$$\Phi_{21} = \begin{bmatrix} \sqrt{\tau_M} PA & \sqrt{\tau_M} PB\overline{\Xi} K & 0 & \sqrt{\tau_M} PB\overline{\Xi} K \end{bmatrix},$$

$$\Phi_{31} = \begin{bmatrix} 0 & \Pi^1 & 0 & 0 \\ 0 & 0 & 0 & \Pi^1 \end{bmatrix}, \quad \Pi^1 = \begin{bmatrix} \sigma_1 \sqrt{\tau_M} PBC_1 K \\ \sigma_2 \sqrt{\tau_M} PBC_2 K \\ \vdots \\ \sigma_m \sqrt{\tau_M} PBC_m K \end{bmatrix}, \tag{46}$$

$$\Phi_{33} = \operatorname{diag}\{\underbrace{-PR^{-1}P, \ldots, -PR^{-1}P}_{2m}\}.$$

Due to $(R - \varepsilon^{-1}P) R^{-1} (R - \varepsilon^{-1}P) \geq 0$, we have

$$-PR^{-1}P \leq -2\varepsilon P + \varepsilon^2 R. \tag{47}$$

Substituting $-PR^{-1}P$ with $-2\varepsilon P + \varepsilon^2 R$ into (45), we obtain

$$\begin{bmatrix} \Sigma_{11} + \Gamma + \Gamma^T & * & * & * \\ \Phi_{21} & -2\varepsilon P + \varepsilon^2 R & * & * \\ \Phi_{31} & 0 & \widehat{\Phi}_{33} & * \\ \Sigma_{41}(s) & 0 & 0 & -R \end{bmatrix} < 0, \quad s = 1, 2, \tag{48}$$

where $\widehat{\Phi}_{33} = \operatorname{diag}\{\underbrace{-2\varepsilon P + \varepsilon^2 R, \ldots, -2\varepsilon P + \varepsilon^2 R}_{2m}\}$.

Denoting $X = P^{-1}$, $\widetilde{Q} = XQX$, $\widetilde{R} = XRX$, $\widetilde{N} = XNX$, $\widetilde{M} = XMX$, $\widetilde{\Omega} = X\Omega X$, and $Y = KX$, then pre- and postmultiplying (48) with $\operatorname{diag}\{\underbrace{X, \ldots, X}_{2m+5}, I\}$, (43) can be obtained. □

Remark 12. Theorem 11 shows that, for given ρ and ε, we can obtain the feedback gain K by solving a set of LMIs in (43); on the other hand, using Theorem 11, for the preselected Ω and the feedback gain K, event-triggered parameter ρ can be obtained. Therefore, we can use Theorem 11 to codesign the the feedback gain K and the event-triggered parameter.

Remark 13. Notice that (43) includes the information transmission delay, and we can use (43) to obtain the feedback gain and the event-triggered parameter, which can be used to guarantee the required performance even though the transmission delay exists.

4. Simulation Examples

Example 14. To illustrate the effectiveness and application of the proposed method, we consider an inverted pendulum on top of a moving cart. The plants linearized state is depicted as the following system [21]:

$$\dot{x}(t) = \begin{bmatrix} 0 & 1 & 0 & 0 \\ 0 & 0 & \dfrac{-mg}{M} & 0 \\ 0 & 0 & 0 & 1 \\ 0 & 0 & \dfrac{g}{l} & 0 \end{bmatrix} x(t) + \begin{bmatrix} 0 \\ \dfrac{1}{M} \\ 0 \\ \dfrac{-1}{Ml} \end{bmatrix} u(t), \quad (49)$$

where $x^T = [x_1\ x_2\ x_3\ x_4] = [y\ \dot{y}\ \theta\ \dot{\theta}]$ is the system state, y is the carts position, and θ is the pendulum bob's angle with respect to the vertical. $M = 10\,\text{kg}$ is the cart mass, $m = 1\,\text{kg}$ is the mass of the pendulum bob, $l = 3\,\text{m}$ is the length of the pendulum arm, and $g = 10\,\text{m/s}^2$ is gravitational acceleration. The initial state is the vector $x_0 = [0.98\ \ 0\ \ 0.2\ \ 0]$.

In the following, we will give two cases. Case 1 is used to show how the upper bound of τ_M varies along the values of ρ, under given feedback gain K as (50), when there is no failure in the actuator. In Case 2, we consider that the actuators have probabilistic failure rates; firstly, we give an example to design both the feedback and the trigger parameters, and the upper bound of τ_M and the release interval for $t \in [0, 40]$ are also derived; secondly, we suppose that the feedback gain K is given as (50) and study how the upper bound of τ_M varies along the values of ρ.

Case 1. When $\Xi = 1$, that is, there is no failure in the actuator, let

$$K = [2\ \ 12\ \ 378\ \ 210] \quad (50)$$

which is the same as the one in [22].

For given $\sigma_1 = 0$, $\varepsilon = 0.53$, and $\rho = 0.2$, by using Theorem 10, we can have $\tau_M = 0.1270$ and

$$\Omega = \begin{bmatrix} 0.0001 & 0.0003 & 0.0112 & 0.0062 \\ 0.0003 & 0.0023 & 0.0666 & 0.0376 \\ 0.0112 & 0.0666 & 2.1100 & 1.1662 \\ 0.0062 & 0.0376 & 1.1662 & 0.6510 \end{bmatrix}. \quad (51)$$

Also, we can obtain Table 1 and Figure 2, which describe the upper bound of τ_M varying along the values of ρ and the state response of (17).

With feedback gain K as (50), from Table 1, we can see that when $\rho = 0.53$ and $\sigma = 0.2$, the upper bound of τ_M is 0.1270. Suppose that $\tau_k = 0$; since $\tau_M = h + \overline{\tau}$, it can be known that the maximum sampling period is 0.1270. Moreover, under our event-trigger scheme, the maximum release period is 2.3.

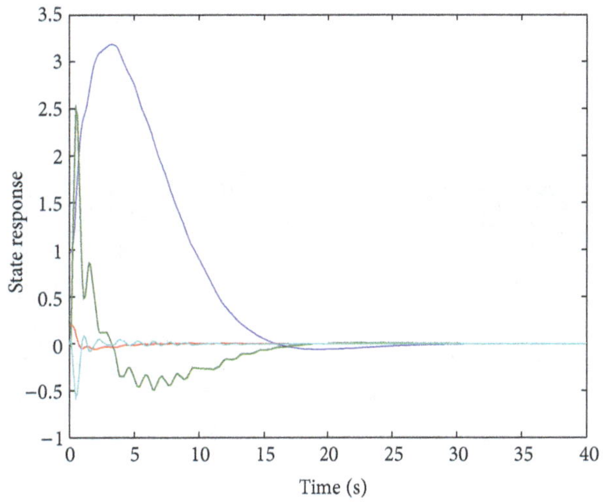

Figure 2: The state response of (49) with the feedback gain (50) for given $\Xi = 1$, $\sigma_1 = 0$, $\varepsilon = 0.53$, and $\rho = 0.2$.

Table 1: The computation results of the upper bound τ_M for given $\Xi = 1$, $\sigma = 0$ and $\varepsilon = 0.53$.

ρ	0	0.05	0.1	0.15	0.2
The upper bound of τ_M	0.1987	0.1582	0.1445	0.1349	0.1270

It can be seen that under the same conditions, our scheme can provide a larger release interval than some existing ones in [21, 22].

Case 2. When $\overline{\Xi} = 0.6$ and $\sigma_1 = 0.02$, that is, the actuators have probabilistic failure rate, setting $\varepsilon = 0.53$ and $\rho = 0.2$, by using Theorem 11, we can obtain the feedback gain K of (4) as follows:

$$K = [5.0696\ \ 23.2085\ \ 594.9876\ \ 331.9409],$$
$$\Omega = \begin{bmatrix} 162.1817 & 72.523 & -13.8855 & -11.9519 \\ 72.5232 & 525.4131 & -67.738 & -32.7170 \\ -13.8855 & -67.7382 & 52.8001 & -70.5906 \\ -11.9519 & -32.7170 & -70.5906 & 132.2203 \end{bmatrix}. \quad (52)$$

For given $\overline{\Xi} = 0.6$, $\sigma = 0.02$, and $\varepsilon = 0.53$, under different values of ρ, the upper bound of τ_M and the release interval for $t \in [0, 40]$ are given in Table 2. Figures 3 and 4, respectively, represent the release instants and release interval and the probabilistic actuator failures, when $\overline{\Xi} = 0.6$, $\sigma = 0.02$, $\varepsilon = 0.53$, $\rho = 0.1$, and $\tau_M = 0.2399$.

When K is given as (50), let $\overline{\Xi} = 0.6$, $\overline{\Xi} = 1$, $\sigma_1 = 0$, and $\varepsilon = 0.53$, and the upper bound of τ_M varies along the values of ρ, which is shown in Table 3.

Remark 15. For given ρ and ε, we can obtain the upper bound of τ_M. we can see that the larger ρ, the larger average release period; Thus, the load of network communication delay is reduced and the transmission delay is decreased. This fact can be illustrated by Tables 1, 2 and 3.

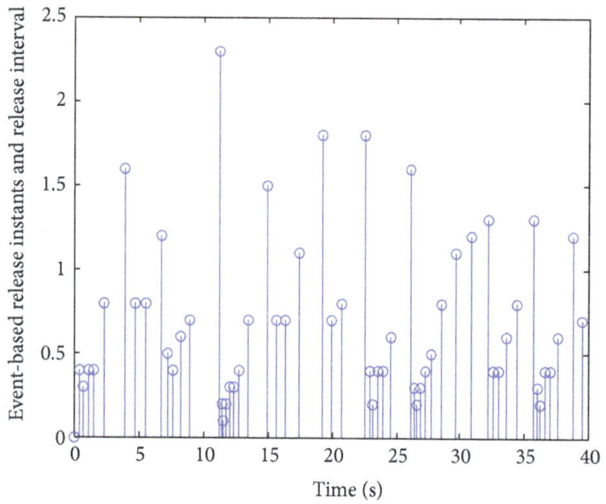

FIGURE 3: The release instants and release interval for given $\overline{\Xi} = 0.6$, $\sigma = 0.02$, and $\varepsilon = 0.53$.

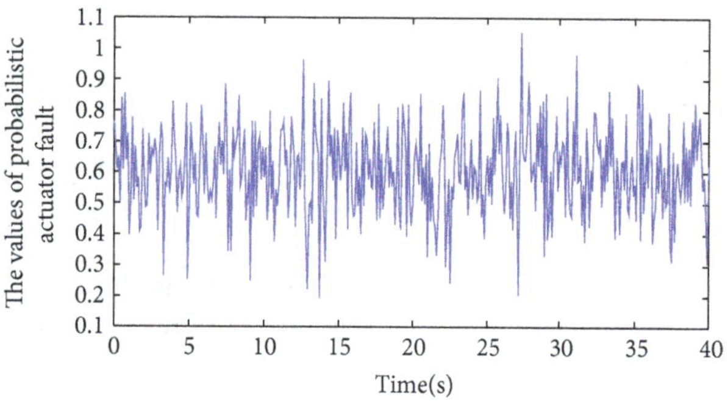

FIGURE 4: The probabilistic actuator failures.

TABLE 2: Some computation results for given $\overline{\Xi} = 0.6$, $\sigma = 0.02$, and $\varepsilon = 0.53$.

ρ	0	0.1	0.2	0.3	0.4
The upper bound of τ_M	0.3556	0.2399	0.1789	0.1394	0.1085
The maximum release interval for $t \in [0, 40]$	0.1	2.3	1.1	1.2	2.3
The total trigger times for $t \in [0, 40]$	401	57	78	111	122

TABLE 3: The computation results of the upper bound τ_M for given $\overline{\Xi} = 0.6$, $\sigma = 0.02$, and $\varepsilon = 0.53$.

ρ	0	0.05	0.1	0.15	0.2
The upper bound of τ_M	0.2963	0.2197	0.1479	0.0546	0

5. Conclusion

In order to save the communication network bandwidth, a novel event triggering scheme is used to determine when to transmit the sampled state information. Under this event-triggered scheme, this paper considers networked systems with probabilistic actuators faults. In terms of different failure rates and the measurements distortion of every actuator, a new probabilistic actuator fault model for event-triggered networked control systems is proposed. By using Lyapunov functional method, criteria for the EMSS and criteria for codesigning both the feedback and the trigger parameters are derived in the form of linear matrix inequalities. A simulation example is given to illustrate that our event-triggered scheme can lead to a larger release period than some existing works.

Acknowledgments

This work is partly supported by the National Natural Science Foundation of China (No. 11226240), the Natural Science Foundation of Jiangsu Province of China (no. BK2012469), the Natural Science Foundation of the Jiangsu Higher Education Institutions of China (no. 12 KJD120001), the Young Teacher Supporting Foundation of Shanghai (no. ZZGJD12036), and a project funded by the Priority Academic Program Development of Jiangsu Higher Education Institutions (PAPD).

References

[1] J. P. Hespanha, P. Naghshtabrizi, and Y. Xu, "A survey of recent results in networked control systems," *Proceedings of the IEEE*, vol. 95, no. 1, pp. 138–172, 2007.

[2] T. Yang, "Networked control system: a brief survey," *IEE Proceedings Control Theory and Applications*, vol. 153, no. 4, pp. 403–412, 2006.

[3] Q.-L. Han, "A discrete delay decomposition approach to stability of linear retarded and neutral systems," *Automatica*, vol. 45, no. 2, pp. 517–524, 2009.

[4] H. Gao, X. Meng, and T. Chen, "Stabilization of networked control systems with a new delay characterization," *IEEE Transactions on Automatic Control*, vol. 53, no. 9, pp. 2142–2148, 2008.

[5] X. Wang and M. D. Lemmon, "Event-triggering in distributed networked control systems," *IEEE Transactions on Automatic Control*, vol. 56, no. 3, pp. 586–601, 2011.

[6] P. Tabuada, "Event-triggered real-time scheduling of stabilizing control tasks," *IEEE Transactions on Automatic Control*, vol. 52, no. 9, pp. 1680–1685, 2007.

[7] V. Nguyen and Y. Suh, "Networked estimation for event-based sampling systems with packet dropouts," *Sensors*, vol. 9, no. 4, pp. 3078–3089, 2009.

[8] A. Eqtami, D. V. Dimarogonas, and K. J. Kyriakopoulos, "Event-triggered control for discrete-time systems," in *Proceedings of the American Control Conference (ACC '10)*, pp. 4719–4724, July 2010.

[9] X. Wang and M. Lemmon, "On event design in event-triggered feedback systems," *Automatica*, vol. 47, no. 10, pp. 2319–2322, 2011.

[10] D. Yue, J. Lam, and D. Ho, "Reliable H_∞ control of uncertain descriptor systems with multiple time delays," *IEE Proceedings Control Theory and Applications*, vol. 150, no. 6, pp. 557–564, 2003.

[11] D. Ye and G.-H. Yang, "Delay-dependent adaptive reliable H_∞ control of linear time-varying delay systems," *International Journal of Robust and Nonlinear Control*, vol. 19, no. 4, pp. 462–479, 2009.

[12] E. Tian, D. Yue, and C. Peng, "Reliable control for networked control systems with probabilistic sensors and actuators faults," *IET Control Theory & Applications*, vol. 4, no. 8, pp. 1478–1488, 2010.

[13] E. Tian, D. Yue, and C. Peng, "Reliable control for networked control systems with probabilistic actuator fault and random delays," *Journal of the Franklin Institute*, vol. 347, no. 10, pp. 1907–1926, 2010.

[14] F. O. Hounkpevi and E. E. Yaz, "Robust minimum variance linear state estimators for multiple sensors with different failure rates," *Automatica*, vol. 43, no. 7, pp. 1274–1280, 2007.

[15] X. He, Z. Wang, and D. Zhou, "Robust filtering for time-delay systems with probabilistic sensor faults," *IEEE Signal Processing Letters*, vol. 16, no. 5, pp. 442–445, 2009.

[16] G. Wei, Z. Wang, and H. Shu, "Robust filtering with stochastic nonlinearities and multiple missing measurements," *Automatica*, vol. 45, no. 3, pp. 836–841, 2009.

[17] Y. Y. Wang, L. Xie, and C. E. de Souza, "Robust control of a class of uncertain nonlinear systems," *Systems & Control Letters*, vol. 19, no. 2, pp. 139–149, 1992.

[18] E. Tian, D. Yue, and Y. Zhang, "Delay-dependent robust H_∞ control for T-S fuzzy system with interval time-varying delay," *Fuzzy Sets and Systems*, vol. 160, no. 12, pp. 1708–1719, 2009.

[19] D. Yue, Q.-L. Han, and J. Lam, "Network-based robust H_∞ control of systems with uncertainty," *Automatica*, vol. 41, no. 6, pp. 999–1007, 2005.

[20] Y. He, M. Wu, J.-H. She, and G.-P. Liu, "Parameter-dependent Lyapunov functional for stability of time-delay systems with polytopic-type uncertainties," *IEEE Transactions on Automatic Control*, vol. 49, no. 5, pp. 828–832, 2004.

[21] X. Wang, *Event-triggering in cyber-physical systems [Ph.D. dissertation]*, University of Notre Dame, 2009.

[22] X. Wang and M. D. Lemmon, "Self-triggered feedback control systems with finite-gain $\mathscr{L}_2$ stability," *IEEE Transactions on Automatic Control*, vol. 54, no. 3, pp. 452–467, 2009.

14

Probabilistic Value-Centric Optimization Design for Fractionated Spacecrafts Based on Unscented Transformation

Ming Xu,[1] Jinlong Wang,[1] Ang Zhang,[1] and Shengli Liu[2]

[1] *Department of Aerospace Engineering, School of Astronautics, Beihang University, Beijing 100191, China*
[2] *DFH Satellite Co., Ltd., Beijing 100094, China*

Correspondence should be addressed to Ming Xu; xuming@buaa.edu.cn

Academic Editor: Suiyang Khoo

Fractionated spacecrafts are of particular interest for pointing-intensive missions because of their ability to decouple physically the satellite bus and some imaging payloads, which possess a lesser lifecycle cost than a comparable monolithic spacecraft. Considering the probabilistic uncertainties during the mission lifecycle, the cost assessment or architecture optimization is essentially a stochastic problem. Thus, this research seeks to quantitatively assess different spacecraft architecture strategies for remote-sensing missions. A dynamical lifecycle simulation and parametric models are developed to evaluate the lifecycle costs, while the mass, propellant usage, and some other constraints on spacecraft are assessed using nonparametric, physics-based computer models. Compared with the traditional Monte Carlo simulation to produce uncertain distributions during the lifecycle, the unscented transformation is employed to reduce the computational overhead, just as it does in improving the extended Kalman filter. Furthermore, the genetic algorithm is applied to optimize the fractionated architecture based on the probabilistic value-centric assessments developed in this paper.

1. Introduction

The Defense Advanced Research Projects Agency (DARPA) implemented the Future, Fast, Flexible, Fractionated, Free-Flying (F6) program, aiming to demonstrate the concept of the fractionated architecture [1]. DARPA conducted extensive researches for the technological performances and economic advantages and invested Lockheed Martin Company (LM), Northrop Grumman Corporation (NG), Orbital Sciences Corporation (OSC), and Boeing Company (BC) for fractionated spacecrafts based on value-centric design methodologies (VCDM) in first phase of the F6 program [2]. This innovative design method for spacecraft has drawn great attentions from the astronautical community.

Considering the application of fractionated modules into responsive space, Richards et al. [3] analyzed the internal relations between the four aspects of technological, organizational, economic, and political supports. Brown and Eremenko [4, 5] summarized the achievements of VCDM in the F6 program from the viewpoint of the relationship between the innovative value-centered design standard and traditional monolithic design standard and then evaluated the concept of the fractionated spacecrafts compared with the traditional spacecraft architecture [5]. Mathieu and Weigel [6] evaluated the advantages and costs of the fractionated architecture in the fields of the attributes, strategies, and models. O'Neill [7] developed the semianalytic tool PIVOT for the model frame, risk, and the net present value and then optimized the PIVOT tool involved in the second phase of the F6 program. O'Neil and Mankins [8] achieved the conclusion that the fractionated spacecrafts were better on the lifecycle cost than a traditional spacecraft through the dynamical simulation and parameter models on the quality of the spacecraft [9]. Lafleur and Saleh [10, 11] developed another design tool GT-FAST, which instanced specific analysis on the F6 program for handling input, model, and attributes. Yao et al. [12, 13] proposed the multidisciplinary optimization about the uncertainty for the spacecraft conceptual design and verified the feasibility and effectiveness of this optimization. Daniels et al. [14] presented a heuristics-based decision model using a Monte Carlo simulation to produce value distributions for satellite operator decision sets and a multi-stage decision process utilizing a dynamic programming algorithm to find value optimal decisions. Compared to the traditional measure metric on spacecraft cost, Collopy [15] derived a new metric rigorously from the view of probability, which places a focus on improving the probability of success rather than on reducing the cost.

One element necessary in enabling a probabilistic, value-centric analysis of such fractionated architecture is a systematic method for sizing and costing many candidate architectures that arise in aerospace engineering. In this paper, a dynamical lifecycle simulation and parametric models are developed to evaluate the lifecycle costs, while the mass, propellant usage, and some other constraints on spacecraft are assessed using nonparametric, physics-based computer models. The lifecycle is divided into three phases, that is, module development, launching, and on-orbit control. Furthermore, the genetic algorithm is applied to optimize the fractionated architecture based on the probabilistic value-centric assessments developed in this paper. To accelerate the optimization, one of the new techniques is to employ the graphic processing unit (GPU) accelerated genetic algorithm based on the Compute Unified Device Architecture (CUDA) to improve the searching efficiency because the assessments in any generation are parallel computerized. Another is to introduce the unscented transformation to reduce the amount of computations rather than the Monte Carlo simulation, just as it does in improving the extended Kalman filter (EKF) [16, 17].

2. Cost, Value, and Assessment Indicators

The cost assessment for a spacecraft design scheme includes module development costs, launch and operation costs, and risk costs from uncertainties during the lifecycle. All the types of the above costs are quantified by the monetary unit, for example, US dollar. The statistical sum of all the costs from the preparation phase to the end of the lifecycle constitutes the total cost of the spacecraft design scheme.

There are several valuation standards in the value-centric design methodologies to measure the practical spacecraft scheme by the quantitative assessments of the cost and value of fractionated spacecrafts. Quantitative assessment of the value gives a measurement indicator, as well as an evaluation methodology to provide information for the task of decision making. Considering the probabilistic uncertainties during the mission lifecycle, the value is acquired at the end of the lifecycle of the spacecraft from the quantitative criteria to quantify the value of spacecraft.

To study the cost and value of the trade-offs, the net present value (NPV) is usually used in the financial evaluation of the business assets [10, 11]. It refers to the investment scheme generating net cash flow to the capital cost for the discount rate discount and original investment present value variance. Therefore, the net present value NPV is used to judge the fractionated spacecraft design in this paper. The higher the value of NPV is, the better the spacecraft design scheme is, which means that the input cost is low but the output value is high. Hence, the input-output ratio could be weighted by the NPV value, which is formulized as follows [5]:

$$\text{NPV} = \frac{N}{\left(1 + D_R\right)^{T_{\text{yearloop}}-1}}, \quad (1)$$

where D_R is the discount rate, N is the free cash flow, and T_{yearloop} is the spacecraft lifecycle (unit: year).

Table 1: Parameter settings for fractionated components.

Components names	Weight (kg)	Power (W)	Cost (M$)	TRL	FIT
EO	40	15	15	9	5000
24/7 Comm	4	25	5	5	5000
HBD	10	25	2	9	5000
SSR	8	100	2	7	6000
MDP	8	18	1	6	5000
AIS	5	15	0.5	8	3500

3. Spacecraft Models

3.1. Fractionated Spacecraft Architecture. The spacecraft architecture modeling is a prerequisite to assess the cost and value of the spacecraft design scheme. But the fractionated spacecrafts are quite different from traditional spacecraft because of the different design principles on the two types of architectures. Common architecture level definitions on the fractionated spacecraft architecture modeling are listed from low to high order, as "component," "module," and "cluster" [12].

Component: it is the smallest unit in the fractionated spacecraft architecture modeling. The four companies modeled the components in the first phase of the DARPA mission, mainly including the following components: (1) payload component: the valuable payloads are the feasible assessment to increase the total value of the fractionated spacecraft for the purpose of using the least cost to achieve the more value; (2) measurement and control component: it realizes the continuous measurement and control through the relay module in the architecture constellation; (3) mission data processing component: usually the on-board computer is used for processing complex data; (4) digital communication component: it is responsible for communication between the space and ground, as well as responsible for the data download and upload; (5) data storage component: it is storing large amounts of data and preparing for data transmission, that is, solid-state drive (SSR).

Module: it consists of several components to achieve independent and free flight on orbit. In addition to payload and other functional components, it includes other associated components of the power supply and thermal control.

Cluster: it is composed of multiple fractionated modules in formation flight to complete certain missions.

Using six separable components in this paper, the parameters are shown in Table 1 [18], where FIT represents the number of failures in 1000 hours and the characterization of component reliability parameters, the Earth observing payload (EO) is employed throughout this paper to meet the imaging missions, 24/7 Comm is the digital component, HBD is the high-bandwidth downloading communication component, SSR is the solid hard drive component, MDP is the mission data processing component, and AIS is the automatic identification component.

The spacecraft scheme includes the design of spacecraft architecture and the selection of launch vehicles [19]. To inject fractionated spacecrafts into space, six alternative categories for the launch vehicle are provided in this paper, that is, Minotaur I, Athena I, Taurus 2210, Taurus 3110, Minotaur IV, and Athena II [1], which have different launch capacities and reliabilities, and different launch vehicle costs. Therefore, the launch vehicle should be selected based upon the aerospace mission and the fractionated architecture.

The combination of separated components, the cluster segmentations, and the selection of launch vehicle impact the assessed cost and value of the fractionated spacecrafts. As a result, the optimization of the cluster segmentation and its separated components should be taken to get a higher input-output ratio.

3.2. Spacecraft Architecture Design Models

3.2.1. Cost Models. There are three parts of the fractionated spacecraft's cost: module development cost, launch and operation cost, and risk cost, where commercial insurance for spacecraft is involved in launch and operation cost, and the ground equipment cost and software development cost are not accounted in this paper. Risk cost includes the cost caused by time delay during the module development and the failure due to launch vehicle and on-orbit maintenance, which are considered to be governed by uncertainties.

According to the recyclability, the cost can be divided into nonrecyclable and recyclable types. Inherited from this attribution, the separated components in the cluster would be divided into the recyclable and nonrecyclable ones. This classification is primarily proposed for some reusable equipment and design schemes for the specified mission. In this case, the second cost is the result from the combination of the first cost and the learning curve rate.

(1) Module Development Costs. Different evaluations arise from the development costs with different components, or different development cycles and technology readiness levels (TRL) for the same components. For any spacecraft architecture scheme, the recyclable and nonrecyclable costs are calculated, respectively. For the specified type of modules, the cost on the module manufactured at the first time is higher than others because of its low TRL; however, the cost on the module built at the second or subsequent time is lower due to the cost of the recyclable components.

Considering the batch production, the cost of the unit module is estimated as follows:

$$\begin{aligned} C_{\text{mod}_{i1}} &= C_{\text{mod}_{i1}}^{\text{NRE}} + C_{\text{mod}_{i1}}^{\text{REC}}, \\ C_{\text{mod}_{iQ}} &= C_{\text{mod}_{i1}}^{\text{REC}} \cdot Q^{(\ln L_R)/\ln 2}, \end{aligned} \tag{2}$$

where $C_{\text{mod}_{i1}}$ is the cost of the first module of the ith type module including nonrecyclable cost $C_{\text{mod}_{i1}}^{\text{NRE}}$ and recyclable cost $C_{\text{mod}_{i1}}^{\text{REC}}$, $C_{\text{mod}_{iQ}}$ is the cost of the Qth module of the ith type module, and L_R is the learning curve rate of the module development and production. The nonrecyclable cost and recyclable cost are calculated by the components and satellite platform, the cost calculation formula with reference to the cost of small satellite model (abbr. SSCM07), which was developed by Mahr [20] for the Aerospace Cooperation and has potential applications in estimating the manufacturing cost empirically [10, 11, 13, 21].

SSCM07 is good at estimating the production cost; however, it is powerless in estimating other types of costs (such as operation, risk, and inflation cost) and their dynamic and uncertain evolutions during the spacecraft's lifecycle. Thus, we employed SSCM07 to estimate the production costs for fractionated modules and then developed a probabilistic method to measure all the uncertain costs during the lifecycle dynamically.

Taking inflation into account, the cost of the unit module is estimated as follows:

$$C_{\text{mod}_{iQ}}^{\text{inflated}} = C_{\text{mod}_{iQ}} \cdot \left(1 + R_{\text{inflated}}\right)^{T_{\text{lateryear}}}, \tag{3}$$

where R_{inflated} is the year inflation rate and $T_{\text{lateryear}}$ is expressed as module development time (unit: year). All module's development cost is as follows:

$$C_M = \sum_{i=1}^{N_{\text{type}}} \left(\sum_{j=1}^{N_{\text{mod}_i}^{\text{type}}} C_{\text{mod}_{ij}}^{\text{inflated}} \right), \tag{4}$$

where N_{type} is the number of the spacecraft module's type and $N_{\text{mod}_i}^{\text{type}}$ is the number of certain type module.

(2) Launching and Operational Costs. The selection of a launch vehicle scheme is determined according to the module properties. Therefore, the cost of the launch vehicle is estimated by the design scheme of the spacecraft architecture. The requirement from the spacecraft and the capacity of the launch vehicle, respectively, restrict the feasible combinations between them, so the number of selecting launch vehicles is less than the number of modules. Thus, the launch cost is listed as follows:

$$C_{\text{launch}} = \sum_{i=1}^{N_{\text{launch}}} C_{\text{launch}}^{i}, \tag{5}$$

where N_{launch} is the total expected number of launch vehicles and C_{launch}^{i} is the cost of the ith vehicle. Owning to the huge cost of launch vehicles, it is necessary to reduce the number of launch vehicles and to load as many modules as possible in every vehicle.

When the fractionated spacecrafts are on orbit, operating each module will generate the operational cost. In case of the annual operational cost C_i^{ops} for each module being equal to

2 M$, the total operational cost C_{ops} during the spacecraft's lifecycle is as follows:

$$C_{ops} = \sum_{i=1}^{N_{type}} \left(\sum_{j=1}^{N_{mod_i}^{type}} C_i^{ops} \cdot T_{ij} \right), \tag{6}$$

where T_{ij} is the total orbiting time of the ith type module on the jth year. Then the launch and operational cost of fractionated spacecrafts during the lifecycle is as follows:

$$C_{lo} = C_{launch} + C_{ops}. \tag{7}$$

(3) Risk Costs. The uncertainties, such as task delay, launching failure, or on-orbit failure, exist in all the phases of spacecraft development, launching, and on-orbit operation, which are required to be maintained with extra costs. The risks are measured by the risk cost C_{risk} as the criterion of the robustness and reliability in spacecraft architecture design, which are allocated by launching failure and on-orbit failure in this paper.

Launching failure: this type of failure comes from the reliability of launch vehicles less than 100%. If a launch vehicle happens to fail, all the modules carried by this vehicle need to be redeveloped and then launched. Hence, the risk cost of launching failure can be formulized as follows:

$$C_{launch}^{failure} = \sum_{i \in I_{launch}^{failure}}^{N_{launch}} C_{launch}^{i} + \sum_{i \in I_{launch}^{failure}}^{N_{launch}} \left(\sum_{j \in I_{mod_i}^{launchfail}}^{N_{mod_{ij}}^{launchfail}} C_{mod_{jk}}^{inflated} \right), \tag{8}$$

where $I_{launch}^{failure}$ is the identification number of the failing launch vehicle, $N_{mod_{ij}}^{launchfail}$ and $I_{mod_i}^{launchfail}$ indicate the quantity and type of the modules carried by the failing vehicle "i," respectively.

On-orbit operation: if a module fails on-orbit, another module of the same type is expected to be launched to replace this faulty module. Thus, the cost of on-orbit operation is essentially the development and launching costs for the new module, which is listed as shown:

$$C_{ops}^{failure} = \sum_{i=1}^{N_{ops}^{failure}} C_{launch}^{i} + \sum_{i \in I_{launch}^{failure}}^{N_{launch}} \left(\sum_{j \in I_{mod_i}^{opsfail}}^{N_{mod_{ij}}^{opsfail}} C_{mod_{jk}}^{inflated} \right), \tag{9}$$

where $N_{ops}^{failure}$ is the identification number of the module failing on-orbit, $N_{mod_{ij}}^{opsfail}$ and $i_{mod_i}^{opsfail}$ indicate the quantity and type of the module failing on-orbit, respectively.

Therefore, the risk cost can be considered as the failure cost during the launching or operational phase, as

$$C_{risk} = C_{launch}^{failure} + C_{ops}^{failure}. \tag{10}$$

According to the above sections presenting the costs when developing, launching, and operating a module and the risk costs due to the failures in these phases, the total cost of the fractionated architecture can be expressed as following:

$$C = C_M + C_{lo} + C_{risk}. \tag{11}$$

3.2.2. Value Models. The revenue model gives a finical rule measuring the spacecraft gains, which are considered as the data products achieved from the payload modules and their communication links. Consequently, some empirical weight factors are introduced to monetize the imaging data to model the benefit value of a spacecraft. When the data have better resolution and positioning accuracy, a larger factor is weighted to price them; on the other hand, a smaller weight factor is expected.

The effectiveness of the communication link affects the value from the imaging data as well, depending upon the type of link hardware (low or high speed transmission), the frequency, and duration of the spacecraft passing through the data receiving stations.

In this paper, the payloads are defined to employ both the low-rate downlink (abbr. LR), high-rate downlink (abbr. HR), and space-ground interlink (abbr. SG) to implement the digital transmissions. Thus, the benefit value of the fractionated spacecrafts can be derived from the following equation:

$$R_{i+j} = \delta_i \left(N_i^{LR} + N_i^{HR} + N_i^{SG} \right) + \delta_j \left(N_j^{LR} + N_j^{HR} + N_j^{SG} \right), \tag{12}$$

where δ_i and δ_j are the weight factors of the payloads, respectively, and N_k^{LR}, N_k^{HR}, and N_k^{SG}, $k = i, j$, are, respectively, the amount of valid data obtained by the link between space and ground.

3.2.3. Uncertainty Models. A systematic study on the investing cost and economic benefit assessments on the spacecraft architectures is implemented in this paper. Considering the probabilistic uncertainties during the mission lifecycle, the cost assessment or architecture optimization is essentially a stochastic problem. The uncertainties such as financial inflation, launching failure, and on-orbit operating failure are required to be maintained by extra costs, which are formulized by (3) in Section 3.2.1(1) and (8), (9), and (10) in Section 3.2.1(3).

In the numerical assessments, all the uncertainties are modeled by the stochastic noises. Generally, the upper and lower bounds of these uncertainties are determined according to the experience in previous work or technology readiness level (TRL), which are formulized as following:

$$C = F\left(C_s, \delta_{max}, \delta_{min}\right), \tag{13}$$

where C_s is the mean value and δ_{max} and δ_{min} are the upper and lower bounds, respectively. The annual inflation rate is assumed as the Gaussian noise approximately with its mean and variance assigned to 3.5% and 0.5%, respectively, according to the historical consumer price index (CPI) of China. The reliability of launch vehicles is set as P_{launch} = 95.1% according to the handbook of the vehicle type of Minotaur IV, which means the launching failure occurs when a random number from the uniform distribution $U(0, 1)$ is larger than the reliability P_{launch}. The on-orbit operating failure is dependent on the mean time to failure θ of the components, which is modeled by the exponential distribution as $f(t) = \lambda e^{-\lambda \cdot t}$, where λ is referred as aging rate equal to $1/\theta$ and is set as 5% in this paper.

For a specified design scheme, the traditional Monte Carlo simulation is employed by the cost and value assessment to cope with the uncertainties. More than 10000 scenarios are created by the Monte Carlo method to simulate all the combinations of the noises mentioned above. Generally, the Monte Carlo simulation costs a mass of computation to generate all the situations ergodicly. Therefore, this statistical method is feasible just for evaluating a specified design scheme within all kinds of scenarios; however, it is not available for refining the optimal one from many candidate schemes. For example, the best design scheme is expected to be selected from 5000 candidates, and then the optimization needs to be evaluated 5×10^7 (=5000×10000) times. The heavy computation makes the optimizing iteration very difficult or impossible.

Thus, it is necessary to introduce the unscented transformation (UT) firstly proposed by Julier and Uhlmann [16] to reduce the amount of computations rather than the Monte Carlo simulation. The most common use of UT is in the nonlinear projection of mean and covariance estimates in the context of nonlinear extensions of the Kalman filter [17]. The principal advantage of the approach is that the nonlinear function is fully exploited, as opposed to the extended Kalman filter which replaces it with a linear one. One immediate advantage is that the UT can be applied with any given function whereas linearization may not be possible for functions that are not differentiable. A practical advantage is that the UT can be easier to implement because it avoids the need to derive and implement a linearizing Jacobian matrix.

The mean and covariance of financial inflation, launching failure, and on-orbit operating failure will be exactly encoded by sigma points and then be propagated by the assessment procedure to each point. Hence, the mean and covariance of the transformed set of points then represent the desired transformed estimate.

The unscented transformation is defined as the application of a given function to any partial characterization of an otherwise unknown distribution, but its most common use is for the case in which only the mean and covariance are given. For the fractionated architecture with six modules, the random uncertainties (denoted by $\mathbf{x} \in \mathbf{R}^6$) from the annual inflation rate, launching failure, and on-orbit operating failure have the mean $\overline{\mathbf{x}}$ and variance $\mathbf{P}$. To yield the statistics of some combination of cost and value $\mathbf{y}$ propagated through the nonlinear mapping defined by the cost and value assessment procedure $\mathbf{y} = \mathbf{h}(\mathbf{x})$, a set of 13 sigma points $\{\overline{\mathbf{x}}_i, i = 0, 1, \ldots, 12\}$ are calculated using the following general selection scheme [16]:

$$\mathbf{x}^{(0)} = \overline{\mathbf{x}},$$
$$\tilde{\mathbf{x}}^{(i)} = \overline{\mathbf{x}} + \left(\sqrt{6 \cdot \mathbf{P}}\right)^T, \quad i = 1, 2, \ldots, 6, \tag{14}$$
$$\tilde{\mathbf{x}}^{(i)} = \overline{\mathbf{x}} - \left(\sqrt{6 \cdot \mathbf{P}}\right)^T, \quad i = 7, 8, \ldots, 12.$$

Once the sigma points are calculated from the prior statistics as shown above, they are propagated through the nonlinear function $\mathbf{y}^{(i)} = \mathbf{h}[\mathbf{x}^{(i)}]$. Thus, the mean y_u and variance $\mathbf{P}_u$ of $\mathbf{y}$ approximated using a weighted sample mean and covariance of the posterior sigma-points are as follows:

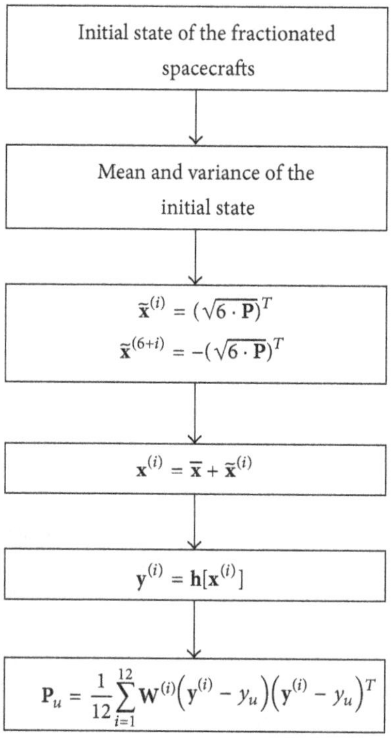

FIGURE 1: The process of applying UT to assess the cost and value for the fractionated architecture.

$$y_u = \frac{1}{12}\sum_{i=1}^{12} \mathbf{W}^{(i)}\mathbf{y}^{(i)},$$
$$\mathbf{P}_u = \frac{1}{12}\sum_{i=1}^{12} \mathbf{W}^{(i)}\left(\mathbf{y}^{(i)} - y_u\right)\left(\mathbf{y}^{(i)} - y_u\right)^T, \tag{15}$$

where the weight coefficients $\mathbf{W}^{(i)}$ are restricted by the normalization condition $\sum_{i=0}^{12} \mathbf{W}^{(i)} = 1$. The process of applying the unscented transformation to calculate the statistical result of the cost and value assessment for the fractionated architecture is presented in Figure 1.

Compared with the traditional Monte Carlo simulation, this approach characterizes a probability distribution only in terms of few set of statistics. Furthermore, considering the ideal case without the uncertainties, the cost and benefit value (including the net present value NPV) will be the same as the single result achieved from the initial state of the fractionated spacecrafts.

4. Cost and Value Assessments

4.1. Cost and Value Assessment Procedure. Cost and value assessment for the fractionated architecture is quite dependent on all the stages divided in the whole lifecycle, that is, module development, launching, and on-orbit operation. Within the cost and value accumulated in each stage, the total

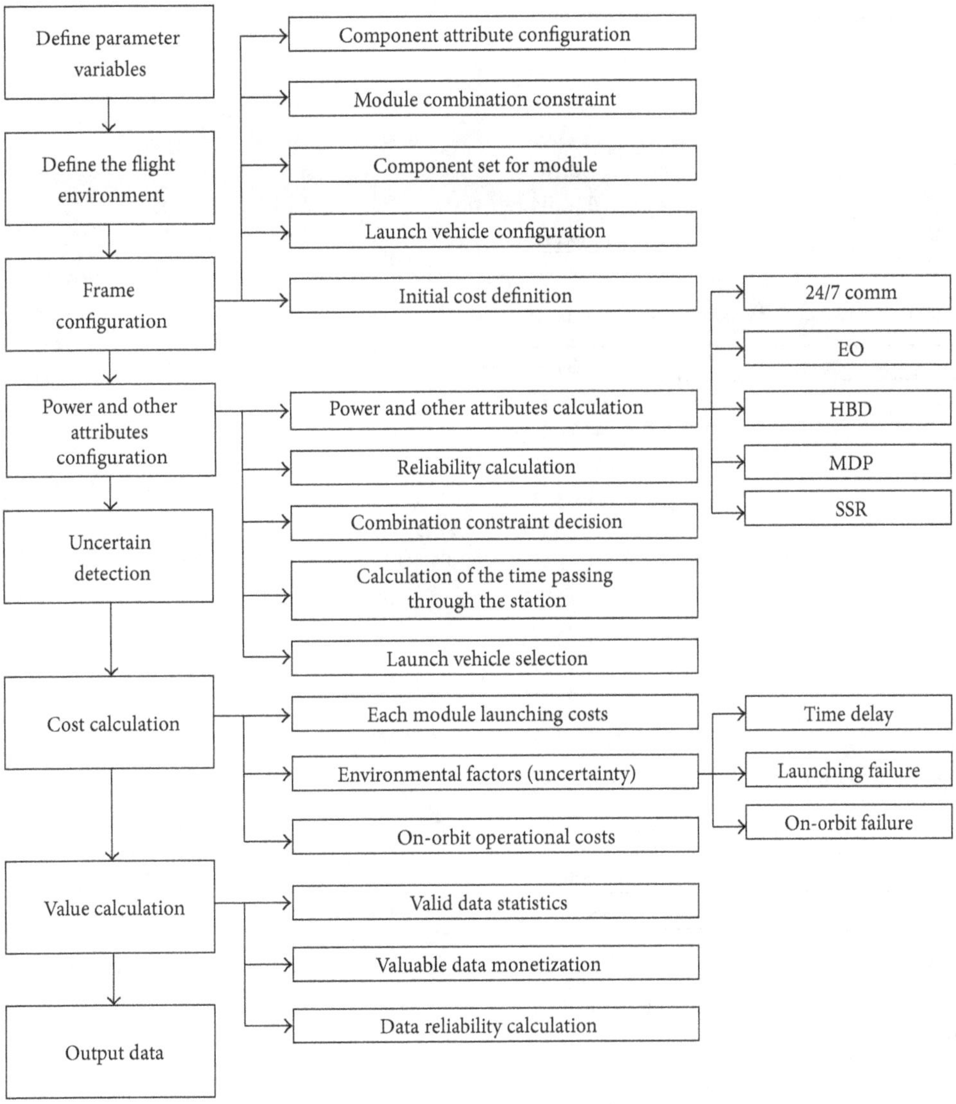

FIGURE 2: The flowchart of cost and value assessment for fractionated architecture.

cost and value will be yielded at the end of the lifecycle. The flowchart of the cost and value assessment is presented in Figure 2 for the fractionated architecture.

The main work is to define some relevant parameters for the fractionated spacecrafts in its early development stage, including setting the uncertainties due to the time delay and choosing the functional components assembled into the modules and their launch vehicles. The cost settlement at this stage should take the inflation rate and others into consideration. Moreover, no value is yielded in this stage because the spacecraft has not been injected into space and no imaging data are produced at the module development stage.

In the launch phase, the certain cost comes from a series of launch vehicles adapted by the developed modules, and the uncertain cost originates from the launching failure which requires more costs in redeveloping and relaunching modules. In addition, no value is yielded in this stage as well.

After injected into space successfully, the spacecraft is expected to be debugged on-orbit for about one month, and then some functional components (like on-board camera) will be put into use and able to produce valuable data, so that the spacecraft starts to benefit from this stage. Furthermore, the debugging stage costs the on-orbit operation a lot.

The uncertain cost in the operational stage results from a module failing on-orbit. In this case, another module is launched to replace this faulty module after redeveloped in accordance with the existing template. During the replacement, the faulty module and its successor have no contributions to the benefit value.

Therefore, the cost and value of the fractionated architecture can be obtained through the whole lifecycle according to the assessment flowchart listed in Figure 2.

4.2. Simulation of Cost and Value Assessment. In this paper, the nominal orbit is set as a sun-synchronous orbit with the orbital altitude of 500 km and the local time at descending node of 10:30 AM. The aerospace mission starts from 1st of January, and the development period of a module is assumed

TABLE 2: The net present value NPV in five-year lifecycle.

Time	NPV
The first year	−105.1077
The second year	−144.0231
The third year	118.7044
The fourth year	132.8142
The fifth year	14.1471

TABLE 3: The net present value NPV in ten-year lifecycle.

Time	NPV
The first year	−95.1603
The second year	−131.6379
The third year	55.4421
The fourth year	62.4822
The fifth year	51.2209
The sixth year	96.6323
The seventh year	62.1255
The eighth year	109.7646
The ninth year	119.1036
The tenth years	119.1036

TABLE 4: Average cost and benefit (unit: M$) in five-year lifecycle and their standard variances.

	Average cost	Cost standard variance	Average benefit	Benefit standard variance
Value	264.0081	32.1762	265.5474	59.9580

TABLE 5: Average cost and benefit (unit: M$) in ten-year lifecycle and their standard variances.

	Average cost	Cost standard variance	Average benefit	Benefit standard variance
Value	696.8454	37.8953	813.8701	88.1731

as two years. All the modules are launched by the vehicle type of Minotaur IV with its reliability of 95.1%. The simulation cases are implemented in this paper with the lifecycles of five and ten years, respectively.

Considering the cost and value uncertainties from schedule change, launching failure, on-orbit failure, commercial insurance, module replacement, and other factors, the net present values counted numerically in the lifecycles of five and ten years are shown in Tables 2 and 3, and the average costs and benefits are shown in Tables 4 and 5. Then the assessments in the lifecycles of five and ten years are classified and labeled on the phase plane of cost and benefit values shown in Figures 3 and 4, where the two error ellipses are yielded by the unscented transformation with the confidence levels of 0.5 and 0.68, respectively, and the scattered points are plotted by the Monte Carlo method.

The numerical simulation indicates that the cost and benefit values quite depend on the mission lifecycle. The longer lifecycle increases the risk of on-orbit failure and the cost of module replacement and produces more valuable imaging data which raises the benefit values, which accounts for why the cost and value created in the lifecycle of ten years are more than five years. Moreover, most of the scattered points located inside the error ellipses validate the feasibility of the unscented transformation. Therefore, the assessments of many candidate schemes will be implemented in Section 5 by the unscented transformation to accelerate the optimization procedure.

However, there exists no relationship between the lifecycle of the fractionated spacecrafts and the net present value, which is accumulated by all the uncertainties during the lifecycle. Furthermore, the points far away from the confidence ellipses may happen with a very low probability, so that they are often ignored in aerospace engineering.

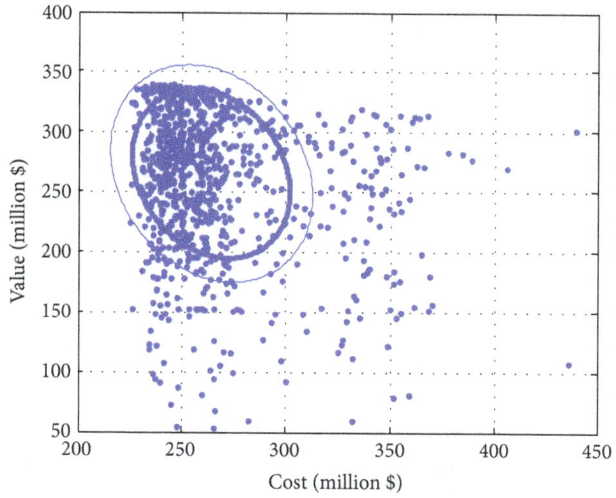

FIGURE 3: The assessments in five-year lifecycle labeled on the phase plane of cost and benefit values.

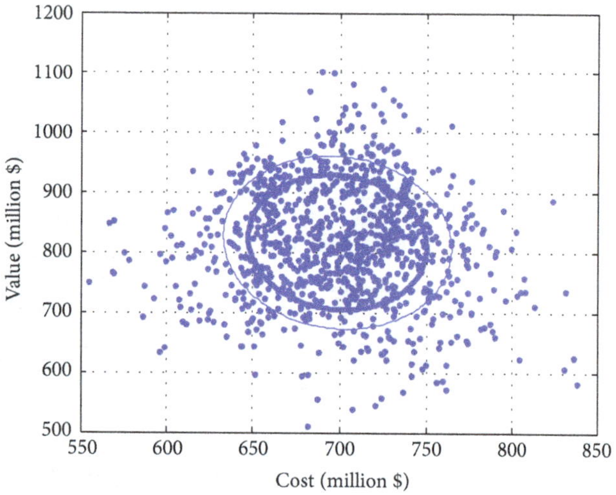

FIGURE 4: The assessments in ten-year lifecycle labeled on the phase plane of cost and benefit values.

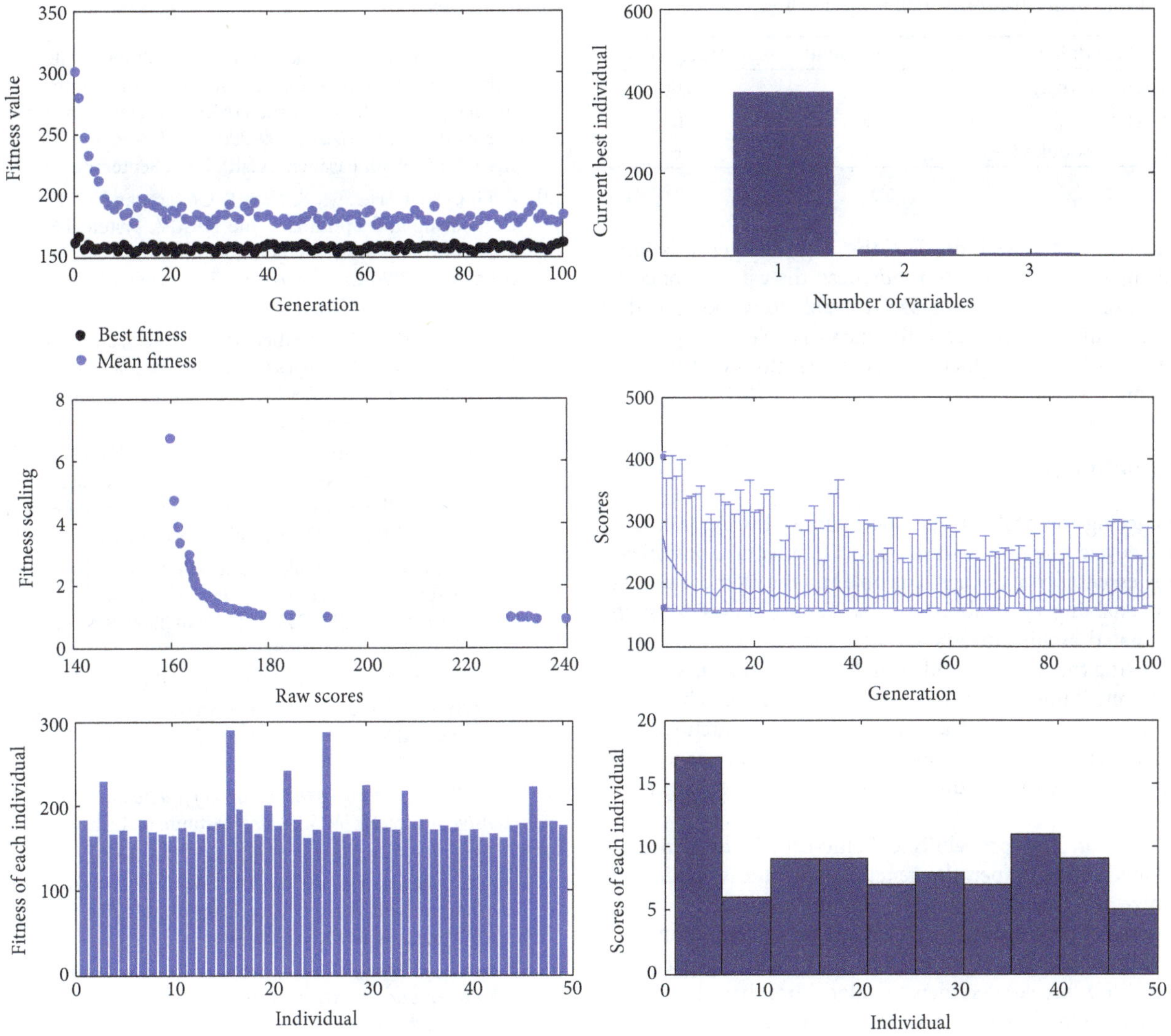

FIGURE 5: The optimization results by genetic algorithm.

5. Probabilistic Value-Centric Optimization for Fractionated Spacecraft Architecture

The value-centric assessment methodology can customize several criterions, such as the mean and variance of the cost, the benefit income, and the net present value NPV. Therefore, the optimization is essentially solving a multiobjective problem in Pareto's sense [13, 22].

For this multidisciplinary optimization problem, there exist analytic gradient functions to guide the refinement of iteration procedure of numerical optimization. Thus, the heuristic method, such as the genetic algorithm, is quite good at refining the optimal design scheme. Compared with the traditional algorithm, the graphic processing unit (GPU) accelerated genetic algorithm based on the Compute Unified Device Architecture (CUDA) is employed to improve the searching efficiency because the cost and value assessments in any generation are parallel computerized.

In this paper, the fractionated spacecrafts with five-years lifecycle are chosen to optimize their architecture. The objective function for this case is to maximize the net present value NPV with the optimized variables of orbital altitude, local time at descending node (LTDN), and the number of modules. The inequality constraints for the three variables are listed in Table 6.

For the genetic initialization, the population number within each generation is set as 100. The evolution after the fifty generations can refine the maximum net present value NPV, shown in Figure 5. In the optimal design scheme, the orbital altitude is 573.7150 km, the local time at descending node is 11:23 AM, and the number of modules employed by the fractionated architecture is six. Ideally, if the design models are accurate enough, the maximum NPV could be reached through the engineering practice.

According to the numerical simulations, the optimal design scheme of the fractionated architecture depends on

TABLE 6: Inequality constraints for the three variables.

Optimal conditions	Lower bound	Upper bound
Orbit altitude (km)	400	700
LTDN (hour)	9	13
Number of modules (—)	1	6

the assessment criterions. For the high-budget missions requiring more reliability or robustness, the variance of cost, income, or net present value is preferable to working as the objective function; however, the mean is selected by some low-budget missions which pay more attentions on the cost-performance ratio.

6. Conclusion

Considering the probabilistic uncertainties during the mission lifecycle, the cost assessment or architecture optimization is essentially a stochastic problem. One element necessary in enabling a probabilistic, value-centric analysis of such fractionated architecture is a systematic method for sizing and costing the many candidate architectures that arise. One of the contributions in this paper is to quantitatively assess the impacts of various fractionated spacecraft architecture strategies on the lifecycle cost, mass, propellant usage, and mission lifetime of pointing-intensive, remote-sensing mission spacecraft.

Based on the probabilistic value-centric assessments developed in this paper, the genetic algorithm is applied to optimize the fractionated spacecraft architecture from the viewpoint of probability. To accelerate the optimization, the second contribution is to employ the graphic processing unit (GPU) accelerated genetic algorithm based on the Compute Unified Device Architecture (CUDA) to improve the searching efficiency because the assessments in any generation are parallel computerized. Furthermore, another is to introduce the unscented transformation to reduce the amount of computations rather than the Monte Carlo method in stochastic simulation.

Finally, for future work, the models can ultimately be developed on a level of confidence such that the results of the surplus value model can be analyzed in detail, in particular to establish the design objective functions for components. This can then provide a platform for optimization to generate the best design scheme for different aerospace missions.

Acknowledgments

The research is supported by the National Natural Science Foundation of China (11172020), the National High Technology Research and Development Program of China (863 Program: 2012AA120601), Talent Foundation supported by the Fundamental Research Funds for the Central Universities, Aerospace Science and Technology Innovation Foundation of China Aerospace Science Corporation, and Innovation Fund of China Academy of Space Technology. The authors have no conflict of interests with the mentioned trademark(s).

References

[1] O. C. Brown, P. Eremenko, and P. D. Collopy, "Value-centric design methodologies for fractionated spacecraft: progress summary from phase 1 of the DARPA system F6 program," in *Proceedings of the AIAA Space 2009 Conference and Exposition*, AIAA 2009-6540, Pasadena, Calif, USA, September 2009.

[2] M. G. O'Neill, H. Yue, S. Nag, P. Grogan, and O. L. deWeck, "Comparing and optimizing the DARPA system F6 program value-centric design methodologies," in *Proceedings of the AIAA Space 2010 Conference and Exposition*, AIAA 2010-8828, Anaheim, Calif, USA, 2010.

[3] M. G. Richards, Z. Szajnfarber, M. G. O'Neill, and A. L. Weigel, "Implementation challenges for responsive space architectures," in *Proceedings of the AIAA Responsive Space Conference*, AIAA-RS7-2009-2004, Los Angeles, Calif, USA, April 2009.

[4] O. Brown and P. Eremenko, "Application of value-centric design to space architectures: the case of fractionated spacecraft," in *Proceedings of the AIAA Space 2008 Conference and Exposition*, AIAA 2009-6540, San Diego, Calif, USA, September 2008.

[5] O. Brown and P. Eremenko, "The value proposition for fractionated space architectures," in *Proceedings of the Space 2006 Conference*, AIAA 2006-7506, pp. 2788–2809, Reston, Va, USA, September 2006.

[6] C. Mathieu and A. L. Weigel, "Assessing the flexibility provided by fractionated spacecraft," in *Proceedings of the Space 2005 Conference*, AIAA 2005-6700, Long Beach, Calif, USA, August-September 2005.

[7] M. G. O'Neill, *Assessing the impacts of fractionation on pointing-intensive spacecraft [M.S. thesis]*, Institute of Technology, 2010.

[8] D. A. O'Neil and J. C. Mankins, "The advanced technology lifecycle analysis system (ATLAS)," in *Proceedings of the 55th International Astronautical Congress*, IAC-04-IAA.3.6.3.01, Vancouver, Canada, October 2004.

[9] O. Brown, A. Long, N. Shah, and P. Eremenko, "System lifecycle cost under uncertainty as a design metric encompassing the value of architectural flexibility," in *Proceedings of the AIAA Space 2007 Conference*, AIAA 2007-6023, pp. 216–229, Long Beach, Calif, USA, September 2007.

[10] J. M. Lafleur and J. H. Saleh, "GT-FAST: a point design tool for rapid fractionated spacecraft sizing and synthesis," in *Proceedings of the AIAA Space 2009 Conference and Exposition*, AIAA 2009-6563, Pasadena, Calif, USA, September 2009.

[11] J. M. Lafleur and J. H. Saleh, "Exploring the F6 fractionated spacecraft trade space with GT-FAST," in *Proceedings of the AIAA Space 2009 Conference and Exposition*, AIAA 2009-6802, Pasadena, Calif, USA, September 2009.

[12] W. Yao, X. Q. Chen, and Y. Zhao, "Based on the uncertainty of MDO satellite overall optimization design," *Chinese Journal of Aeronautics*, vol. 30, no. 5, pp. 68–76, 2009.

[13] W. Yao, X. Q. Chen, W. C. Luo, M. von Tooren, and J. Guo, "Review of uncertainty-based multidisciplinary design optimization methods for aerospace vehicles," *Progress in Aerospace Sciences*, vol. 47, pp. 450–479, 2011.

[14] M. Daniels, J. Irvine, B. Tracey, W. Schram, and M. E. Paté-Cornell, "Probabilistic simulation of multi-stage decisions for operation of a fractionated satellite mission," in *Proceedings of the IEEE Aerospace Conference (AERO '11)*, pp. 1–16, March 2011.

[15] P. Collopy, "Value of the probability of success," in *Proceedings of the AIAA Space 2008 Conference and Exposition*, AIAA 2008-7868, San Diego, Calif, USA, September 2008.

[16] S. J. Julier and J. K. Uhlmann, "Unscented filtering and nonlinear estimation," *Proceedings of the IEEE*, vol. 92, no. 3, pp. 401–422, 2004.

[17] R. van der Merwe, E. A. Wan, and S. I. Julier, "Sigma-point kalman filters for nonlinear estimation and sensor-fusion: applications to integrated navigation," in *Proceedings of the AIAA Guidance, Navigation, and Control Conference*, AIAA 2004-5120, pp. 1735–1764, Providence, RI, USA, August 2004.

[18] P. D. Collopy, "Economic-based distributed optimal design," in *Proceedings of the AIAA Space 2001 Conference and Exposition*, AIAA 2001-4675, Albuquerque, NM, USA, 2001.

[19] NASA, *NASA Systems Engineering Handbook*, SP-610S, NASA, 1995.

[20] E. M. Mahr, *Small Satellite Cost Model 2007 (SSCM07) User's Manual*, ATR-2007(8617)-5, The Aerospace Corporation, 2007.

[21] A. Braukhane, "Lessons learned from one-week concurrent engineering study approach," in *Proceedings of the 17th International Conference on Concurrent Enterprising (ICE '01)*, pp. 1–10, Bremen, Germany, 2011.

[22] R. Hassan and W. Crossley, "Spacecraft reliability-based design optimization under uncertainty including discrete variables," *Journal of Spacecraft and Rockets*, vol. 45, no. 2, pp. 394–405, 2008.

15

Observer-Based Decentralized Control for Uncertain Interconnected Systems of Neutral Type

Heli Hu,[1] Dan Zhao,[2] and Qingling Zhang[3]

[1] *Key Laboratory of Manufacturing Industrial Integrated Automation, Shenyang University, Shenyang 110044, China*
[2] *Department of Fundamental Teaching, Shenyang Institute of Engineering, Shenyang 110136, China*
[3] *Institute of Systems Science, Northeastern University, Shenyang, Liaoning 110004, China*

Correspondence should be addressed to Heli Hu; huheli2002@yahoo.com.cn

Academic Editor: Rongni Yang

The observer-based decentralized control problem is investigated for a class of uncertain interconnected systems of neutral type. Using the singular value decomposition approach, a full-order observer is designed to guarantee the asymptotic stability of the error dynamic system. A novel mathematical technique is developed to solve this design problem. Sufficient condition for uncertain interconnected systems of neutral type to be asymptotic stable is established based on the singular value decomposition method. Furthermore, the desired gains of observer and controller are obtained by the explicit expressions in terms of some free parameters. Finally, an illustrative example is used to demonstrate the proposed approach, and the corresponding simulation results are given to elucidate the effectiveness.

1. Introduction

Nowadays the systems have become more and more large, and for the interconnected systems decentralized control has obvious advantage that overcomes the limitations of the traditional centralized control requiring sufficiently large communication bandwidth to exchange information between the subsystems. Therefore, the decentralized control scheme [1–11] only making use of local information available is very popular among the researchers and the engineers. In [1], an algorithm formulated within the convex optimization framework is proposed to investigate the strict dissipativity of the linear interconnected systems. A decentralized structure of dissipative state-feedback controllers is designed. In [2, 3], a decentralized adaptive output-feedback stabilizer and a decentralized L_1 adaptive controller are proposed to stabilize a class of large-scale nonlinear systems, respectively. Different from the constant delays involved in the considered system without a priori knowledge of subsystem high-frequency-gain signs in [2], the interconnected nonlinearities and unmodeled dynamics are included in the considered systems in [3]. Based on the concepts of dynamic graphs and dynamic adjacency matrix, a modeling method of a complex dynamic interconnected system is considered in [4]. In [5], a decentralized dynamic output feedback based on linear controller is proposed to robust stabilize a class of nonlinear interconnected systems coupled by nonlinear interconnections that are unknown and quadratically bounded. However, neither constant delay nor time-varying delays are considered in [4, 5].

On the other hand, the neutral system is the general form of delay system that contains the same highest order derivatives for the state vector $x(t)$, at both time t and past time(s) $t_s \le t$. Many models of practical systems can be described by functional differential equation of neutral type [12]. Physical examples for neutral system have distributed networks [13], population ecology [14], heat exchangers, robots in contact with rigid environments [15], and so forth. In recent years, the stability analysis and robust control problems of neutral delay systems have been considered extensively (see, e.g., [16–23]). In [16–18], the stability problems of neutral systems are investigated. The difference among them is that Balasubramaniam et al. [16] focus on the stability of the neutral systems with both constant and time-varying delays using the method of nonuniformly dividing the whole delay interval into multiple segments, Rakkiyappan et al. [17] study the stability of the neutral systems with interval time-varying delays and nonlinear perturbations using the method of a new Lyapunov functional approach, and Nian et al. [18] deal with the stability of neutral systems with only a constant delay using the method of state matrix decomposition. However, the design problems of the control law have not been considered in them.

Furthermore, Ma et al. [19] develop a control method for neutral systems with a single input and some restrictions on the system matrices using a differential-difference inequality and the transformation technique. Han et al. [20] utilize a discretized Lyapunov-Krasovskii functional approach to investigate the stability of the linear neutral systems with small and nonsmall discrete delays. But its discrete application to control design yields nonlinear conditions, which may not be easily computable. In [21], the robust adaptive stabilization problem is investigated for neutral time-delay systems with uncertainties, and an adaptive scheme is introduced to estimate the bounds on uncertainties. But the matched condition is required for the disturbance vector of the considered system. It is worth pointing out that many researches on the neutral systems are often restricted to the stability analysis without controller or static state feedback control schemes.

Nowadays, observer control is also an attractive topic [24–28]. In [24], two controllers based on state feedback and observer output feedback are designed for networked systems with discrete and distributed delays subject to quantization and packet dropout. Also, a compensation scheme is proposed to deal with the effect of random packet dropout through communication network. In [25], Su et al. introduce a new model transformation for considered discrete-time T-S fuzzy systems to realize the design of dynamic output-feedback controller. Utilizing an approximation for time-varying delay state, a new comparison model is proposed. Furthermore, the l_2-l_∞ filtering problem for a class of discrete-time T-S fuzzy systems with time-varying delays is studied in [26]. The anticipated full- and reduced-order filter design is cast into a convex optimization problem, which can be efficiently solved by standard numerical algorithms. In [27], an improved fuzzy observer design that has some advantages, such as the less conservatism and the satisfactory multiple performance required by the fault detection, is presented. Moreover, Messaoud et al. [28] propose a new design of functional unknown input observer for nonlinear systems. In particular, necessary and sufficient conditions for the existence of the anticipated observer are presented.

To the best of the authors' knowledge, the observer-based decentralized control problems of uncertain interconnected systems with neutral type have not yet been investigated, which motivates the present study. In this paper, the asymptotic stabilization of a class of uncertain neutral interconnected systems with time-varying delays in state, control input, and interconnections is made. In framework of linear matrix equalities, the design of the observer and controller is formulated. Sufficiently the influence of interconnections on the system performance is taken into account, and interconnections are dealt with by flexible techniques. These strategies allow one to obtain less conservative stabilization conditions. A numerical example and the corresponding simulation results are given to illustrate the effectiveness of the proposed method of decentralized controller based on observer.

The remainder of the paper is organized as follows. The observer-based decentralized control problem formulation is described in Section 2. In Section 3, the desired gains of observer and controller are obtained by the explicit expressions in terms of some free parameters. A numerical example and the corresponding simulation results are presented in Section 4. The conclusion is provided in Section 5.

2. Problem Formulation

Consider the following uncertain neutral interconnected systems composed of N subsystems:

$$\begin{aligned}
&\dot{x}_i(t) - A_{i\eta_i}\dot{x}_i(t-\eta_i(t)) \\
&\quad = \left[A_i + \Delta A_i(t)\right] x_i(t) \\
&\qquad + \left[A_{i\sigma_i} + \Delta A_{i\sigma_i}(t)\right] x_i(t-\sigma_i(t)) \\
&\qquad + \left[B_i + \Delta B_i(t)\right] u_i(t) \\
&\qquad + \sum_{j=1, j\neq i}^{N} \left[A_{ij} + \Delta A_{ij}\right] x_j\left(t - \tau_{ij}(t)\right), \\
&y_i(t) = C_i x_i(t) + D_i u_i(t), \\
&x_i(t) = \phi_i(t), \quad t \in [-l, 0], \; i = 1, 2, \ldots, N,
\end{aligned} \tag{1}$$

where $x_i(t) \in \Re^{n_i}$, $u_i(t) \in \Re^{m_i}$, and $y_i(t) \in \Re^{p_i}$ are the state, control input, and measurement output of the ith subsystem, respectively. A_i, $A_{i\sigma_i}$, $A_{i\eta_i}$, B_i, A_{ij}, C_i, and D_i are known constant matrices of appropriate dimensions. $\phi_i(t)$ is the initial condition. $\sigma_i(t)$, $\eta_i(t)$, and $\tau_{ij}(t)$ are the time-varying delays. Assume that there exist constants f_{i0}, g_{i0}, l_{i0}, f_i, g_i, l_i, and l satisfying

$$\begin{gathered}
0 \le \sigma_i(t) \le f_{i0}, \qquad 0 \le \eta_i(t) \le g_{i0}, \qquad 0 \le \tau_{ij}(t) \le l_{i0}, \\
\dot{\sigma}_i(t) \le f_i < 1, \qquad \dot{\eta}_i(t) \le g_i < 1, \qquad \dot{\tau}_{ij}(t) \le l_i < 1, \\
l = \max\{f_{i0}, g_{i0}, l_{i0}\}, \quad i, j = 1, 2 \ldots, N, \; j \neq i.
\end{gathered} \tag{2}$$

Time-varying parametric uncertainties $\Delta A_i(t)$, $\Delta A_{i\sigma_i}(t)$, $\Delta B_i(t)$, and $\Delta A_{ij}(t)$ are assumed to satisfy

$$\begin{aligned}
&\left[\Delta A_i(t) \;\; \Delta A_{i\sigma_i}(t) \;\; \Delta B_i(t) \;\; \Delta A_{ij}(t)\right] \\
&\quad = M_i F_i(t) \left[N_{i1} \;\; N_{i\sigma_i} \;\; N_{i2} \;\; E_{ij}\right],
\end{aligned} \tag{3}$$

where matrices M_i, N_{i1}, $N_{i\sigma_i}$, N_{i2}, and E_{ij} are constant matrices of appropriate dimensions and $F_i(t)$ is the unknown matrix function satisfying

$$F_i^T(t) F_i(t) \le I, \quad \forall t \ge 0. \tag{4}$$

The following assumptions are made on the considered system (1).

Assumption 1. Suppose that the matrix C_i has full row rank (i.e., $\text{rank}(C_i) = p_i$). Then the singular value decomposition of C_i presents as follows:

$$C_i = U_i [S_i \ \ 0] V_i^T, \tag{5}$$

where $S_i \in \mathfrak{R}^{p_i \times p_i}$ is a diagonal matrix with positive elements in a decreasing order, $0 \in \mathfrak{R}^{p_i \times (n_i - p_i)}$ is a zero matrix, and $U_i \in \mathfrak{R}^{p_i \times p_i}$ and $V_i \in \mathfrak{R}^{n_i \times n_i}$ are unitary matrices.

Assumption 2. The matrix $A_{i\eta_i} \neq 0$ and $\|A_{i\eta_i}\| < 1$.

Consider the following observer-based decentralized control for system (1):

$$\begin{aligned}
&\dot{\hat{x}}_i(t) - A_{i\eta_i}\dot{\hat{x}}_i(t - \eta_i(t)) \\
&\quad = A_i \hat{x}_i(t) + B_i u_i(t) + A_{i\sigma_i}\hat{x}_i(t - \sigma_i(t)) \\
&\quad + \sum_{j=1, j \neq i}^{N} A_{ij}\hat{x}_j(t - \tau_{ij}(t)) + L_i(y_i(t) - \hat{y}_i(t)), \\
&\hat{y}_i(t) = C_i \hat{x}_i(t) + D_i u_i(t), \\
&u_i(t) = -K_i \hat{x}_i(t),
\end{aligned} \tag{6}$$

where $\hat{x}_i(t) \in \mathfrak{R}^{n_i}$ and $\hat{y}_i(t) \in \mathfrak{R}^{p_i}$ are the state and output vectors of the observer and $K_i \in \mathfrak{R}^{m_i \times n_i}$ and $L_i \in \mathfrak{R}^{n_i \times p_i}$ are the controller gain and observer gain to be designed. System (1) with the observer-based control (6) can be rewritten as

$$\begin{aligned}
\begin{bmatrix} \dot{\hat{x}}_i(t) \\ \dot{e}_i(t) \end{bmatrix} &= \begin{bmatrix} A_i - B_i K_i & L_i C_i \\ 0 & A_i - L_i C_i \end{bmatrix} \begin{bmatrix} \hat{x}_i(t) \\ e_i(t) \end{bmatrix} \\
&+ \begin{bmatrix} A_{i\sigma_i} & 0 \\ 0 & A_{i\sigma_i} \end{bmatrix} \begin{bmatrix} \hat{x}_i(t - \sigma_i(t)) \\ e_i(t - \sigma_i(t)) \end{bmatrix} \\
&+ \begin{bmatrix} A_{i\eta_i} & 0 \\ 0 & A_{i\eta_i} \end{bmatrix} \begin{bmatrix} \dot{\hat{x}}_i(t - \eta_i(t)) \\ \dot{e}_i(t - \eta_i(t)) \end{bmatrix} \\
&+ \begin{bmatrix} \sum_{j=1, j \neq i}^{N} A_{ij}\hat{x}_j(t - \tau_{ij}(t)) \\ \sum_{j=1, j \neq i}^{N} A_{ij} e_j(t - \tau_{ij}(t)) \end{bmatrix} \\
&+ \begin{bmatrix} 0 \\ M_i \end{bmatrix} \Delta_i(t),
\end{aligned} \tag{7}$$

where the signal $e_i(t) = x_i(t) - \hat{x}_i(t)$ is defined as the estimated error of system, $\Delta_i(t) = F_i(t)[(N_{i1} - N_{i2}K_i)\hat{x}_i(t) + N_{i1}e_i + N_{i\sigma_i}\hat{x}_{i\sigma_i} + N_{i\sigma_i}e_{i\sigma_i} + \sum_{j=1, j \neq i}^{N} E_{ij}\hat{x}_j(t - \tau_{ij}(t)) + \sum_{j=1, j \neq i}^{N} E_{ij}e_j(t - \tau_{ij}(t))]$, and the uncertainty $\Delta_i(t)$ satisfies the following quadratic inequality:

$$\Delta_i^T(t)\Delta_i(t) \le \xi_i^T(t)\Theta_i^T\Theta_i\xi_i(t), \tag{8}$$

where

$$\begin{aligned}
\xi_i^T(t) = \big[&\hat{x}_i^T(t) \ \ e_i^T(t) \ \ \hat{x}_i^T(t - \sigma_i(t)) \ \ e_i^T(t - \sigma_i(t)) \\
&\dot{\hat{x}}_i^T(t - \eta_i(t)) \ \ \dot{e}_i^T(t - \eta_i(t)) \ \ \hat{x}_1^T(t - \tau_{i1}(t)) \\
&\cdots \ \hat{x}_{i-1}^T(t - \tau_{i\ i-1}(t)) \ \ \hat{x}_{i+1}^T(t - \tau_{i\ i+1}(t)) \\
&\cdots \ \hat{x}_N^T(t - \tau_{iN}(t)) \ \ e_1^T(t - \tau_{i1}(t)) \\
&\cdots \ e_{i-1}^T(t - \tau_{i\ i-1}(t)) \ \ e_{i+1}^T(t - \tau_{i\ i+1}(t)) \\
&\cdots \ e_N^T(t - \tau_{iN}(t)) \big], \\
\Theta_i = \big[&(N_{i1} - N_{i2}K_i) \ \ N_{i1} \ \ N_{i\sigma_i} \ \ N_{i\sigma_i} \ \ 0 \ \ 0 \ \ E_{i1} \\
&\cdots \ E_{i\ i-1} \ \ E_{i\ i+1} \ \cdots \ E_{iN} \ \ E_{i1} \\
&\cdots \ E_{i\ i-1} \ \ E_{i\ i+1} \ \cdots \ E_{iN} \big].
\end{aligned} \tag{9}$$

Definition 3. Consider system (1) with the observer-based control (6). System (1) is said to be robustly stabilizable by the observer-based control (6), if the closed-loop system (7) satisfying Assumptions 1 and 2 is asymptotically stable.

Lemma 4 (see [29]). *For a given $C_i \in \mathfrak{R}^{p_i \times n_i}$ with $\text{rank}(C_i) = p_i$, assume that $Q_{i2} \in \mathfrak{R}^{n_i \times n_i}$ is a symmetric matrix; then there exists a matrix $\hat{Q}_{i2} \in \mathfrak{R}^{p_i \times p_i}$ such that $C_i Q_{i2} = \hat{Q}_{i2} C_i$ if and only if*

$$Q_{i2} = V_i \begin{bmatrix} \hat{X}_{i1} & 0 \\ 0 & \hat{X}_{i2} \end{bmatrix} V_i^T, \tag{10}$$

where $\hat{X}_{i1} \in \mathfrak{R}^{p_i \times p_i}$ and $\hat{X}_{i2} \in \mathfrak{R}^{(n_i - p_i) \times (n_i - p_i)}$.

3. Main Result

Theorem 5. *Consider uncertain neutral interconnected systems (1) with (2), (3), and (6). If there exists a solution $Q_{ik} > 0$ $(k = 1, 3, 4, 5, 6)$, $\hat{X}_{i1} > 0$, $\hat{X}_{i2} > 0$, $\overline{W}_{ji} > 0$, $\overline{G}_{ji} > 0$, $W_{ij} > 0$, $G_{ij} > 0$, R_{i1}, and R_{i2} such that the following inequalities hold:*

$$\Omega_i < 0, \tag{11}$$

then system (1) is robustly stabilizable by the observer-based control (6) with $K_i = R_{i1}Q_{i1}^{-1}$ *and* $L_i = R_{i2}U_iS_i\widehat{X}_{i1}^{-1}S_i^{-1}U_i^T$, where

$$
\Omega_i = \begin{bmatrix}
\Omega^i_{11} & \Omega^i_{12} & \Omega^i_{13} & 0 & \Omega^i_{15} & 0 & \Omega^i_{17} & 0 & 0 & Q_{i1} & 0 & \Omega^i_{1\ 12} & 0 & \Omega^i_{1\ 14} \\
* & \Omega^i_{22} & 0 & \Omega^i_{24} & 0 & \Omega^i_{26} & 0 & \Omega^i_{28} & M_i & 0 & \Omega^i_{2\ 11} & \Omega^i_{2\ 12} & \Omega^i_{2\ 13} & \Omega^i_{2\ 14} \\
* & * & -Q_{i3} & 0 & 0 & 0 & 0 & 0 & 0 & 0 & 0 & \Omega^i_{3\ 12} & 0 & \Omega^i_{314} \\
* & * & * & -Q_{i4} & 0 & 0 & 0 & 0 & 0 & 0 & 0 & 0 & \Omega^i_{4\ 13} & \Omega^i_{4\ 14} \\
* & * & * & * & -Q_{i5} & 0 & 0 & 0 & 0 & 0 & 0 & \Omega^i_{5\ 12} & 0 & 0 \\
* & * & * & * & * & -Q_{i6} & 0 & 0 & 0 & 0 & 0 & 0 & \Omega^i_{6\ 13} & 0 \\
* & * & * & * & * & * & \Omega^i_{77} & 0 & 0 & 0 & 0 & \Omega^i_{7\ 12} & 0 & \Omega^i_{7\ 14} \\
* & * & * & * & * & * & * & \Omega^i_{88} & 0 & 0 & 0 & 0 & \Omega^i_{8\ 13} & \Omega^i_{8\ 14} \\
* & * & * & * & * & * & * & * & -I & 0 & 0 & 0 & M_i^T & 0 \\
* & * & * & * & * & * & * & * & * & \Omega^i_{10\ 10} & 0 & 0 & 0 & 0 \\
* & * & * & * & * & * & * & * & * & * & \Omega^i_{11\ 11} & 0 & 0 & 0 \\
* & * & * & * & * & * & * & * & * & * & * & \Omega^i_{12\ 12} & 0 & 0 \\
* & * & * & * & * & * & * & * & * & * & * & * & \Omega^i_{13\ 13} & 0 \\
* & * & * & * & * & * & * & * & * & * & * & * & * & I
\end{bmatrix},
$$

$$\Omega^i_{11} = A_iQ_{i1} + Q_{i1}A_i^T - B_iR_{i1} - R_{i1}^TB_i^T + \sum_{j=1,j\neq i}^{N} \frac{1}{1-l_j}\overline{W}_{ji}, \qquad \Omega^i_{12} = R_{i2}C_i, \qquad \Omega^i_{13} = A_{i\sigma_i}Q_{i3},$$

$$\Omega^i_{15} = A_{i\eta_i}Q_{i5}, \qquad \Omega^i_{1\ 12} = Q_{i1}A_i^T - R_{i1}^TB_i^T, \qquad \Omega^i_{17} = \begin{bmatrix} A_{i1} & \cdots & A_{i\ i-1} & A_{i\ i+1} & \cdots & A_{iN} \end{bmatrix},$$

$$\Omega^i_{1\ 14} = Q_{i1}N_{i1}^T - R_{i1}^TN_{i2}^T, \qquad \Omega^i_{24} = A_{i\sigma_i}Q_{i4}, \qquad \Omega^i_{22} = A_iQ_{i2} + Q_{i2}A_i^T - R_{i2}C_i - C_i^TR_{i2}^T + \sum_{j=1,j\neq i}^{N} \frac{1}{1-l_j}\overline{G}_{ji},$$

$$\Omega^i_{26} = A_{i\eta_i}Q_{i6}, \qquad \Omega^i_{28} = \Omega^i_{17}, \qquad \Omega^i_{2\ 11} = Q_{i2}, \qquad \Omega^i_{2\ 12} = C_i^TR_{i2}^T,$$

$$\Omega^i_{2\ 14} = Q_{i2}N_{i1}^T, \qquad \Omega^i_{2\ 13} = Q_{i2}A_i^T - C_i^TR_{i2}^T, \qquad \Omega^i_{3\ 12} = Q_{i3}A_{i\sigma_i}^T, \qquad \Omega^i_{3\ 14} = Q_{i3}N_{i\sigma_i}^T,$$

$$\Omega^i_{4\ 13} = Q_{i4}A_{i\sigma_i}^T, \qquad \Omega^i_{6\ 13} = Q_{i6}A_{i\eta_i}^T, \qquad \Omega^i_{4\ 14} = Q_{i4}N_{i\sigma_i}^T, \qquad \Omega^i_{5\ 12} = Q_{i5}A_{i\eta_i}^T,$$

$$\Omega^i_{77} = \mathrm{diag}\left\{-W_{i1},\ldots,-W_{i\ i-1},-W_{i\ i+1},\ldots,-W_{iN}\right\}, \qquad \Omega^i_{7\ 12} = \left(\Omega^i_{17}\right)^T, \qquad \Omega^i_{8\ 13} = \Omega^i_{7\ 12},$$

$$\Omega^i_{7\ 14} = \begin{bmatrix} E_{i1} & \cdots & E_{i\ i-1} & E_{i\ i+1} & \cdots & E_{iN} \end{bmatrix}^T, \qquad \Omega^i_{8\ 14} = \Omega^i_{7\ 14},$$

$$\Omega^i_{88} = \mathrm{diag}\left\{-G_{i1},\ldots,-G_{i\ i-1},-G_{i\ i+1},\ldots,-G_{iN}\right\}, \qquad \Omega^i_{10\ 10} = -\left(1-f_i\right)Q_{i3}, \qquad \Omega^i_{11\ 11} = -\left(1-f_i\right)Q_{i4},$$

$$\Omega^i_{12\ 12} = -\left(1-g_i\right)Q_{i5}, \qquad \Omega^i_{13\ 13} = -\left(1-g_i\right)Q_{i6}, \qquad \overline{W}_{ji} = Q_{i1}W_{ji}Q_{i1}, \qquad \overline{G}_{ji} = Q_{i1}G_{ji}Q_{i1},$$

$$Q_{i2} = V_i\begin{bmatrix} \widehat{X}_{i1} & 0 \\ 0 & \widehat{X}_{i2} \end{bmatrix}V_i^T, \quad i,j = 1,2\ldots,N, \; j\neq i. \tag{12}$$

Proof. According to Schur complement, inequality (11) is equivalent to the following inequality:

$$\Gamma_{i1} = \begin{bmatrix} \Omega_{11}^{i} & \Omega_{12}^{i} & \Omega_{13}^{i} & 0 & \Omega_{15}^{i} & 0 & \Omega_{17}^{i} & 0 & 0 \\ * & \Omega_{22}^{i} & 0 & \Omega_{24}^{i} & 0 & \Omega_{26}^{i} & 0 & \Omega_{28}^{i} & M_i \\ * & * & -Q_{i3} & 0 & 0 & 0 & 0 & 0 & 0 \\ * & * & * & -Q_{i4} & 0 & 0 & 0 & 0 & 0 \\ * & * & * & * & -Q_{i5} & 0 & 0 & 0 & 0 \\ * & * & * & * & * & -Q_{i6} & 0 & 0 & 0 \\ * & * & * & * & * & * & \Omega_{77}^{i} & 0 & 0 \\ * & * & * & * & * & * & * & \Omega_{88}^{i} & 0 \\ * & * & * & * & * & * & * & * & -I \end{bmatrix}$$

$$+ \Upsilon_{i1}^{T}\left(\frac{1}{1-f_i}Q_{i3}^{-1}\right)\Upsilon_{i1} + \Upsilon_{i2}^{T}\left(\frac{1}{1-f_i}Q_{i4}^{-1}\right)\Upsilon_{i2} + \Upsilon_{i3}^{T}\left(\frac{1}{1-g_i}Q_{i5}^{-1}\right)\Upsilon_{i3} + \Upsilon_{i4}^{T}\left(\frac{1}{1-g_i}Q_{i6}^{-1}\right)\Upsilon_{i4} + \Upsilon_{i5}^{T}\Upsilon_{i5} < 0, \tag{13}$$

where

$$\begin{aligned} \Upsilon_{i1} &= \left[(Q_{i1})^{T}\ 0\ 0\ 0\ 0\ 0\ 0\ 0\ 0\right], \qquad \Upsilon_{i2} = \left[0\ \left(\Omega_{2\ 11}^{i}\right)^{T}\ 0\ 0\ 0\ 0\ 0\ 0\ 0\right], \\ \Upsilon_{i3} &= \left[\left(\Omega_{1\ 12}^{i}\right)^{T}\ \left(\Omega_{2\ 12}^{i}\right)^{T}\ \left(\Omega_{3\ 12}^{i}\right)^{T}\ 0\ \left(\Omega_{5\ 12}^{i}\right)^{T}\ 0\ \left(\Omega_{7\ 12}^{i}\right)^{T}\ 0\ 0\right], \\ \Upsilon_{i4} &= \left[0\ \left(\Omega_{2\ 13}^{i}\right)^{T}\ 0\ \left(\Omega_{4\ 13}^{i}\right)^{T}\ 0\ \left(\Omega_{6\ 13}^{i}\right)^{T}\ 0\ \left(\Omega_{8\ 13}^{i}\right)^{T}\ M_i\right], \\ \Upsilon_{i5} &= \left[\left(\Omega_{1\ 14}^{i}\right)^{T}\ \left(\Omega_{2\ 14}^{i}\right)^{T}\ \left(\Omega_{3\ 14}^{i}\right)^{T}\ \left(\Omega_{4\ 14}^{i}\right)^{T}\ 0\ 0\ \left(\Omega_{7\ 14}^{i}\right)^{T}\ \left(\Omega_{8\ 14}^{i}\right)^{T}\ 0\right]. \end{aligned} \tag{14}$$

By Lemma 4, the conditions $Q_{i2} = V_i \left[\begin{smallmatrix} \widehat{X}_{i1} & 0 \\ 0 & \widehat{X}_{i2} \end{smallmatrix}\right] V_i^T$ and $\widehat{Q}_{i2} = U_i S_i \widehat{X}_{i1} S_i^{-1} U_i^T$ imply the condition $C_i Q_{i2} = \widehat{Q}_{i2} C_i$. In view of $K_i = R_{i1} Q_{i1}^{-1}$, $L_i = R_{i2} U_i S_i \widehat{X}_{i1}^{-1} S_i^{-1} U_i^T$, $\overline{W}_{ji} = Q_{i1} W_{ji} Q_{i1}$, and $\overline{G}_{ji} = Q_{i1} G_{ji} Q_{i1}$, the matrix inequality (13) is equivalent to

$$\Gamma_{i2} = \begin{bmatrix} \Xi_{11}^{i} & \Xi_{12}^{i} & \Xi_{13}^{i} & 0 & \Xi_{15}^{i} & 0 & \Omega_{17}^{i} & 0 & 0 \\ * & \Xi_{22}^{i} & 0 & \Xi_{24}^{i} & 0 & \Xi_{26}^{i} & 0 & \Omega_{18}^{i} & M_i \\ * & * & -Q_{i3} & 0 & 0 & 0 & 0 & 0 & 0 \\ * & * & * & -Q_{i4} & 0 & 0 & 0 & 0 & 0 \\ * & * & * & * & -Q_{i5} & 0 & 0 & 0 & 0 \\ * & * & * & * & * & -Q_{i6} & 0 & 0 & 0 \\ * & * & * & * & * & * & \Omega_{77}^{i} & 0 & 0 \\ * & * & * & * & * & * & * & \Omega_{88}^{i} & 0 \\ * & * & * & * & * & * & * & * & -I \end{bmatrix} + \Lambda_i^{T}\left(\frac{1}{1-g_i}Q_{i5}^{-1}\right)\Lambda_i + \widehat{\Lambda}_i^{T}\left(\frac{1}{1-g_i}Q_{i6}^{-1}\right)\widehat{\Lambda}_i + \overline{\Lambda}_i^{T}\overline{\Lambda}_i$$

$$+ \begin{bmatrix} Q_{i1} \\ 0 \\ 0 \\ 0 \\ 0 \\ 0 \\ 0 \\ 0 \\ 0 \end{bmatrix} \left(\frac{1}{1-f_i}Q_{i3}^{-1}\right) \begin{bmatrix} Q_{i1} \\ 0 \\ 0 \\ 0 \\ 0 \\ 0 \\ 0 \\ 0 \\ 0 \end{bmatrix}^{T} + \begin{bmatrix} 0 \\ Q_{i2} \\ 0 \\ 0 \\ 0 \\ 0 \\ 0 \\ 0 \\ 0 \end{bmatrix} \left(\frac{1}{1-f_i}Q_{i4}^{-1}\right) \begin{bmatrix} 0 \\ Q_{i2} \\ 0 \\ 0 \\ 0 \\ 0 \\ 0 \\ 0 \\ 0 \end{bmatrix}^{T} < 0, \tag{15}$$

where

$$\Lambda_i = \left[(A_i - B_i K_i) Q_{i1} \ \ L_i C_i Q_{i2} \ \ A_{i\sigma_i} Q_{i3} \ \ 0 \ \ A_{i\eta_i} Q_{i5} \ \ 0 \ \ \Omega^i_{17} \ \ 0 \ \ 0 \right],$$
$$\widehat{\Lambda}_i = \left[0 \ \ (A_i - L_i C_i) Q_{i2} \ \ 0 \ \ A_{i\sigma_i} Q_{i4} \ \ 0 \ \ A_{i\eta_i} Q_{i6} \ \ 0 \ \ \Omega^i_{17} \ \ M_i \right],$$
$$\overline{\Lambda}_i = \left[(N_{i1} - N_{i2} K_i) Q_{i1} \ \ N_{i1} Q_{i2} \ \ N_{i\sigma_i} Q_{i3} \ \ N_{i\sigma_i} Q_{i4} \ \ 0 \ \ 0 \ \ (\Omega^i_{7\ 14})^T \ \ (\Omega^i_{8\ 14})^T \ \ 0 \right],$$
$$\Xi^i_{11} = Q_{i1}(A_i - B_i K_i)^T + (A_i - B_i K_i) Q_{i1} + \sum_{j=1, j\neq i}^{N} \frac{1}{1 - l_j} \overline{W}_{ji}, \qquad \Xi^i_{12} = L_i C_i Q_{i2}, \qquad \Xi^i_{13} = A_{i\sigma_i} Q_{i3},$$
$$\Xi^i_{15} = A_{i\eta_i} Q_{i5}, \qquad \Xi^i_{22} = Q_{i2}(A_i - L_i C_i)^T + (A_i - L_i C_i) Q_{i2} + \sum_{j=1, j\neq i}^{N} \frac{1}{1 - l_j} \overline{G}_{ji}, \qquad \Xi^i_{24} = A_{i\sigma_i} Q_{i4},$$
$$\Xi^i_{26} = A_{i\eta_i} Q_{i6}. \tag{16}$$

Pre- and postmultiplying matrix Γ_{i2} in (15) by Π_i^T and Π_i, where

$$\Pi_i = \operatorname{diag}\left(Q_{i1}^{-1}, Q_{i2}^{-1}, Q_{i3}^{-1}, Q_{i4}^{-1}, Q_{i5}^{-1}, Q_{i6}^{-1}, I, I, I \right) = \operatorname{diag}\left(P_{i1}, P_{i2}, P_{i3}, P_{i4}, P_{i5}, P_{i6}, I, I, I \right) > 0, \tag{17}$$

we obtain

$$\Gamma_{i3} = \begin{bmatrix} \Psi^i_{11} & \Psi^i_{12} & \Psi^i_{13} & 0 & \Psi^i_{15} & 0 & \Psi^i_{17} & 0 & 0 \\ * & \Psi^i_{22} & 0 & \Psi^i_{24} & 0 & \Psi^i_{26} & 0 & \Psi^i_{28} & P_{i2} M_i \\ * & * & -Q_{i3} & 0 & 0 & 0 & 0 & 0 & 0 \\ * & * & * & -Q_{i4} & 0 & 0 & 0 & 0 & 0 \\ * & * & * & * & -Q_{i5} & 0 & 0 & 0 & 0 \\ * & * & * & * & * & -Q_{i6} & 0 & 0 & 0 \\ * & * & * & * & * & * & \Psi^i_{77} & 0 & 0 \\ * & * & * & * & * & * & * & \Psi^i_{88} & 0 \\ * & * & * & * & * & * & * & * & -I \end{bmatrix}$$
$$+ \Phi_i^T \left(\frac{1}{1 - g_i} P_{i5} \right) \Phi_i + \widehat{\Phi}_i^T \left(\frac{1}{1 - g_i} P_{i6} \right) \widehat{\Phi}_i + \overline{\Phi}_i^T \overline{\Phi}_i < 0, \tag{18}$$

where

$$\Phi_i = \left[A_i - B_i K_i \ \ L_i C_i \ \ A_{i\sigma_i} \ \ 0 \ \ A_{i\eta_i} \ \ 0 \ \ \Omega^i_{17} \ \ 0 \ \ 0 \right],$$
$$\widehat{\Phi}_i = \left[0 \ \ A_i - L_i C_i \ \ 0 \ \ A_{i\sigma_i} \ \ 0 \ \ A_{i\eta_i} \ \ 0 \ \ \Omega^i_{17} \ \ M_i \right],$$
$$\overline{\Phi}_i = \left[N_{i1} - N_{i2} K_i \ \ N_{i1} \ \ N_{i\sigma_i} \ \ N_{i\sigma_i} \ \ 0 \ \ 0 \ \ (\Phi^i_{7\ 14})^T \ \ (\Phi^i_{8\ 14})^T \ \ 0 \right],$$
$$\Psi^i_{11} = (A_i - B_i K_i)^T P_{i1} + P_{i1}(A_i - B_i K_i) + \sum_{j=1, j\neq i}^{N} \frac{1}{1 - l_j} \overline{W}_{ji} + \frac{1}{1 - f_i} P_{i3},$$
$$\Psi^i_{12} = P_{i1} L_i C_i, \qquad \Psi^i_{13} = P_{i1} A_{i\sigma_i},$$
$$\Psi^i_{15} = P_{i1} A_{i\eta_i}, \qquad \Psi^i_{17} = P_{i1} \Omega^i_{17},$$
$$\Psi^i_{24} = P_{i2} A_{i\sigma_i}, \qquad \Psi^i_{26} = P_{i2} A_{i\eta_i},$$
$$\Psi^i_{22} = (A_i - L_i C_i)^T P_{i2} + P_{i2}(A_i - L_i C_i) + \sum_{j=1, j\neq i}^{N} \frac{1}{1 - l_j} \overline{G}_{ji} + \frac{1}{1 - f_i} P_{i4},$$
$$\Psi^i_{28} = P_{i2} \Omega^i_{17}. \tag{19}$$

Construct the following Lyapunov functional candidate:

$$V(\widehat{x}_t, e_t) = \sum_{i=1}^{N} \Bigg\{ \widehat{x}_i^T(t) P_{i1} \widehat{x}_i(t) + e_i^T(t) P_{i2} e_i(t) + \frac{1}{1 - f_i} \times \int_{t-\sigma_i(t)}^{t} \left[\widehat{x}_i^T(s) P_{i3} \widehat{x}_i(s) + e_i^T(s) P_{i4} e_i(s) \right] ds$$

$$+\frac{1}{1-g_i}\times\int_{t-\eta_i(t)}^{t}\left[\dot{\hat{x}}_i^T(s)P_{i5}\dot{\hat{x}}_i(s)+\dot{e}_i^T(s)P_{i6}\dot{e}_i(s)\right]ds+\frac{1}{1-l_i}\times\sum_{j=1,j\neq i}^{N}\int_{t-\tau_{ij}(t)}^{t}\left[\hat{x}_j^T(s)W_{ij}\hat{x}_j(s)+e_j^T(s)G_{ij}e_j(s)\right]ds\Bigg\}. \quad (20)$$

For the system (1), the following structural identity holds:

$$\sum_{i=1}^{N}\frac{1}{1-l_i}\sum_{j=1,j\neq i}^{N}x_j^T(t)W_{ij}x_j(t)=\sum_{i=1}^{N}x_i^T(t)\sum_{j=1,j\neq i}^{N}\frac{1}{1-l_j}W_{ji}x_i(t). \quad (21)$$

By some simple derivations, the time derivative of $V(\hat{x}_t, e_t)$ along the trajectories of (7) satisfies the following inequality:

$$\dot{V}(\hat{x}_t, e_t)\le\sum_{i=1}^{N}\zeta_i^T(t)\Gamma_{i3}\zeta_i(t), \quad (22)$$

where

$$\zeta_i^T(t)=\left[\xi_i^T(t)\;\;\Delta_i^T(t)\right]. \quad (23)$$

According to conditions (18) and (22), one can obtain that, for all $\zeta_i\neq 0$,

$$\dot{V}(\hat{x}_t, e_t)<0. \quad (24)$$

In addition, Assumption 2 guarantees system (1) is Lipschitzian in the term $\dot{x}_i(t-\eta_i(t))$ with Lipschitz constant less than 1 [12]. Therefore, by Definition 3 with conditions (20) and (24), system (7) is asymptotically stable and system (1) is robustly stabilizable by observer-based control (6). This completes the proof. □

Remark 6. One of the distinctive features of this paper is that the N inequalities are included in (11), and only one is LMI due to the existence of the interconnected matrix variables $\overline{W}_{ji}=Q_{i1}W_{ji}Q_{i1}$, W_{ij}, $\overline{G}_{ji}=Q_{i2}G_{ji}Q_{i2}$, and G_{ij} among N inequalities. When $i=1$, the corresponding equality $\Omega_1<0$ in (11) is an LMI. Under the solvable condition of $\Omega_1<0$, we can apply Schur complement formula to the second inequality $\Omega_2<0$ in order to obtain the solvable LMI. The process is repeated until the last inequality. Thus, the controller gain K_i and observer gain L_i can be obtained by finding feasible set to $\Omega_i<0$ with feasp in [30].

4. Illustrative Example

Consider system (1) composed of two three-order subsystems with the following parameters:

$$A_1=\begin{bmatrix}-7.1&1.2&2.5\\-3.8&-2.1&0.4\\2.1&-3.1&-5.4\end{bmatrix},$$

$$A_{1\sigma_1}=\begin{bmatrix}-0.2&-0.1&-0.2\\0.2&0.1&0.1\\0.1&-0.2&-0.1\end{bmatrix},$$

$$B_1=\begin{bmatrix}-2.9&-1.3\\-0.3&2.5\\2.7&-2.2\end{bmatrix},\quad A_{1\eta_1}=\begin{bmatrix}-0.2&0.1&0.2\\0.2&-0.3&0.2\\0.1&-0.2&-0.1\end{bmatrix},$$

$$M_1=\begin{bmatrix}-0.2&-0.1&-0.4\\0.6&-0.2&0.2\\-0.1&-0.3&0.3\end{bmatrix},\quad N_{12}=\begin{bmatrix}-0.1&0.3\\0.2&-0.5\\0.1&0.2\end{bmatrix},$$

$$A_{12}=\begin{bmatrix}-2.2&-1.5&1.1\\-1.1&1.3&0.2\\0.3&1.1&-0.5\end{bmatrix},\quad N_{11}=\begin{bmatrix}0.2&0.7&0.3\\0.2&-0.5&-0.2\\0.5&-0.1&-0.1\end{bmatrix},$$

$$D_1=\begin{bmatrix}-1.4&0.4\\1.1&-1.1\end{bmatrix},\quad N_{1\sigma_1}=\begin{bmatrix}0.1&-0.1&-0.3\\0.2&-0.8&-0.2\\-0.5&0.1&0.1\end{bmatrix},$$

$$E_{12}=\begin{bmatrix}0.3&0.4&0.2\\-0.3&-0.3&-0.1\\-0.1&0.2&0.1\end{bmatrix},\quad \sigma_1(t)=0.1(2+\sin(t)),$$

$$C_1=\begin{bmatrix}-2.3&3.9&1.2\\0.2&2.3&-2.1\end{bmatrix},\quad \eta_1(t)=0.2(1+\cos(t)),$$

$$\tau_{12}(t)=0.1(1+\cos(t)),\quad A_2=\begin{bmatrix}-7.6&2.1&1.3\\-2.1&-1.5&-2.1\\1.2&-0.5&-5.3\end{bmatrix},$$

$$A_{2\sigma_2}=\begin{bmatrix}0.1&-0.1&-0.2\\-0.2&0.3&0.1\\0.2&0.3&0.2\end{bmatrix},\quad B_2=\begin{bmatrix}-2.1&1.8\\-4.1&0.2\\2.1&-2.1\end{bmatrix},$$

$$A_{2\eta_2}=\begin{bmatrix}0.1&0.1&-0.1\\0.3&0.2&0.4\\0.1&0.2&0.1\end{bmatrix},\quad A_{21}=\begin{bmatrix}-0.1&0.3&-0.1\\0.1&-0.2&0.2\\0.1&-0.1&0.1\end{bmatrix},$$

$$M_2=\begin{bmatrix}0.5&-0.1&0.3\\-0.2&-0.1&0.1\\-0.2&0.3&0.1\end{bmatrix},\quad N_{21}=\begin{bmatrix}-0.1&0.2&0.1\\-0.1&-0.3&0.5\\-0.2&0.1&-0.5\end{bmatrix},$$

$$N_{2\sigma_2}=\begin{bmatrix}0.2&-0.1&0.1\\0.3&0.2&0.2\\-0.2&0.1&-0.1\end{bmatrix},\quad N_{22}=\begin{bmatrix}0.3&0.1\\-0.2&-0.2\\0.1&0.3\end{bmatrix},$$

$$E_{21}=\begin{bmatrix}-0.2&0.2&-0.1\\0.1&0.1&0.2\\-0.1&-0.2&0.1\end{bmatrix},\quad C_2=\begin{bmatrix}-4.5&0.1&0.1\\0.1&0.3&-0.2\end{bmatrix},$$

$$D_2=\begin{bmatrix}0.17&0.31\\-0.12&0.23\end{bmatrix},\quad \sigma_2(t)=0.2(1+\cos(t)),$$

$$\eta_2(t) = 0.1(2+\cos(t)), \qquad \tau_{21}(t) = 0.2(2+\sin(t)). \tag{25}$$

Applying Matlab toolbox to solving inequality (11), we obtain the following results:

$$Q_{11} = \begin{bmatrix} 1.4865 & 0.7955 & 0.4231 \\ 0.7955 & 2.2754 & -1.5140 \\ 0.4231 & -1.5140 & 5.2053 \end{bmatrix},$$

$$Q_{13} = \begin{bmatrix} 1.4324 & 0.2646 & 0.8637 \\ 0.2646 & 1.1728 & -1.3599 \\ 0.8637 & -1.3599 & 5.0693 \end{bmatrix},$$

$$Q_{14} = \begin{bmatrix} 1.4285 & 0.2413 & 0.6513 \\ 0.2413 & 1.0183 & -1.1782 \\ 0.6513 & -1.1782 & 4.2542 \end{bmatrix},$$

$$Q_{15} = \begin{bmatrix} 47.9235 & 6.6463 & -2.8148 \\ 6.6463 & 44.1028 & 1.5746 \\ -2.8148 & 1.5746 & 43.3414 \end{bmatrix},$$

$$Q_{16} = \begin{bmatrix} 40.1902 & 9.0430 & -2.8492 \\ 9.0430 & 39.5447 & 2.1458 \\ -2.8492 & 2.1458 & 41.1207 \end{bmatrix},$$

$$K_1 = \begin{bmatrix} 1.4531 & -2.1903 & -0.7468 \\ -1.4570 & 2.4554 & 0.6993 \end{bmatrix},$$

$$\widehat{Q}_{12} = \begin{bmatrix} 0.5747 & 0.2102 \\ -0.7420 & 3.6344 \end{bmatrix},$$

$$\widehat{X}_{11} = \begin{bmatrix} 0.8267 & -0.7426 \\ -0.7426 & 3.3824 \end{bmatrix}, \qquad \widehat{X}_{12} = 2.0431,$$

$$\overline{W}_{21} = \begin{bmatrix} 6.6256 & 0.8468 & -2.8166 \\ 0.8468 & 3.7842 & -1.9443 \\ -2.8166 & -1.9443 & 7.7792 \end{bmatrix},$$

$$W_{21} = \begin{bmatrix} 8.0860 & -4.9568 & -2.7975 \\ -4.9568 & 3.9197 & 1.9259 \\ -2.7975 & 1.9259 & 1.2428 \end{bmatrix},$$

$$\overline{G}_{21} = \begin{bmatrix} 3.6329 & 1.1769 & -1.7276 \\ 1.1769 & 3.4642 & -1.9414 \\ -1.7276 & -1.9414 & 7.6914 \end{bmatrix},$$

$$G_{21} = \begin{bmatrix} 2.2006 & -2.2199 & -1.0134 \\ -2.2199 & 4.2063 & 1.6203 \\ -1.0134 & 1.6203 & 1.4176 \end{bmatrix},$$

$$R_{11} = \begin{bmatrix} -5.3350 & 3.3676 & -0.9899 \\ -1.5843 & 4.0258 & 0.3958 \end{bmatrix},$$

$$R_{12} = \begin{bmatrix} -1.2497 & -0.6324 \\ 1.4433 & 0.6711 \\ 0.6835 & -0.8005 \end{bmatrix},$$

$$W_{12} = \begin{bmatrix} 32.5716 & -0.1560 & 0.7872 \\ -0.1560 & 33.0680 & 1.2248 \\ 0.7872 & 1.2248 & 32.6264 \end{bmatrix},$$

$$G_{12} = \begin{bmatrix} 33.1699 & 0.4384 & -0.0119 \\ 0.4384 & 33.4629 & 0.6883 \\ -0.0119 & 0.6883 & 33.2668 \end{bmatrix},$$

$$L_1 = \begin{bmatrix} -2.2325 & -0.0449 \\ 2.5588 & 0.0366 \\ 0.8420 & -0.2690 \end{bmatrix},$$

$$Q_{21} = \begin{bmatrix} 0.2476 & -0.0193 & -0.1246 \\ -0.0193 & 0.1848 & -0.0060 \\ -0.1246 & -0.0060 & 0.2346 \end{bmatrix},$$

$$\widehat{Q}_{22} = \begin{bmatrix} 0.1239 & -0.0000 \\ -0.0000 & 0.1239 \end{bmatrix},$$

$$\widehat{X}_{21} = \begin{bmatrix} 0.1239 & -0.0005 \\ -0.0005 & 0.1197 \end{bmatrix},$$

$$Q_{23} = \begin{bmatrix} 1.3638 & 0.0541 & -2.6550 \\ 0.0541 & 2.8900 & -0.9487 \\ -2.6550 & -0.9487 & 11.4203 \end{bmatrix},$$

$$R_{22} = \begin{bmatrix} -0.0336 & 1.5912 \\ 0.0797 & 5.7463 \\ 0.0145 & -2.5460 \end{bmatrix},$$

$$Q_{24} = \begin{bmatrix} 1.8054 & 1.0651 & -3.0558 \\ 1.0651 & 1.7394 & -3.5831 \\ -3.0558 & -3.5831 & 9.2685 \end{bmatrix},$$

$$K_2 = \begin{bmatrix} 0.5773 & -8.9195 & -3.6070 \\ 3.8724 & -6.5449 & -12.5895 \end{bmatrix},$$

$$Q_{25} = \begin{bmatrix} 416.8897 & -302.9290 & -158.6713 \\ -302.9290 & 231.3825 & 118.1184 \\ -158.6713 & 118.1184 & 79.4590 \end{bmatrix},$$

$$\widehat{X}_{22} = 0.1013,$$

$$Q_{26} = \begin{bmatrix} 139.5613 & -92.2506 & -33.9039 \\ -92.2506 & 66.7926 & 18.6639 \\ -33.9039 & 18.6639 & 12.5276 \end{bmatrix},$$

$$R_{21} = \begin{bmatrix} 0.7642 & -1.6381 & -0.8647 \\ 2.6539 & -1.2087 & -3.3972 \end{bmatrix},$$

$$L_2 = \begin{bmatrix} -0.2712 & 12.8424 \\ 0.6434 & 46.3766 \\ 0.1172 & -20.5476 \end{bmatrix}. \tag{26}$$

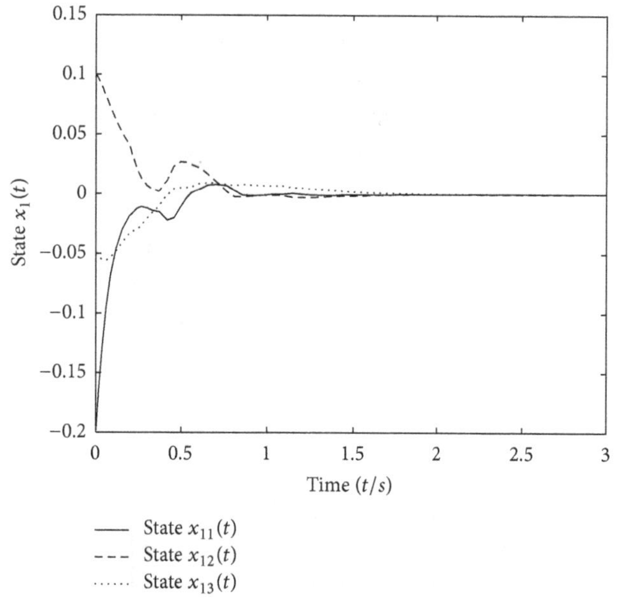

FIGURE 1: State responses of the first closed-loop subsystem.

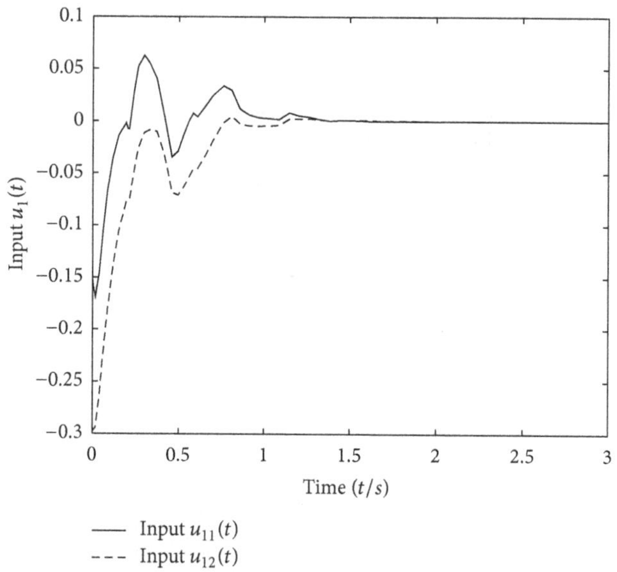

FIGURE 3: The control signals of the first subsystem.

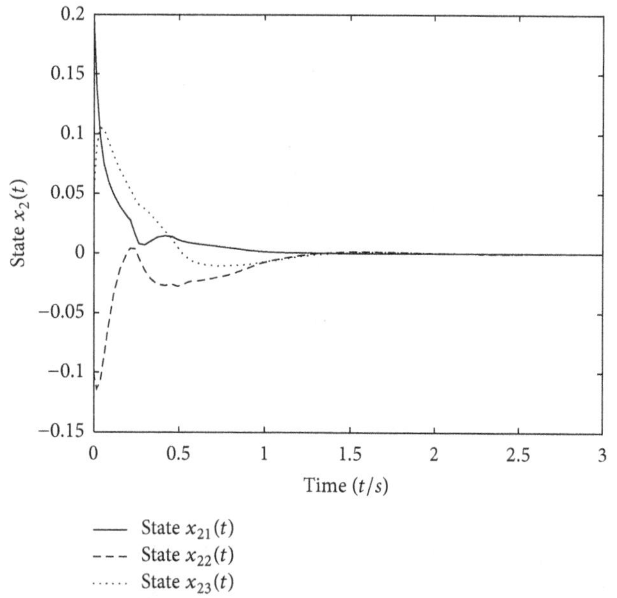

FIGURE 2: State responses of the second closed-loop subsystem.

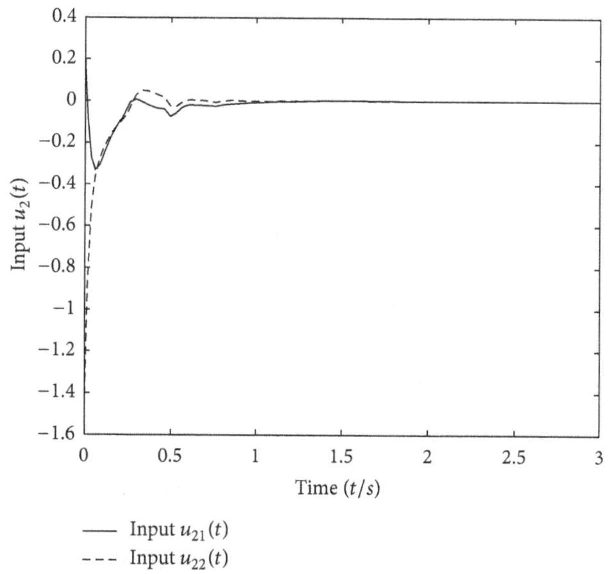

FIGURE 4: The control signals of the second subsystem.

Under the following initial condition

$$\phi_1(t) = \begin{bmatrix} -0.2e^{2t} \\ 0.1 \\ 0.4t - 0.05 \end{bmatrix}, \qquad \phi_2(t) = \begin{bmatrix} 0.2e^{t} \\ 0.2t - 0.1 \\ 0.05 \end{bmatrix}, \quad (27)$$

the simulation results are shown in Figures 1, 2, 3, 4, 5, 6, 7, 8, 9, 10, 11, and 12. With the observer-based control applied, the state responses of the closed-loop system are depicted in Figures 1-2. The control signals based on the observer are rather smooth in Figures 3 and 4. The output signals of the closed-loop system are shown in Figures 5 and 6. The state responses and the output signals of the observer system (6) are presented in Figures 7–10, respectively. The state responds of the error dynamic system (7) are shown in Figures 11 and 12.

The simulations results indicate that the designed observer can stabilize the error dynamic system and estimate the states of the interconnected systems of neutral type.

Remark 7. For this example, $N = 2$. When $i = 1$, we can obtain $Q_{11}, Q_{12}, \overline{W}_{21}, W_{12}, \overline{G}_{21}$, and G_{12} by solving LMI $\Omega_1 < 0$ in (11). Furthermore, according to $\overline{W}_{21} = Q_{11}W_{21}Q_{11}$ and $\overline{G}_{21} = Q_{12}G_{21}Q_{12}$, we have W_{21} and G_{21}. When $i = 2$, inequality $\Omega_2 < 0$ in (11) is not LMI because of the existence of the interconnection matrix $\overline{W}_{12} = Q_{21}W_{12}Q_{21}$ and $\overline{G}_{12} = Q_{22}G_{12}Q_{22}$ (here, W_{12} and G_{12} are constant matrices coming

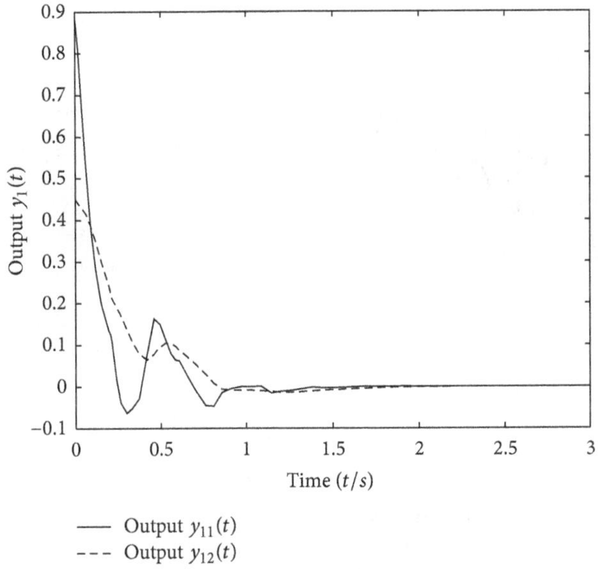

FIGURE 5: Output responses of the first subsystem.

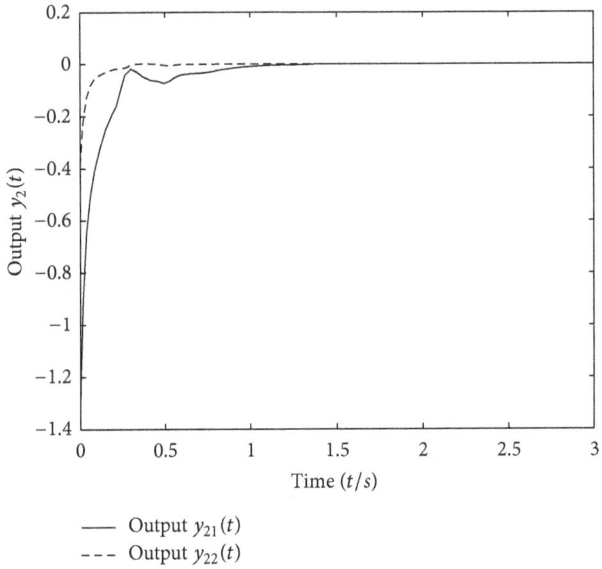

FIGURE 6: Output responses of the second subsystem.

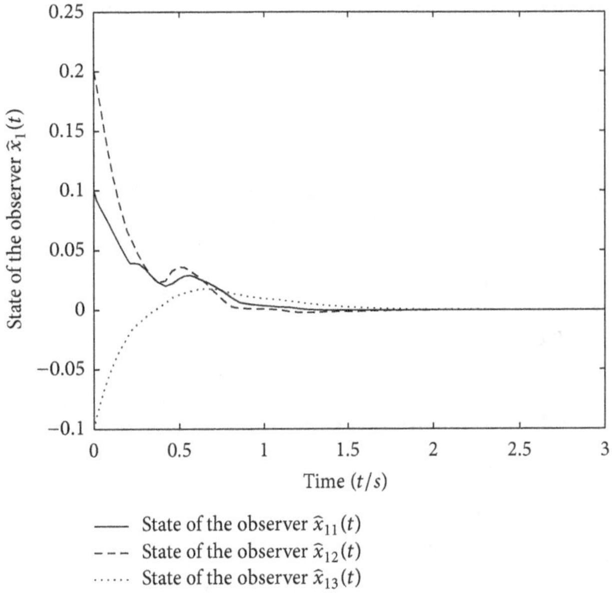

FIGURE 7: State responses of the first observer subsystem.

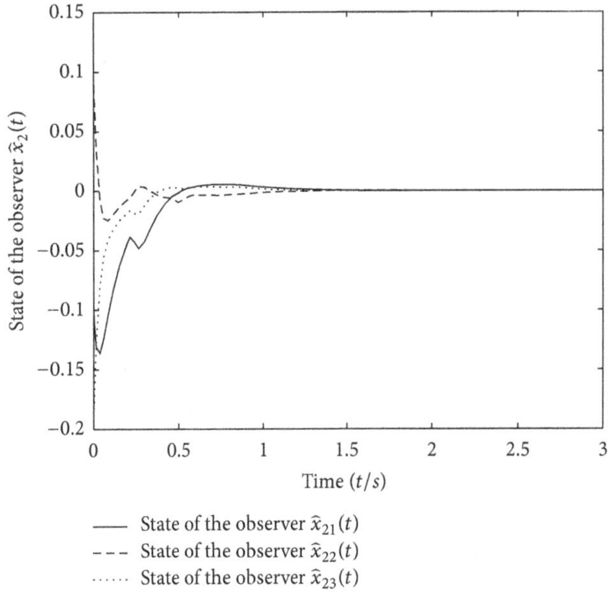

FIGURE 8: State responses of the second observer subsystem.

from $\Omega_1 < 0$). However, the transformed inequality is an LMI after applying Schur complement to $\Omega_2 < 0$. Hence, the controller gain K_2 and observer gain L_2 can be obtained by finding feasible set to the transformed inequality of $\Omega_2 < 0$.

5. Conclusion

The observer-based decentralized control problem of uncertain interconnected systems of neutral type is complex and challenging. In framework of Lyapunov stability theory, a novel mathematical technique to deal with the parametric disturbances is developed to obtain the sufficient conditions of existing anticipated controller and observer. The sufficient conditions are the coupled LMIs and depend on not only the sizes of delays, but also the information of derivatives. The numerical example and the corresponding simulation results elucidate that the results obtained in this paper are effective.

Acknowledgments

This research is supported by the Natural Science Foundation of China (no. 61104106), the Science Foundation of Department of Education of Liaoning Province (no. L2012422), and

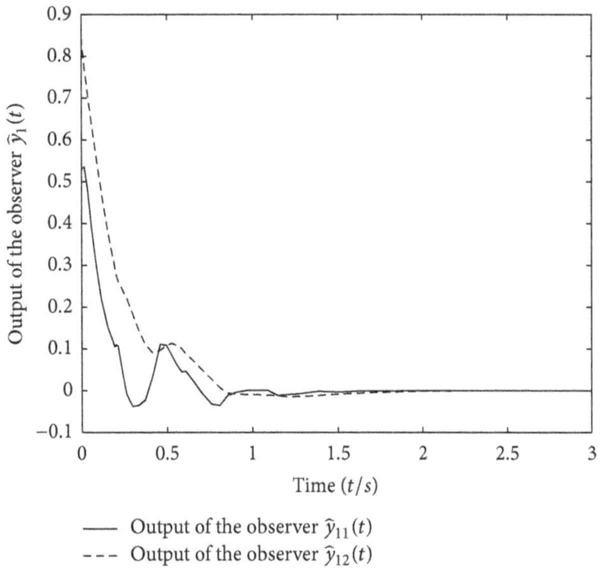

Figure 9: Output responses of the first observer subsystem.

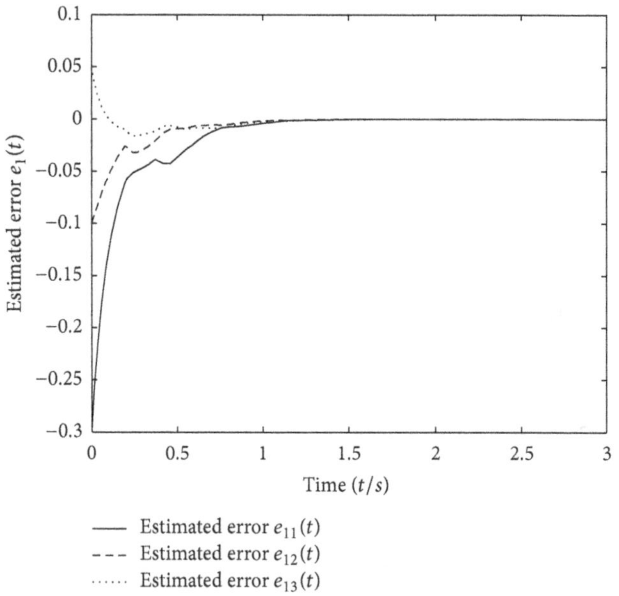

Figure 11: State responses of the first error subsystem.

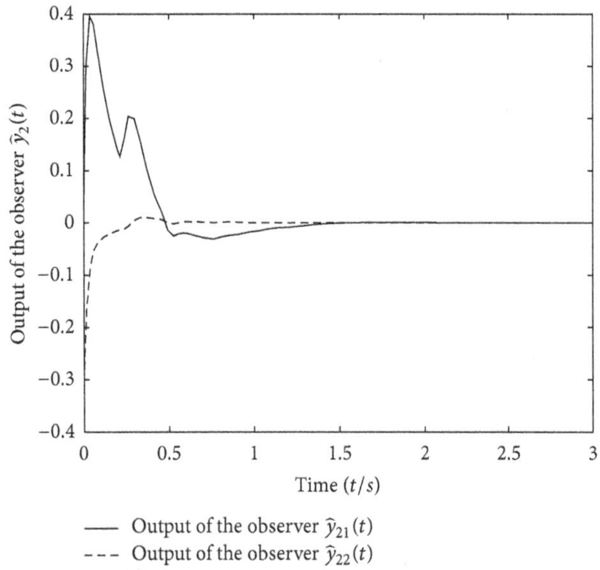

Figure 10: Output responses of the second observer subsystem.

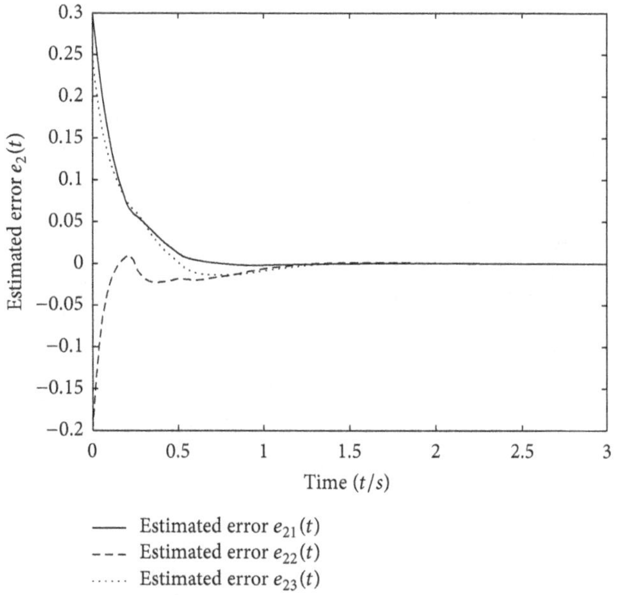

Figure 12: State responses of the second error subsystem.

the Doctorial Science Foundation of Shenyang University (no. 20212340).

References

[1] M. S. Mahmoud and Y. Xia, "A generalized approach to stabilization of linear interconnected time-delay systems," *Asian Journal of Control*, vol. 14, no. 6, pp. 1539–1552, 2012.

[2] X. Ye, "Decentralized adaptive stabilization of large-scale nonlinear time-delay systems with unknown high-frequency-gain signs," *IEEE Transactions on Automatic Control*, vol. 56, no. 6, pp. 1473–1478, 2011.

[3] S. J. Yoo, N. Hovakimyan, and C. Cao, "Decentralised L_1 adaptive control for large-scale non-linear systems with interconnected unmodelled dynamics," *IET Control Theory & Applications*, vol. 4, no. 10, pp. 1972–1988, 2010.

[4] X. Ouyang and X. Chen, "Modeling and decentralized control of complex dynamic interconnected systems," *International Journal of Information & Systems Sciences*, vol. 5, no. 2, pp. 248–259, 2009.

[5] K. Kalsi, J. Lian, and S. H. Źak, "Decentralized dynamic output feedback control of nonlinear interconnected systems," *IEEE Transactions on Automatic Control*, vol. 55, no. 8, pp. 1964–1970, 2010.

[6] M. S. Mahmoud, "Decentralized stabilization of interconnected systems with time-varying delays," *IEEE Transactions on Automatic Control*, vol. 54, no. 11, pp. 2663–2668, 2009.

[7] G. Zhai, N. Chen, and W. Gui, "A study on decentralized H_∞ feedback control systems with local quantizers," *Kybernetika*, vol. 45, no. 1, pp. 137–150, 2009.

[8] S. K. Yee and J. V. Milanović, "Fuzzy logic controller for decentralized stabilization of multimachine power systems," *IEEE Transactions on Fuzzy Systems*, vol. 16, no. 4, pp. 971–981, 2008.

[9] R. Garcia-Hernandez, E. N. Sanchez, E. Bayro-Corrochano, M. A. Llama, and J. A. Ruz-Hernandez, "Real-time decentralized neural backstepping control: application to a two DOF robot manipulator," *International Journal of Innovative Computing, Information and Control*, vol. 7, no. 2, pp. 965–976, 2011.

[10] Y. Xu and S. Hu, "The existence and uniqueness of the solution for neutral stochastic functional differential equations with infinite delay in abstract space," *Acta Applicandae Mathematicae*, vol. 110, no. 2, pp. 627–638, 2010.

[11] H. Mukaidani, "Decentralized stochastic guaranteed cost control for uncertain nonlinear large-scale interconnected systems under gain perturbations," in *Proceedings of the American Control Conference (ACC '09)*, pp. 5097–5102, St. Louis, Mo, USA, June 2009.

[12] V. Kolmanoskii and A. Myshkis, *Applied Theory of Functional Differential Equations*, Kluwer Academic Publishers, Dodrecht, The Netherlands, 1992.

[13] R. K. Brayton, "Bifurcation of periodic solutions in a nonlinear difference-differential equations of neutral type," *Quarterly of Applied Mathematics*, vol. 24, pp. 215–224, 1966.

[14] Y. Kuang, *Delay Differential Equations with Applications in Population Dynamics*, vol. 191 of *Mathematics in Science and Engineering*, Academic Press, Boston, Mass, USA, 1993.

[15] S. Niculescu, *Delay Effects on Stability: A Robust Control Approach*, vol. 269 of *Lecture Notes in Control and Information Sciences*, Springer, Berlin, Germany, 2001, A robust control approach.

[16] P. Balasubramaniam, R. Krishnasamy, and R. Rakkiyappan, "Delay-dependent stability of neutral systems with time-varying delays using delay-decomposition approach," *Applied Mathematical Modelling*, vol. 36, no. 5, pp. 2253–2261, 2012.

[17] R. Rakkiyappan, P. Balasubramaniam, and R. Krishnasamy, "Delay dependent stability analysis of neutral systems with mixed time-varying delays and nonlinear perturbations," *Journal of Computational and Applied Mathematics*, vol. 235, no. 8, pp. 2147–2156, 2011.

[18] X. Nian, H. Pan, W. Gui, and H. Wang, "New stability analysis for linear neutral system via state matrix decomposition," *Applied Mathematics and Computation*, vol. 215, no. 5, pp. 1830–1837, 2009.

[19] W. B. Ma, N. Adachi, and T. Amemiya, "Delay-independent stabilization of uncertain linear systems of neutral type," *Journal of Optimization Theory and Applications*, vol. 84, no. 2, pp. 393–405, 1995.

[20] Q. Han, X. Yu, and K. Gu, "On computing the maximum time-delay bound for stability of linear neutral systems," *IEEE Transactions on Automatic Control*, vol. 49, no. 12, pp. 2281–2285, 2004.

[21] K. Moezzi and A. G. Aghdam, "Adaptive robust control of uncertain neutral time-delay systems," in *Proceedings of the American Control Conference (ACC '08)*, pp. 5162–5167, Seattle, Wash, USA, June 2008.

[22] K. Zhang, B. Jiang, and A. Shumsky, "A new criterion of fault estimation for neutral delay systems using adaptive observer," *Acta Automatica Sinica*, vol. 35, no. 1, pp. 85–91, 2009.

[23] L. Huang and X. Mao, "Delay-dependent exponential stability of neutral stochastic delay systems," *IEEE Transactions on Automatic Control*, vol. 54, no. 1, pp. 147–152, 2009.

[24] R. Yang, P. Shi, G.-P. Liu, and H. Gao, "Network-based feedback control for systems with mixed delays based on quantization and dropout compensation," *Automatica*, vol. 47, no. 12, pp. 2805–2809, 2011.

[25] X. Su, P. Shi, L. Wu, and Y.-D. Song, "A novel control design on discrete-time Takagi- Sugeno fuzzy systems with time-Varying delays," *IEEE Transactions on Fuzzy Systems*, 2012.

[26] X. Su, P. Shi, L. Wu, and Y. -D. Song, "A novel approach to filter design for T-S fuzzy discrete-time systems with time-varying delay," *IEEE Transactions on Fuzzy Systems*, vol. 20, no. 6, pp. 1114–1129, 2012.

[27] D. Zhang, H. Wang, B. Lu, and Z. Wang, "LMI-based fault detection fuzzy observer design with multiple performance constraints for a class of non-linear systems: comparative study," *International Journal of Innovative Computing, Information and Control*, vol. 8, no. 1 B, pp. 633–645, 2012.

[28] R. B. Messaoud, N. Zanzouri, and M. Ksouri, "Local feedback unknown input observer for nonlinear systems," *International Journal of Innovative Computing, Information and Control*, vol. 8, no. 2, pp. 1145–1154, 2012.

[29] D. W. C. Ho and G. Lu, "Robust stabilization for a class of discrete-time non-linear systems via output feedback: the unified LMI approach," *International Journal of Control*, vol. 76, no. 2, pp. 105–115, 2003.

[30] P. Gahinet, A. Nemirovski, A. J. Laub, and M. Chilali, *LMI Control Toolbox User's Guide*, The Math Works, Natick, Mass, USA, 1995.

16

Piecewise-Smooth Support Vector Machine for Classification

Qing Wu and Wenqing Wang

School of Automation, Xi'an University of Posts and Telecommunications, Xi'an 710121, China

Correspondence should be addressed to Qing Wu; xidianwq@yahoo.com.cn

Academic Editor: Jun Zhao

Support vector machine (SVM) has been applied very successfully in a variety of classification systems. We attempt to solve the primal programming problems of SVM by converting them into smooth unconstrained minimization problems. In this paper, a new twice continuously differentiable piecewise-smooth function is proposed to approximate the plus function, and it issues a piecewise-smooth support vector machine (PWSSVM). The novel method can efficiently handle large-scale and high dimensional problems. The theoretical analysis demonstrates its advantages in efficiency and precision over other smooth functions. PWSSVM is solved using the fast Newton-Armijo algorithm. Experimental results are given to show the training speed and classification performance of our approach.

1. Introduction

In the last several years, support vector machine (SVM) has become one of the most promising learning machines because of its high generalization performance and wide applicability for classification, forecasting, and estimation in small-sample cases [1–6]. In addition, SVMs have surpassed the performance of artificial neural networks in many areas such as text categorization, speech recognition, and bioinformatics [7–11].

Basically, the main idea behind SVM is the construction of an optimal hyper plane, which has been used widely in classification [5, 8–16]. It can be formulated into an unconstrained optimization problem [17–21], but the objective function is nonsmooth. To overcome this disadvantage, Lee and Mangasarian used the integral of the sigmoid function to get a smooth SVM(SSVM) model in 2001 [17]. It is a very important and significant result to SVM since many famous algorithms can be used to solve it. In 2005, Yuan et al. proposed two polynomial functions, namely, the smooth quadratic polynomial function and the smooth forth polynomial function, and got QPSSVM and FPSSVM models [20, 21]. Xiong et al. derived an important recursive formula and a new class of smoothing functions using the technique of interpolation functions in [20]. In 2007, Yuan et al. used a three-order spline function to smooth the objective function of unconstrained optimization problem of SVM and obtained TSSVM model [21]. However, the efficiency or the precision of the algorithms was limited.

A natural problem is whether there is another smooth function to get a more efficient smooth SVM than existing works. In this paper, we introduce a piecewise function to smooth SVM and obtain a novel piecewise smooth support vector machine (PWSSVM). Theoretical analyses show that approximation accuracy of the piecewise smooth function to the plus function is higher than the available. Rough set theory is used to prove the global convergence of PWSSVM and the upper bound of convergence is proposed. The fast Newton-Armijo algorithm [22, 23] is employed to train the PWSSVM. Our new method is implemented in batches and can efficiently handle large-scale and high dimensional problems. Numerical experiments confirm the theoretical results and demonstrate that PWSSVM is more effective than the previous smooth support vector machine models.

The paper is organized as follows. In Section 2, we state the pattern classification and describe the PWSSVM. The approximation performance of smooth functions to the plus function is compared in Section 3. The convergence performance of PWSSVM is given in Section 4. The Newton-Armijo algorithm is applied to train PWSSVM in Section 5. Section 6 shows numerical comparisons. Finally, a brief conclusion of this paper is made.

In this paper, unless otherwise stated, all vectors are column vectors. For a vector x in the n-dimensional real space R^n, the plus function x_+ is defined as $(x_+)_i = \max(0, x_i)$, $i = 1, \ldots, n$. The scalar (inner) product of two vectors x, y in the n-dimensional real space will be denoted by $x \cdot y$ and the p-norm will be denoted by $\|\cdot\|_p$. For a matrix $A \in R^{m \times n}$, A_i is the ith row of A which is a row vector in R^n. A column vector of ones of n dimension will be denoted by e. If φ is a real valued function defined in the n-dimensional real space R^n, the gradient of φ is denoted by $\nabla\varphi(x)$ which is a row vector in R^n and $n \times n$ the Hessian matrix of φ at x is denoted by $\nabla^2\varphi(x)$.

2. Piecewise-Smooth Support Vector Machine

In this paper, let us consider a binary classification problem with m training samples in the n-dimensional real space R^n. It is represented by the $m \times n$ matrix A, according to membership of each point A_i in the class 1 or −1 as specified by a given $m \times m$ diagonal matrix D with 1 or −1 along its diagonal. For this problem, the standard SVM with a linear kernel is given by the following quadratic program with parameter $v > 0$

$$\begin{aligned} \min_{(w,\gamma,y)\in R^{n+1+m}} \quad & ve^{\mathrm{T}}y + \frac{1}{2}w^{\mathrm{T}}w \\ \text{s.t.} \quad & D(Aw - e\gamma) + y \geq e \quad y \geq 0, \end{aligned} \tag{1}$$

where e is a vector of ones, w is the normal to the bounding plane, and b is the distance of the bounding plane to the origin. The linear separating plane is defined as follows:

$$P = \left\{x_i \mid x_i \in R^n, w^{\mathrm{T}}x_i = b\right\}. \tag{2}$$

The first term in the objective function of (1) is the 1-norm of the slack variable y with weight v. Replace the first term with the 2-norm vector y. Add $(1/2)\gamma^{\mathrm{T}}\gamma$ to the objective function which induces strong convexity but has little or no effect on the problem. SVM model is replaced by the following problem:

$$\begin{aligned} \min_{(w,\gamma,y)\in R^{n+1+m}} \quad & \frac{v}{2}y^{\mathrm{T}}y + \frac{1}{2}\left(w^{\mathrm{T}}w + \gamma^{\mathrm{T}}\gamma\right) \\ \text{s.t.} \quad & D(Aw - e\gamma) + y \geq e \quad y \geq 0. \end{aligned} \tag{3}$$

Let $y = (e - D(Aw - e\gamma))_+$, where $(\cdot)_+$ replaces negative components of a vector by zeros, then we can convert the SVM problem (3) into the following unconstrained optimization problem:

$$\min_{(w,\gamma)\in R^{n+1}} \frac{1}{2}v\left\|(e - D(Aw - e\gamma))_+\right\|^2 + \frac{1}{2}\left(w^{\mathrm{T}}w + \gamma^{\mathrm{T}}\gamma\right). \tag{4}$$

This is a strongly convex minimization problem and it has a unique solution. Let $\max\{0, x\} = x_+$. The function $(x)_+$ is a continuous but nonsmooth function. Therefore, many optimization algorithms based on derivative and gradient cannot solve the problem (4) directly.

In 2001, Lee and Mangasarian [17] employed the integral of the sigmoid function $p(x, k)$ to approximate the nondifferentiable function x_+ as follows:

$$p(x, k) = x + \frac{1}{k}\log\left(1 + \varepsilon^{-kx}\right), \quad k > 0, \tag{5}$$

where ε is the base of natural logarithm and k is a smoothing parameter. They got the SSVM model.

In 2005, Yuan et al. [18] presented two polynomial functions as follows:

$$q(x,k) = \begin{cases} x, & x \geq \frac{1}{k}, \\ \frac{k}{4}x^2 + \frac{1}{2}x + \frac{1}{4k}, & -\frac{1}{k} < x < \frac{1}{k}, \quad k > 0, \\ 0, & x \leq -\frac{1}{k}, \end{cases}$$

$$h(x,k) = \begin{cases} x, & x \geq \frac{1}{k}, \\ -\frac{1}{16k}(kx+1)^3(kx-3), & -\frac{1}{k} < x < \frac{1}{k}, \quad k > 0, \\ 0, & x \leq -\frac{1}{k}. \end{cases} \tag{6}$$

Using the above smooth functions to proximate plus function x_+, they got two smooth polynomial support vector machine models (FPSSVM and QPSSVM). The authors also showed that FPSSVM and QPSSVM were more effective than SSVM in [18].

In 2007, Yuan et al. [21] presented a three-order spline function as follows:

$$T(x,k) = \begin{cases} 0, & x < -\frac{1}{k}, \\ \frac{k^2}{6}x^3 + \frac{k}{2}x^2 + \frac{1}{2}x + \frac{1}{6k}, & -\frac{1}{k} \leq x < 0, \\ -\frac{k^2}{6}x^3 + \frac{k}{2}x^2 + \frac{1}{2}x + \frac{1}{6k}, & 0 \leq x \leq \frac{1}{k}, \\ x, & x > \frac{1}{k}. \end{cases} \tag{7}$$

They used the smooth function $T(x, k)$ to approach the plus function and got a new smooth SVM model TSSVM. However, the efficiency or the precision of these algorithms above was limited.

In this paper, we propose a novel smooth function $\varphi(x, k)$ with smoothing parameter $k > 0$ to approximate to the function x_+ as follows:

$$\varphi(x,k) = \begin{cases} 0, & x < -\frac{1}{3k}, \\ \frac{3}{2}k^2\left(x + \frac{1}{3k}\right)^3, & -\frac{1}{3k} \leq x < 0, \\ x + \frac{3}{2}k^2\left(\frac{1}{3k} - x\right)^3, & 0 \leq x \leq \frac{1}{3k}, \\ x, & x > \frac{1}{3k}. \end{cases} \tag{8}$$

The first- and second-order derivatives of $\varphi(x,k)$ are

$$\nabla\varphi(x,k)=\begin{cases}0, & x<-\dfrac{1}{3k},\\ \dfrac{9}{2}k^2\left(x+\dfrac{1}{3k}\right)^2, & -\dfrac{1}{3k}\le x<0,\\ 1-\dfrac{9}{2}k^2\left(\dfrac{1}{3k}-x\right)^2, & 0\le x\le\dfrac{1}{3k},\\ 1, & x>\dfrac{1}{3k},\end{cases}$$
$$\nabla^2\varphi(x,k)=\begin{cases}0, & x<-\dfrac{1}{3k},\\ 9k^2\left(x+\dfrac{1}{3k}\right), & -\dfrac{1}{3k}\le x<0,\\ 9k^2\left(\dfrac{1}{3k}-x\right), & 0\le x\le\dfrac{1}{3k},\\ 0, & x>\dfrac{1}{3k}.\end{cases} \quad (9)$$

The solution of the problem (3) can be obtained by solving the following smooth unconstrained optimization problem with the smoothing parameter k approaching infinity as

$$\min_{(w,\gamma)\in R^{n+1}}\Phi_k(w,\gamma)=\frac{1}{2}\upsilon\left\|\varphi\left(e-D\left(Aw-e\gamma\right),k\right)\right\|^2+\frac{1}{2}\left(w^{\mathrm{T}}w+\gamma^{\mathrm{T}}\gamma\right). \quad (10)$$

Thus, we develop a new smooth approximation for problem (3).

3. Approximation Performance Analysis of Smooth Functions

In this section, we will compare the approximation performance of smooth functions to plus function.

Lemma 1 (see [17]). *$p(x,k)$ is defined as the integral of the sigmoid function in [17], and x_+ is the plus function. The following conclusions can be obtained:*

(1) *$p(x,k)$ is arbitrary rank smooth about x;*

(2) *$p(x,k)\ge x_+$;*

(3) *for $\rho>0$, $|x|<\rho$, $p(x,k)^2-x_+^2\le(\log 2/k)^2+(2\rho/k)\log 2$.*

Lemma 2 (see [18]). *$q(x,k)$ and $h(x,k)$ are defined as (6), and x_+ is plus function. We can obtain the following conclusions:*

(1) *$q(x,k)$ is one rank smooth about x, and $h(x,k)$ is twice rank smooth about x;*

(2) *$q(x,k)\ge x_+$ and $h(x,k)\ge x_+$;*

(3) *for any $x\in R$, $q(x,k)^2-x_+^2\le 1/11k^2$ and $h(x,k)^2-x_+^2\le 1/19k^2$.*

Lemma 3 (see [19]). *$T(x,k)$ is defined as (7), and x_+ is plus function. The following results are easily obtained:*

(1) *$T(x,k)$ is twice rank smooth about x;*

(2) *$T(x,k)\ge x_+$;*

(3) *for any $x\in R$, $T(x,k)^2-x_+^2\le 1/24k^2$.*

Theorem 4. *The piecewise approximation function $\varphi(x,k)$ defined in (8) has the following properties:*

(1) *$\varphi(x,k)$ is twice rank smooth about x;*

(2) *for any $x\in R$, $\varphi(x,k)\ge x_+$,*

(3) *for any $x\in R$, then $\varphi(x,k)^2-x_+^2\le 1/216k^2$.*

Proof. (1) According to the formulas (8) and (9), one can easily obtain the results in (1).

(2) In the following, we verify the fact $\varphi(x,k)\ge(x)_+\ge 0$. (i) The equation $\varphi(x,k)=(x)_+$ holds while $x\in(-\infty,-(1/3k))\cup(1/3k,\infty)$. (ii) Since $\varphi(x,k)$ is a monotone increasing function, we have the following result $\varphi(x,k)-(x)_+=\varphi(x,k)\ge\varphi(-(1/3k),k)=0$ while $x\in[-(1/3k),0)$. (iii) For $x\in[0,1/3k]$, we have $\varphi(x,k)-(x)_+=(3/2)k^2(1/3k-x)^3\ge 0$. Hence, we have the conclusion $\varphi(x,k)\ge(x)_+\ge 0$.

(3) For $x\in(-\infty,-(1/3k))\cup(1/3k,\infty)$, $\varphi(x,k)^2-x_+^2=0$, the inequality in conclusion (3) is satisfied naturally.

For $-(1/3k)\le x\le 0$, since $x_+=0$, $\varphi(x,k)^2-x_+^2=\varphi(x,k)^2$. Because $\varphi(x,k)$ is positive value, continuous, and increasing function for $-(1/3k)\le x\le 0$, we have $\varphi(x,k)^2\le\varphi(0,k)^2=1/324k^2\le 1/216k^2$.

For $0<x\le 1/3k$, let

$$s(x)=\varphi(x,k)^2-x_+^2=\left(x+\frac{3}{2}k^2\left(\frac{1}{3k}-x\right)^3\right)^2-x^2=\frac{9}{4}k^4\left(x-\frac{1}{3k}\right)^6-3k^2x\left(x-\frac{1}{3k}\right)^3. \quad (11)$$

In order to obtain the result, making the variable substitution $t=kx$ (obviously $t\in(0,1/3)$), then we have $s(t)=(3/k^2)[(3/4)(t-(1/3))^6-t(t-(1/3))^3]$. For $t\in(0,1/3)$, the maximum point of $s(t)$ is $t=0.0605$ and $s(t)=\varphi^2(x,k)-(x)_+^2\le g(0.0605)\approx 0.0046/k^2\le 1/216k^2$.

In conclusion, we have $\varphi(x,k)^2-x_+^2\le 1/216k^2$. □

Theorem 5. *Let $\rho=1/k$, and $k>0$. Consider the following.*

(1) *If the smooth function is defined as (5), then by Lemma 1, we have*

$$p(x,k)^2-x_+^2\le\left(\frac{\log 2}{k}\right)^2+\frac{2\rho}{k}\log 2=\left(\log^2 2+2\log 2\right)\frac{1}{k^2}\approx 0.6927\frac{1}{k^2}. \quad (12)$$

(2) *If the smooth functions are defined as (6), by Lemma 2,*

$$q(x,k)^2-x_+^2\le\frac{1}{11k^2}\approx 0.0909\frac{1}{k^2},\qquad h(x,k)^2-x_+^2\le\frac{1}{19k^2}\approx 0.0526\frac{1}{k^2}. \quad (13)$$

(3) *If the smooth function is defined as* (7), *by Lemma 3,*

$$T(x,k)^2 - x_+^2 \le \frac{1}{24k^2} \approx 0.0417\frac{1}{k^2}. \quad (14)$$

(4) *If the smooth function is defined as* (8), *by Theorem 4,*

$$\varphi(x,k)^2 - x_+^2 \le \frac{1}{216k^2} \approx 0.0046\frac{1}{k^2}. \quad (15)$$

Theorem 5 shows that the proposed piecewise smooth function $\varphi(x,k)$ achieves the best degree of approximation to the plus function x_+. When k is fixed, it is easy to obtain the different smooth capability of the above smooth functions. The smooth performance comparison is given in Figure 1, where we set the smooth parameter $k = 10$ and $\rho = 1/k$.

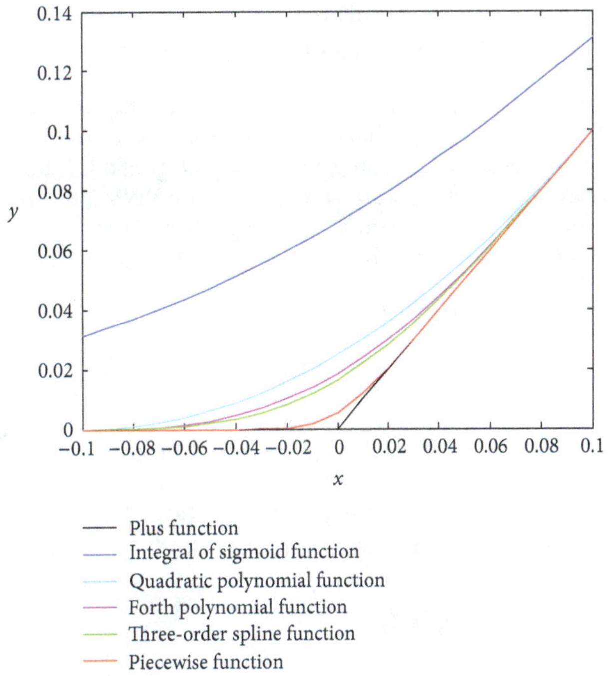

FIGURE 1: Comparison of approximation performance of smooth functions ($k = 10$).

4. Convergence Performance of PWSSVM

In this section, the convergence of PWSSVM will be presented. By using rough set theory, we prove that the solution of PWSSVM can closely approximate the optimal solution of the original model (4) when k goes to infinity. Furthermore, a formula for computing the upper bound of convergence is deduced.

Theorem 6. *Let $A \in R^{m\times n}$ and $b \in R^{m\times 1}$. Define the real-valued functions in the n-dimensional real space R^n as follows:*

$$f(x) = \frac{1}{2}\|(Ax-b)_+\|_2^2 + \frac{1}{2}\|x\|_2^2,$$
$$g(x,k) = \frac{1}{2}\|\varphi(Ax-b,k)\|_2^2 + \frac{1}{2}\|x\|_2^2, \quad (16)$$

where $\varphi(\cdot)$ is defined in (8), *$k > 0$. Then we have the following results:*

(1) *$f(x)$ and $g(x,k)$ are strongly convex functions;*

(2) *there exists a unique solution x^* to $\min_{x\in R^n} f(x)$, and a unique solution x_k^* to $\min_{x\in R^n} g(x,k)$;*

(3) *for any $k > 0$, x_k^* and x^* satisfy the following condition:*

$$\|x_k^* - x^*\|^2 \le \frac{m}{432k^2}; \quad (17)$$

(4) $\lim_{k\to\infty}\|x_k^* - x^*\| = 0$.

Proof. (1) For any $k > 0$, $f(x)$ and $g(x,k)$ are strongly convex functions because $\|\cdot\|^2$ is strong convex function.

(2) Let $L_v(f(x))$ be the level set of $f(x)$ and let $L_v(g(x,k))$ be the level set of $g(x,k)$. Since $x_+ \le \varphi(x,k)$, it is easy to obtain $L_v(g(x,k)) \subset L_v(f(x)) \subset \{x \mid \|x\|_2^2 \le 2v\}$. Therefore, $L_v(f(x))$ and $L_v(g(x,k))$ are compact subsets in R^n. Using the strong convexity property of $f(x)$ and $g(x,k)$ for $k > 0$, there is a unique solution to $\min_{x\in R^n} f(x)$ and $\min_{x\in R^n} g(x,k)$, respectively.

(3) By using the first order optimization condition and considering convex property of $f(x)$ and $g(x,k)$, we have

$$f(x_k^*) - f(x^*) \ge \nabla f(x^*)(x_k^* - x^*) + \frac{1}{2}\|x_k^* - x^*\|_2^2 = \frac{1}{2}\|x_k^* - x^*\|_2^2,$$
$$g(x^*,k) - g(x_k^*,k) \ge \nabla g(x_k^*,k)(x^* - x_k^*) + \frac{1}{2}\|x_k^* - x^*\|_2^2 = \frac{1}{2}\|x_k^* - x^*\|_2^2. \quad (18)$$

Add the two formulas above and notice that $\varphi(x,k) \ge x_+$, and then we have

$$\begin{aligned}\|x_k^* - x^*\|_2^2 &\le f(x_k^*) - f(x^*) + g(x^*,k) - g(x_k^*,k)\\ &= (g(x^*,k) - f(x^*)) - (g(x_k^*,k) - f(x_k^*))\\ &\le g(x^*,k) - f(x^*)\\ &= \frac{1}{2}\|\varphi(Ax^*-b,k)\|_2^2 - \frac{1}{2}\|(Ax^*-b)_+\|_2^2.\end{aligned} \quad (19)$$

According to the third result of Theorem 4, we obtain the conclusion $\|x_k^* - x^*\|^2 \le m/432k^2$.

(4) According to $\|x_k^* - x^*\|^2 \le m/432k^2$, we have $\lim_{k\to\infty} x_k^* = x^*$. □

5. The Newton-Armijo Algorithm for PWSSVM

Following the results of the previous section, one can obtain the twice continuous differentiability of the objective function of problem (10). In order to take advantage of this feature, we use the Newton-Armijo method to train PWSSVM since it is a faster method than the BFGS algorithm [18, 19, 21]. The Newton-Armijo algorithm for problem (10) works as follows.

5.1. Newton-Armijo Algorithm

Step 1 (initialization). Start with any $(w^0, \gamma^0) \in R^{n+1}$, τ and set $i := 0$.

Step 2. Compute $\Phi^i = \Phi_k(w^i, \gamma^i)$ and $g^i = \nabla\Phi_k(w^i, \gamma^i)$.

Step 3. If $\|g^i\|_2 \le \tau$, then stop and accept (w^i, γ^i). Otherwise, compute Newton direction $d^i \in R^{n+1}$ from the following linear system:

$$\nabla^2\Phi\left(w^i, \gamma^i; k^i\right) d^i = -\left(g^i\right)^{\mathrm{T}}, \quad (20)$$

where "T" denotes transpose symbol.

Step 4 (Armijo stepsize). Choose a stepsize $\lambda_i = \max\{1, 1/2, 1/4, \ldots\}$ such that

$$\Phi_k\left(w^i, \gamma^i\right) - \Phi_k\left(\left(w^i, \gamma^i\right) + \lambda_i d^i\right) \ge -\rho\lambda_i g^i d^i, \quad (21)$$

where $\rho \in (0, 1/2)$ and set

$$\left(w^{i+1}, \gamma^{i+1}\right) = \left(w^i, \gamma^i\right) + \lambda_i d^i. \quad (22)$$

Step 5. Replace i by $i + 1$ and go to Step 2.

We need to only solve a linear system of (20) instead of a quadratic program in our smooth approach. Because the objective function is strong convex, it is not difficult to obtain that our Newton-Armijo algorithm for training PWSSVM converges globally to the unique solution [17, 23]. Hence, the start point is not important. In this paper, we always set $(w^0, \gamma^0) = e$, where e denotes a column vector of ones of n dimension.

PWSSVM described above can solve the linear classification problems. In fact, we can extend some of the results in Section 2 to nonlinear PWSSVM with kernel technique as [17]. Furthermore, The Newton-Armijo algorithm can also solve nonlinear PWSSVM successfully.

6. Numerical Experiments

Newton-Armijo cannot be applied to QPSSVM model due to lack of the second-order derivative. In fact, the classification capacity of FPSSVM is slightly better than QPSSVM [18–21]. In our experiment, we do not compare QPSSVM with the other smooth SVM method. To demonstrate the effectiveness and speed of PWSSVM, we compare the performance numerically among SSVM, FPSSVM, TSSVM, and PWSSVM. The four smooth SVMs are all trained by the fast Newton-Armijo algorithm. All experiments are run on Personal Computer with 3.0 GHz and a maximum of 1.99 GB of the memory available. The programs of PWSVM, FPSSVM, and TSSVM are written in the MATLAB language. This computer runs Win7 with MATLAB 7.0.1. The source code of SSVM, "ssvm.m," is obtained from the author's website for the linear problem [24], and "lsvmk.m" for the nonlinear problem. In our experiments, all of the input data and the variables needed in programs are kept in the memory. For SSVM, TSSVM, FPSSVM, and PWSSVM, an optimality tolerance of 10^{-8} is used to determine when to terminate. Gaussian kernel is used in all our experiments.

Table 1: PWSSVM compared with SSVM, FPSSVM, and TSSVM on NDC generated dataset with difference sizes ($C = 10$).

Trains/dimension	Algorithm	Train correctness (%)	Test correctness (%)	Time (s)
2,000,000/10	SSVM	90.86	91.25	**278.97**
	FPSSVM	90.86	91.25	367.46
	TSSVM	90.98	91.33	342.45
	PWSSVM	**91.34**	**91.75**	339.64
2,000,000/20	SSVM	87.64	87.08	**417.64**
	FPSSVM	87.88	88.05	449.28
	TSSVM	87.89	88.05	446.45
	PWSSVM	**88.14**	**88.36**	449.20
10,000/100	SSVM	94.26	93.60	11.17
	FPSSVM	94.78	93.60	11.33
	TSSVM	94.77	93.77	6.24
	PWSSVM	**94.85**	**95.52**	**4.26**
10,000/1000	SSVM	96.67	86.20	56.22
	FPSSVM	96.69	86.14	66.52
	TSSVM	96.69	86.16	26.69
	PWSSVM	**97.94**	**86.23**	**18.73**

The first experiment is used to demonstrate the capability of PWSSVM in solving larger problems. The results in Table 1 are designed to compare the training correctness, the testing correctness, and the training time among the four smooth SVMs on a massively sized dataset. The datasets are created using Musicants NDC Data Generator [25] with different sizes. The test samples are 5% of the training samples. The experiment results show that PWSSVM has the highest training accuracy and testing accuracy. Furthermore, PWSSVM can be used to solve problems more quickly than the other three smooth SVMs when the number of the sample data is relative small.

The second experiment is designed to demonstrate the effectiveness of PWSSVM through the "tried and true" checkerboard dataset [26]. One highly nonlinearly separable but simple example is the checkerboard dataset which has often been used to show the effectiveness of nonlinear kernel methods [17]. The checkerboard dataset is generated by uniformly discretizing the regions $[0, 1] \times [0, 1]$ to $200^2 = 40000$ points and labeling two classes "White" and "Black" spaced by 3×3 grid as Figure 2 shows.

Table 2: PWSSVM compared with SSVM, FPSSVM, and TSSVM on checkerboard dataset with difference sizes ($C = 100$).

Training size	Algorithm	Train correctness (%)	Test correctness (%)	Time (s)
1000	SSVM	99.60	98.28	4.61
	FPSSVM	99.60	98.28	4.35
	TSSVM	99.62	98.28	4.23
	PWSSVM	**99.80**	**98.76**	**4.01**
2000	SSVM	98.84	98.54	11.97
	FPSSVM	99.22	98.59	9.63
	TSSVM	99.22	98.59	**9.35**
	PWSSVM	99.22	**98.68**	9.55
3000	SSVM	98.47	98.88	25.21
	FPSSVM	98.64	98.88	21.58
	TSSVM	98.68	98.90	**17.21**
	PWSSVM	**98.85**	**98.94**	17.36
5000	SSVM	99.34	99.51	47.09
	FPSSVM	99.48	99.51	39.29
	TSSVM	99.52	99.51	38.76
	PWSSVM	**99.79**	**99.65**	**38.23**

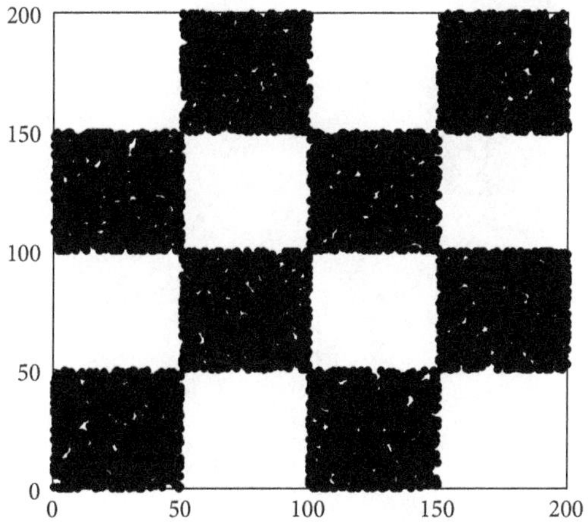

Figure 2: The figure of the checkerboard dataset.

In the first trial of this experiment, the training set contains 1000 points randomly sampled from the checkerboard (for comparison, they are obtained from [17]) which contain 514 "white" samples and 486 "black" samples and the rest 39,000 points are in the testing set. Gaussian kernel function $K(x, y) = \exp(-0.5\|x - y\|^2)$ is used and $C = 100$. Total time for the 1000-point training set using PWSSVM with a Gaussian kernel is 4.01 s. The train accuracy of PWSSVM is 99.80%. The test accuracy of PWSSVM is 98.76% on a 39,000-point test set. TSSVM solves the same problem within 4.23 s, and the train accuracy and the test accuracy are 99.62% and 98.28%. FPSSVM and SSVM obtain the train accuracy of 99.60% within 4.35 s and 4.61 s, respectively. The test accuracy of them are 98.28%.

The rest results are presented in Table 2. The training set is randomly selected from the checkerboard with different sizes. The remaining samples are used as test samples. We compare the classification results of PWSSVM, TSSVM, FPSSVM, and SSVM with the same Gaussian kernel function. The results in Table 2 demonstrate that PWSSVM can solve massive problems quickly, followed by TSSVM, FPSVM, and SSVM in turn. The experimental results show that PWSSVM can obtain the highest train precision and test precision.

7. Conclusions

A novel PWSSVM is proposed in this paper. It only needs to find the unique minima of the unconstrained differentiable convex quadratic function. The proposed method has many advantages over those available, such as good classification performance and less training time cost. The numerical results show that PWSSVM has excellent generalization ability.

Acknowledgments

The authors would like to thank the anonymous reviewers for their valuable comments. This work was supported in part by the National Natural Science Foundation of China under Grants 61100165, 61100231, and 51205309 and the Natural Science Foundation of Shaanxi Province (2010JQ8004, 2012JQ8044).

References

[1] V. N. Vapnik, *The Nature of Statistical Learning Theory*, Springer, New York, NY, USA, 1995.

[2] V. N. Vapnik, *Statistical Learning Theory*, John Wiley & Sons, New York, NY, USA, 1998.

[3] K. R. Mller, A. J. Smola, G. Rtsch, B. Schlkopf, J. Kohlmorgen, and V. Vapnik, "Using support vector machines for time series prediction," in *Advances in Kernel Methods: Support Vector Machine*, B. Scholkopf, J. Burges, and A. Smola, Eds., MIT Press, Cambridge, Mass, USA, 1999.

[4] T. Farooq, A. Guergachi, and S. Krishnan, "Knowledge-based Green's Kernel for support vector regression," *Mathematical Problems in Engineering*, vol. 2010, Article ID 378652, 16 pages, 2010.

[5] J. Zheng and B. L. Lu Bao-Liang, "A support vector machine classifier with automatic confidence and its application to gender classification," *Neurocomputing*, vol. 74, no. 11, pp. 1926–1935, 2011.

[6] J. Y. Zhu, B. Ren, H. X. Zhang, and Z. T. Deng, "Time series prediction via new support vector machines," in *Proceedings of the International Conference on Machine Learning and Cybernetics (ICMLC '02)*, pp. 364–366, November 2002.

[7] J. Ramana and D. Gupta, "LipocalinPred: a SVM-based method for prediction of lipocalins," *BMC Bioinformatics*, vol. 10, p. 445, 2009.

[8] T. Joachims, "Text categorization with support vector machines: learning with many relevant features," in *Proceedings of the 10th European Conference on Machine Learning (ECML '98)*, pp. 137–142, Springer, Heidelberg, Germany, 1998.

[9] P. S. Jaume, M. I. Darío, and D. M. Fernando, "Support vector machines for continuous speech recognition," in *Proceedings of*

the 14th European Signal Processing Conference (EUSIPCO '06), Florence, Italy, September 2006.

[10] V. Bevilacqua, P. Pannarale, M. Abbrescia, and C. Cava, "Comparison of data-merging methods with SVM attribute selection and classification in breast cancer gene expression," *BMC Bioinformatics*, vol. 13, supplement 7, p. S9, 2012.

[11] E. J. Spinosa and A. C. Carvalho, "Support vector machines for novel class detection in bioinformatics," *Genetics and Molecular Research*, vol. 4, no. 3, pp. 608–615, 2005.

[12] A. Nurettin and G. Cüneyt, *An Application of Support Vector Machine in Bioinformatics: Automated Recognition of Epileptiform Patterns in EEG Using SVM Classifier Designed by a Perturbation Method*, vol. 3261 of *Advances in Information Systems Lecture Notes in Computer Science*, 2005.

[13] Y. F. Sun, X. D. Fan, and Y. D. Li, "Identifying splicing sites in eukaryotic RNA: support vector machine approach," *Computers in Biology and Medicine*, vol. 33, no. 1, pp. 17–29, 2003.

[14] H. J. Lin and J. P. Yeh, "Optimal reduction of solutions for support vector machines," *Applied Mathematics and Computation*, vol. 214, no. 2, pp. 329–335, 2009.

[15] A. Christmann and R. Hable, "Consistency of support vector machines using additive kernels for additive models," *Computational Statistics and Data Analysis*, vol. 56, no. 4, pp. 854–873, 2012.

[16] Y. H. Shao and N. Y. Deng, "A coordinate descent margin based-twin support vector machine for classification," *Neural Networks*, vol. 25, pp. 114–121, 2012.

[17] Y.-J. Lee and O. L. Mangasarian, "SSVM: a smooth support vector machine for classification," *Computational Optimization and Applications*, vol. 20, no. 1, pp. 5–22, 2001.

[18] Y. B. Yuan, J. Yan, and C. X. Xu, "Polynomial smooth support vector machine (PSSVM)," *Chinese Journal of Computers*, vol. 28, no. 1, pp. 9–17, 2005.

[19] Y. B. Yuan and T. Z. Huang, *A Polynomial Smooth Support Vector Machine for Classification*, vol. 3584 of *Lecture Note on Artificial Intelligence*, 2005.

[20] J. Z. Xiong, J. L. Hu, H. Q. Yuan, T. M. Hu, and G. M. Li, "Research on a new class of functions for smoothing support vector machines," *Acta Electronica Sinica*, vol. 35, no. 2, pp. 366–370, 2007.

[21] Y. Yuan, W. Fan, and D. Pu, "Spline function smooth support vector machine for classification," *Journal of Industrial and Management Optimization*, vol. 3, no. 3, pp. 529–542, 2007.

[22] D. P. Bertsekas, *Nonlinear Programming*, Athena Scientific, Belmont, Ma, USA, 2nd edition, 1999.

[23] C. Xu and J. Zhang, "A survey of Quasi-Newton equations and Quasi-Newton methods for optimization," *Annals of Operations Research*, vol. 103, no. 1–4, pp. 213–234, 2001.

[24] D. R. Musicant and O. L. Managsarian, "LSVM: Lagrangian support vector machine," 2000, http://www.cs.wisc.edu/dmi/svm/ .

[25] D. R. Musicant, "NDC: normally distributed clustered datasets," 1998, http://www.cs.wisc.edu/~musicant/data/ndc/ .

[26] T. K. Ho and E. M. Kleinberg, "Checkerboard dataset," 1996, http://www.cs.wisc.edu/~musicant/data/ndc/ .

Image Fusion Based on Nonsubsampled Contourlet Transform and Saliency-Motivated Pulse Coupled Neural Networks

Liang Xu, Junping Du, and Qingping Li

Beijing Key Laboratory of Intelligent Telecommunication Software and Multimedia, School of Computer Science, Beijing University of Posts and Telecommunications, Beijing 100876, China

Correspondence should be addressed to Junping Du; junpingdu@126.com

Academic Editor: H. K. Leung

In the nonsubsampled contourlet transform (NSCT) domain, a novel image fusion algorithm based on the visual attention model and pulse coupled neural networks (PCNNs) is proposed. For the fusion of high-pass subbands in NSCT domain, a saliency-motivated PCNN model is proposed. The main idea is that high-pass subband coefficients are combined with their visual saliency maps as input to motivate PCNN. Coefficients with large firing times are employed as the fused high-pass subband coefficients. Low-pass subband coefficients are merged to develop a weighted fusion rule based on firing times of PCNN. The fused image contains abundant detailed contents from source images and preserves effectively the saliency structure while enhancing the image contrast. The algorithm can preserve the completeness and the sharpness of object regions. The fused image is more natural and can satisfy the requirement of human visual system (HVS). Experiments demonstrate that the proposed algorithm yields better performance.

1. Introduction

Due to a tremendous growth in the application of image sensors, image fusion technique has huge potential for growth and has been used successfully in many fields, such as remote sensing, medical imaging, defense surveillance, and computer vision [1–3]. The aim of image fusion is to combine several source images (obtained from different sensors and view points) into a fused image, which contains all important contents from source images and expresses more abundant information in a scene. In practical applications, the direct obtained image is not able to satisfy the requirements because of many factors, for example, the limitations of sensors, varying illumination, occlusions and angles, and so forth. Image fusion technique can solve effectively the problems by taking advantage of multiple-source information producing the fused result which satisfies perception system.

According to the level, image fusion approaches can be generally classified into three types: pixel level, feature level, or decision level [4]. According to whether the fusion methods need the assistant of multiscale transform (MST) tools or not, they can also be categorized into two main classes [5]: MST-based and non-MST-based approaches. A variety of MST tools have been developed for image fusion. The earliest and the most popular MST tools are pyramid [6, 7] and wavelet [8, 9] transform. They are directly constructed by combination of two 1D transforms, so they are not the true 2D transforms. To improve the accuracy of decomposition and reconstruction, the more advanced MST tools have been proposed, such as ridgelets [10], contourlets [11, 12], curvelets [13, 14], and NSCT [15]. The approaches are the true 2D geometric MST tools, which can achieve the decomposition and reconstruction of image signals perfectly and satisfy the requirement of image fusion. In addition, for non-MST-based methods, Piella [16] performs the image fusion by a variational model, and the fused result contains the geometry structure of all the inputs and enhances the contrast for visualization. Ludusan and Lavialle [17] propose a variational approach based on error estimation theory and partial differential equations for concurrent image fusion and denoising of multifocus images.

The combination strategy of the decomposed coefficients is another key step in the MST-based fusion approaches. Fusion strategies can mainly be divided into three categories:

pixel-based, window based, and area based [18]. The simplest pixel-based fusion rule directly selects the fused coefficients using single pixel, but the method is easily influenced by noise. Window based and area based fusion rules take advantage of the local characteristics of neighborhood pixels and, thus, are superior to pixel-based rules [19].

The existing image fusion approaches do not take fully into account the characteristics of HVS, which the HVS tends to focus on the most relevant saliency regions in a scene. According to the visual perception mechanism, the fused image should improve the quality of object areas in a scene. The goal of the proposed algorithm is to preserve the completeness, saliency, and sharpness of object areas and satisfy the requirements of HVS. Consequently, based on NSCT and saliency-motivated PCNN, the paper proposes a novel image fusion algorithm. The visual saliency model and PCNN are two very important tools in image processing. The former is inspired by the behavior and the neuronal architecture of the early primate visual system; the latter is a visual cortex-inspired neural network and characterized by the global coupling and pulse synchronization of neurons. The saliency map produced by the visual saliency model as input to motivate PCNN is used as the fusion rule which can preserve the saliency objects from source images leading to more abundant content contained in a fused image.

The rest of the paper is organized as follows. Section 2 reviews basic NSCT theory in brief. Section 3 presents the proposed image fusion algorithm in detail. Section 4 demonstrates and discusses the experimental results. Section 5 concludes.

2. Nonsubsampled Contourlet Transform

In this section, we briefly review the theory and properties of NSCT, which will be used in the rest of this paper (see [15] for details).

NSCT is a kind of overcomplete transform and is a shift-invariant version of contourlet transform. NSCT has some excellent properties in the process of image decomposition, including shift invariance, multiscale, and multidirection. NSCT is used as the MST tool to provide a better representation of the contours and overcome pseudo-Gibbs phenomena. The main components of the NSCT are a nonsubsampled pyramid filter bank (NSPFB) structure for multiscale decomposition and a nonsubsampled directional filter bank (NSDFB) structure for directional decomposition. The NSCT is displayed in Figure 1.

The multiscale property of the NSCT is achieved by using two-channel nonsubsampled 2-D filter banks (NSFBs), called as NSPFB. The filters for next level are obtained by upsampling the filters of the previous level, by which the multiscale property is obtained without the need for additional filter design. We assume that the NSPFB decomposition is with $J = N$ levels. At the first level, input images are decomposed by the low-pass filter $H_0(z)$ and the corresponding high-pass filter $H_1(z)$, respectively. The ideal passband support of the low-pass filter at the jth level is the region $[-(\pi/2^j), (\pi/2^j)]^2$. The ideal support of the equivalent high-pass filter is the complement of the low-pass filter, that is, the region $[-(\pi/2^{j-1}), (\pi/2^{j-1})]^2 \setminus [-(\pi/2^j), (\pi/2^j)]^2$. The NSDFB, a shift-invariant directional filter bank (DFB), is obtained by eliminating the downsamplers and upsamplers in the DFB. To achieve multidirection decomposition, the NSDFB is iteratively used. All filter banks in the NSDFB tree structure are obtained from a single NSFB with fan filters. Each filter bank in the NSDFB tree has the same computational complexity as that of the building-block NSFB.

Figure 1 shows the NSCT which is constructed by combining the NSPFB and the NSDFB. The two-channel NSFBs in the NSPFB and the NSDFB satisfy the Bezout identity and are invertible, so the NSCT is invertible. The key of NSCT is the filter design problem of the NSPFB and NSDFB. The aim is to design the filters supporting the Bezout identity and obtaining other useful properties. In addition, for a fast implementation, the mapping approach is used to transform the filter into a ladder or lifting structure. More details can be seen in [15].

3. The Proposed Algorithm

In the section, the proposed image fusion algorithm based on NSCT and saliency-motivated PCNN is presented in detail. The main idea is that the visual saliency map is first built on high-pass subband coefficients of the NSCT using the visual attention model (phase spectrum of Fourier transform (PFT) model presented in Section 3.1) and then is combined with source high-pass subband coefficients as input to motivate PCNN. Coefficients with large firing times are employed as the fused high-pass subband coefficients. Low-pass subband coefficients are merged to develop a weighted fusion rule based on firing times of PCNN. PCNN is built in each subband to simulate the biological activity of HVS. The fused image has more natural visual appearance and can satisfy the requirements of HVS. The framework of the proposed algorithm is shown in Figure 2. For the clearness of the presentation, we assume that two registered source images are combined.

The algorithm first decomposes source images into the low-pass subband and high-pass directional subband coefficients by the NSCT. The coarsest subband contains the main energy from source images and denotes the abundant structural information. Therefore, an adaptive weighted average fusion rule based on the firing times of PCNN is developed to merge the low-pass subband. High-pass directional subbands contain the abundant detail contents of source images, so we create a maximum selection fusion principle based on saliency-motivated PCNN for selecting the fused coefficients. The final fused image is reconstructed by applying the inverse NSCT on the merged coefficients.

3.1. Images Decomposition and Saliency-Motivated PCNN. The decomposition of source images employs NSCT presented in Section 2. Input images A and B are decomposed into different scale and direction subbands using NSCT. The subbands $\{C^A_{j_0}(x, y), C^A_{j,l}(x, y)\}$ and $\{C^B_{j_0}(x, y), C^B_{j,l}(x, y)\}$ are

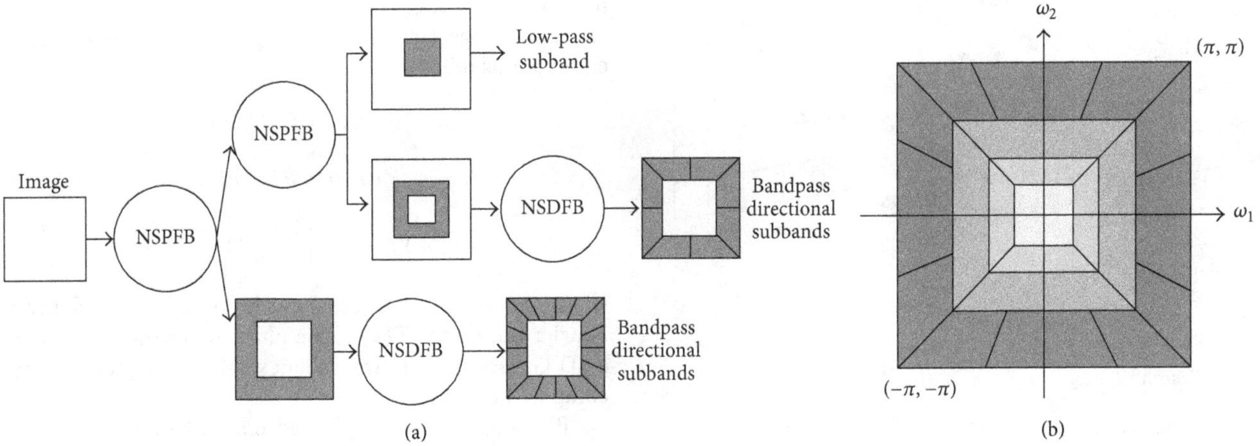

FIGURE 1: Nonsubsampled contourlet transform. (a) NSFB structure that implements the NSCT and (b) idealized frequency partitioning obtained with the proposed structure.

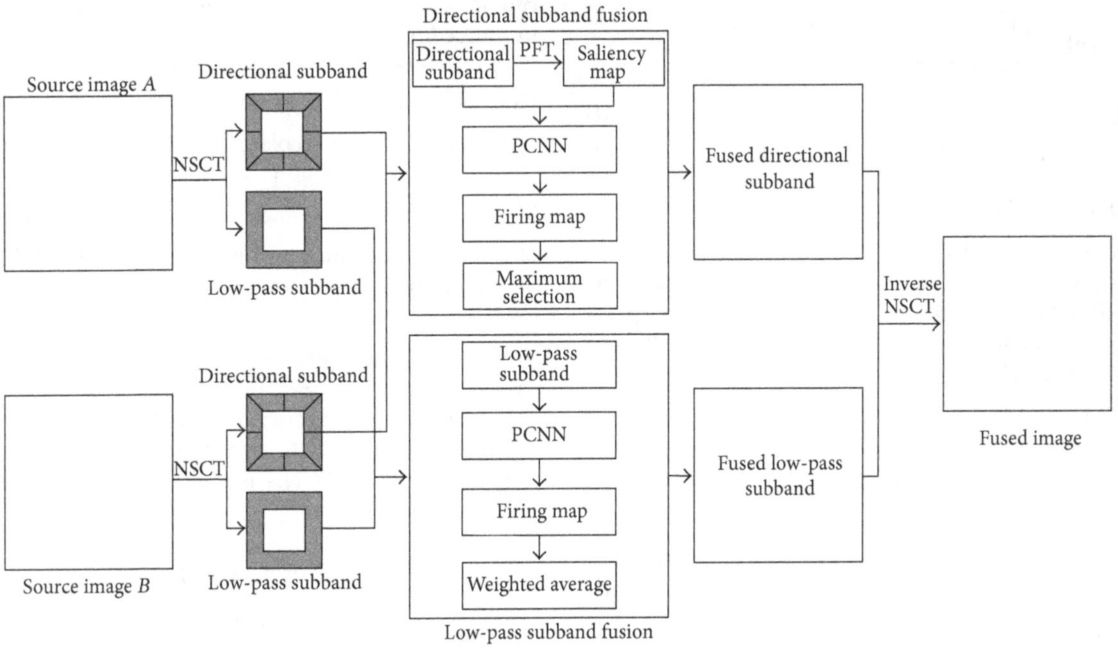

FIGURE 2: Architecture of the proposed algorithm.

obtained, where $C_{j0}(x,y)$ denotes the low-pass subband coefficients of the input images at the coarsest scale and $C_{j,l}(x, y)$ denotes the high-pass directional subband coefficients at the jth scale and in the lth direction.

The following the proposed saliency-motivated PCNN model is discussed. Eckhorn develops a novel biological neural network, called PCNN which is based on the experimental observations of synchronous pulse bursts in cat and monkey visual cortices [20]. PCNN is a feedback network and each PCNN neuron consists of three parts: receptive field, modulation field, and pulse generator [21]. In image processing, PCNN is a single-layer and a two-dimensional connection neural network [22, 23] shown in Figure 3.

In this paper, let $C_{j,l}(x, y)$ denotes the coefficient located at (x, y) in the jth scale at the lth direction. $C_{j,l}(x, y)$ in each subband is inputted to PCNN to motivate the neurons and generate pulse of neurons with (1). Firing times $T_{xy}^{j,l}$ are then computed as in (2):

$$F_{xy}^{j,l}[n] = C_{j,l}(x, y)$$

$$L_{xy}^{j,l}[n] = \exp(-\alpha_L) L_{xy}^{j,l}[n-1] + V_L \sum_{pq} W_{xy,pq}^{j,l} Y_{xy,pq}^{j,l}[n-1]$$

$$U_{xy}^{j,l}[n] = F_{xy}^{j,l}[n] * \left(1 + \beta L_{xy}^{j,l}[n]\right)$$

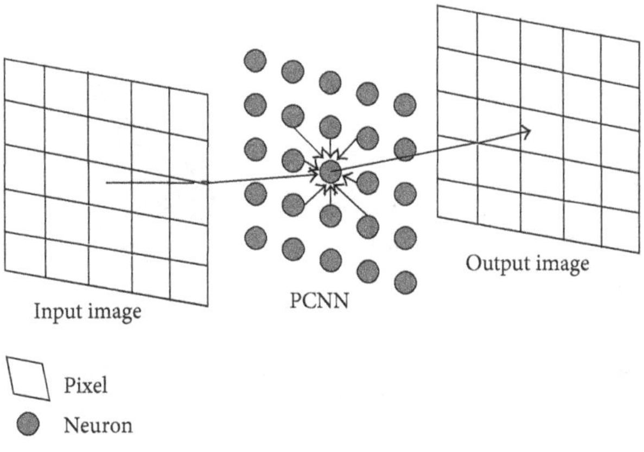

FIGURE 3: Connection model of PCNN neuron.

$$\theta_{xy}^{j,l}[n] = \exp(-\alpha_\theta)\theta_{xy}^{j,l}[n-1] + V_\theta Y_{xy}^{j,l}[n-1]$$

$$Y_{xy}^{j,l}[n] = \begin{cases} 1, & \text{if } U_{xy}^{j,l}[n] > \theta_{xy}^{j,l}[n] \\ 0, & \text{otherwise,} \end{cases} \tag{1}$$

$$T_{xy}^{j,l}[n] = T_{xy}^{j,l}[n-1] + Y_{xy}^{j,l}[n]. \tag{2}$$

In (1), the coefficient $C_{j,l}(x, y)$ is assigned to the feeding input $F_{xy}^{j,l}$. The linking input $L_{xy}^{j,l}$ is equal to the sum of neurons firing times in linking range, where α_L indicates the decay constants and V_L is the amplitude gain. $W_{xy,pq}$ is the weighted coefficient (p and q point out the size of linking range in PCNN). The internal state signal $U_{xy}^{j,l}$ is obtained by modulating $F_{xy}^{j,l}$ and $L_{xy}^{j,l}$, where β is the linking strength. $\theta_{xy}^{j,l}$ is the threshold, where α_θ and V_θ are the decay constants and the amplitude gain, respectively. n denotes the iteration times. If $Y_{xy}^{j,l} = 1$, the neuron will generate a pulse, called one firing. If $Y_{xy}^{j,l} = 0$, the neuron will not generate a pulse. In applications, $T_{xy}^{j,l}[n]$ defined in (2) are often used to indicate the total firing times in n iteration. The firing times are employed to represent image information.

The saliency maps $S_{j,l}^{A}(x, y)$ and $S_{j,l}^{B}(x, y)$ are computed on the high-pass directional subbands $C_{j,l}^{A}(x, y)$ and $C_{j,l}^{B}(x, y)$, which denotes the jth scale and lth direction. The saliency maps are used as the importance indicator of the coefficients for preserving important information of source images.

Phase spectrum of Fourier transform (PFT) proposed in [24] is employed as a saliency detection model for grayscale image. PFT showed that the saliency map can be easily computed by the phase spectrum of an image's Fourier transform when its amplitude spectrum is at nonzero constant value. Only the phase spectrum is used to reconstruct an image which reflects the saliency information of the source image. The implementation of PFT model consists of three steps. An image is first transformed into frequency domain using Fourier transform, and the amplitude and phase spectrums are then obtained. Finally, the saliency map is obtained by inverse Fourier transform on only the phase spectrum. Given an input image $I(x, y)$, three steps have the corresponding equations as follows:

$$F(u, v) = F(I(x, y)) \tag{3}$$

$$P(u, v) = P(F(u, v)) \tag{4}$$

$$S(x, y) = g * \left\| F^{-1}\left\{\exp^{i\cdot P(u,v)}\right\}\right\|^2, \tag{5}$$

where F and F^{-1} denote Fourier transform and inverse Fourier transform. $P(F)$ is the phase spectrum of I and g is a 2D Gaussian filter. The saliency value in location (x, y) is computed using (5).

PFT model is a simple and efficient saliency detection method. An example of the PFT saliency detection is shown in Figure 4. Figures 4(a) and 4(c) are two multifocus source images which show complementary focus point regions. Figures 4(b) and 4(d) are the corresponding saliency maps which indicate different saliency regions of source images. We can observe that the saliency maps present focus point areas in source images.

Consequently, in this paper, the saliency value of high-pass subbands $S_{j,l}^{A/B}(x, y)$ can be computed by (5) in which coefficient $C_{j,l}^{A/B}(x, y)$ replaces $I(x, y)$ as the input. Instead of using PCNN in NSCT domain directly, the product $SC_{j,l}(x, y)$ between coefficient $C_{j,l}^{A/B}(x, y)$ in location (x, y) of high-pass subbands and its saliency value $S_{j,l}^{A/B}(x, y)$ are used to motivate PCNN. $SC_{j,l}(x, y)$ is computed as follows:

$$SC_{j,l}(x, y) = S_{j,l}^{A/B}(x, y) \cdot C_{j,l}^{A/B}(x, y). \tag{6}$$

Then, $SC_{j,l}(x, y)$ is normalized as $SC\_Norm_{j,l}(x, y)$ which is inputted to PCNN to motivate neurons. The proposed saliency-motivated PCNN model is defined in (7) by modifying (1):

$$\begin{aligned} F_{xy}^{j,l}[n] &= SC\_Norm_{j,l}(x, y) \\ L_{xy}^{j,l}[n] &= \exp(-\alpha_L) L_{xy}^{j,l}[n-1] \\ &\quad + V_L \sum_{pq} W_{xy,pq}^{j,l} Y_{xy,pq}^{j,l}[n-1] \\ U_{xy}^{j,l}[n] &= F_{xy}^{j,l}[n] * \left(1 + \beta L_{xy}^{j,l}[n]\right) \\ \theta_{xy}^{j,l}[n] &= \exp(-\alpha_\theta)\theta_{xy}^{j,l}[n-1] + V_\theta Y_{xy}^{j,l}[n-1] \\ Y_{xy}^{j,l}[n] &= \begin{cases} 1, & \text{if } U_{xy}^{j,l}[n] > \theta_{xy}^{j,l}[n] \\ 0, & \text{otherwise.} \end{cases} \end{aligned} \tag{7}$$

3.2. Subband Coefficients Fusion. The high-pass subbands of NSCT decomposition contain abundant detailed information and indicate the saliency components of images, for example, lines, edges, contours, and so forth. In order to preserve the saliency components in the process of image fusion, we propose the fusion rule based on saliency-motivated

Step 0: Given source images A and B.
Step 1: Perform decomposition on source images A and B using NSCT to obtain the high-pass directional subband coefficients and the low-pass subband coefficients.
Step 2: Merge the high-pass directional subbands with (8), (7) and (2).
Step 3: Obtain the fused low-pass subband with (9), (1) and (2).
Step 4: Construct the fused image by applying the inverse NSCT to the fused subband coefficients.

ALGORITHM 1: Image fusion method with saliency-motivated PCNN.

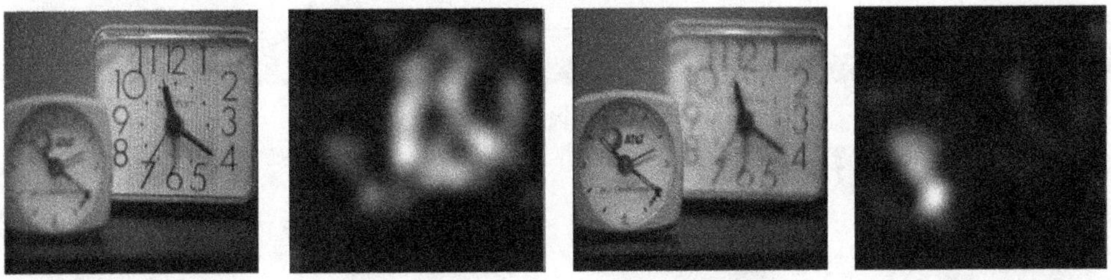

(a) Source image I_1 (b) Saliency map M_1 (c) Source image I_2 (d) Saliency map M_2

FIGURE 4: The results of saliency detection from two complementary input images. (a) and (c) Multifocus source images (b) and (d) saliency maps from PFT.

PCNN for the high-pass subbands. According to the visual attention mechanism, different regions in an image have varying importance for HVS, so the saliency detection is performed on source images to yield saliency maps which indicate the significance level of every pixel in source images. Based on the characteristics, the PFT model is performed on the high-pass subbands to produce the saliency maps, which indicate the importance level of coefficients. And then, the obtained saliency maps are combined with the corresponding high-pass subband coefficients as the input to motivate PCNN. Coefficients with large firing times are selected as the fused coefficients. In addition, the low-pass subband of NSCT decomposition in the coarsest scale contains the main energy of source images and denotes abundant structural information. The fusion rule of the low-pass subband employs a weighted fusion rule based on firing times of PCNN.

The activity maps of high-pass subbands as the criteria of selecting coefficients are presented by the firing map of saliency-motivated PCNN. The activity level indicates the magnitude of coefficients. The coefficients of greater energy carry more important information, so the coefficients of greater activity level are selected as the fused coefficients. Now, according to (7) and (2), the fused coefficients in location (x, y) of high-pass subbands denoted by $F_{j,l}(x, y)$ are defined as follows:

$$F_{j,l}(x,y) = \begin{cases} C_{j,l}^{A}(x,y), & \text{if } T_{A,xy}^{j,l} > T_{B,xy}^{j,l} \\ C_{j,l}^{B}(x,y), & \text{otherwise.} \end{cases} \quad (8)$$

The fused coefficients of low-pass subbands denoted by $F_{j0}(x, y)$ employ a weighted fusion rule based on firing times of PCNN on coefficients $C_{j0}^{A}(x, y)$ and $C_{j0}^{B}(x, y)$, which are defined as follows:

$$F_{j0}(x,y) = \omega * C_{j0}^{A}(x,y) + (1-\omega) * C_{j0}^{B}(x,y)$$
$$\omega = \frac{T_{A,xy}^{j0}}{T_{A,xy}^{j0} + T_{B,xy}^{j0}}, \quad (9)$$

where ω is the weight of coefficients and T_{xy}^{j0} is computed by (1) and (2). Because the low-pass subband at the coarsest scale does not contain the direction, here the symbol l in (1) and (2) is changed to 0. Specifically, $C_{j,l}(x, y)$ in (1) is replaced by $C_{j0}^{A/B}(x, y)$, and $T_{xy}^{j,l}$ in (2) is replaced by T_{xy}^{j0}.

Finally, apply the inverse NSCT to the fused coefficients $\{F_{j0}(x, y), F_{j,l}(x, y)\}$ and then obtain the fused image F. At last, the algorithm description of the proposed image fusion approach is shown in Algorithm 1 for better understanding.

4. Experiments and Analysis

In this section, the proposed image fusion algorithm based on NSCT and saliency-motivated PCNN (named as NSCT-SPCNN) is tested on several sets of images. The goal of the tests is to validate if the proposed algorithm can be used in the real applications and varying surroundings. For comparison, besides the fusion scheme proposed in this paper, another three fusion algorithms, the Laplacian pyramid transform based (LPT), discrete wavelet transform based (DWT), and NSCT-simple based, are used to fuse the same images. All of these use averaging and absolute maximum selection schemes for merging low- and high-pass subband coefficients, respectively. The decomposition level of all of the transforms is three. Extensive experiments with multifocus image fusion and different sensor image fusion

FIGURE 5: "Clock" source images (256 level, size of 256 × 256) and fused images: (a) focus on the right; (b) focus on the left and fused images using (c) LPT, (d) DWT, (e) NSCT-simple, and (f) NSCT-SPCNN methods.

have been performed. Here, three groups of different images were tested to evaluate the performance of the proposed algorithm: a set of multifocus images, a set of multimodal medical images, and a set of artificial out-of-focus images. It is assumed that source images have been registered. The fused results were evaluated using subjective visual inspection and objective assessment tools.

4.1. Visual Analysis. The first experiment uses two multifocus source images and four fused images produced by LPT, DWT, NSCT-simple, and NSCT-SPCNN methods, shown in Figure 5. Figure 5(a) focuses on the right region. Figure 5(b) focuses on the left region. The fused images contain all of focus point regions of source images and expand effectively the depth of a scene. In Figure 5(f), the saliency value associated with coefficients as the input to motivate PCNN is employed to compute the activity level of coefficients. In this way, the algorithm makes sure that the activity level of the saliency pixel is higher, so that the fused image preserves the saliency regions of source images. The images in Figures 5(c)–5(e) are not clear enough and have lower contrast; artifacts were also introduced. The differences among the fused images are very slight, so it is difficult to evaluate the image quality by direct visual inspection. To observe the image quality in more detail, one area in the fused images was magnified.

Figures 6(a)–6(d) show magnified images of the region marked by the boxes in Figures 5(c)–5(f). The performance of the different fusion algorithms can be observed from these magnified images. The images fused using LPT, DWT and NSCT-simple methods (Figures 6(a)–6(c)) have some deformation leading to bend edges. Figure 6(d) has the better visual quality than others with the best visual effect and smoother and sharper edges. This comparison reveals that the NSCT-SPCNN-based fusion approach effectively determines complementary or redundant information between source images. It can preserve all the important information of the source images while avoiding artifacts. In addition, a clearer comparison is made by examining the differences between the fused and source images, shown in Figure 7. Figures 7(a)–7(d) show the difference images between Figures 5(c)–5(f) and Figure 5(a). Observing Figures 7(a)–7(d), we can see that there is little difference between the fused image by NSCT-SPCNN (Figure 5(d)) and the right-focus clock (Figure 5(a)), while a lot of difference between the fused images in Figures 5(c)–5(e) and the right-focus source image in Figure 5(a) can be seen. This further demonstrates that the NSCT-SPCNN-based method is with higher fusion performance.

Figure 8 shows a group of multimodal medical images and images fused using four different fusion algorithms. A set of spatial out-of-focus images are shown in Figure 9,

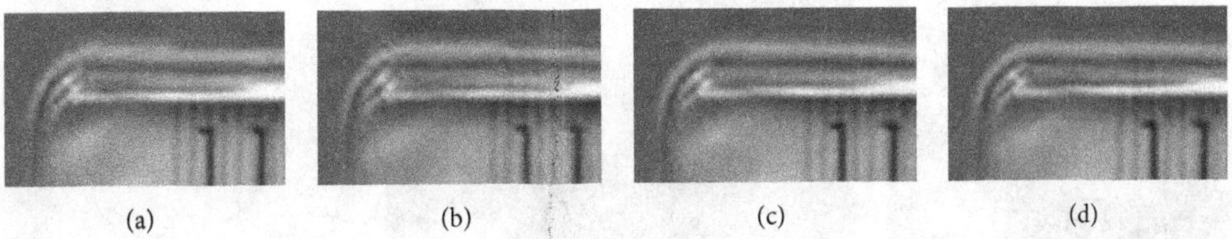

FIGURE 6: Magnified regions from the fused images in Figures 5(c)–5(f) using (a) LPT, (b) DWT, (c) NSCT-simple, and (d) NSCT-SPCNN methods.

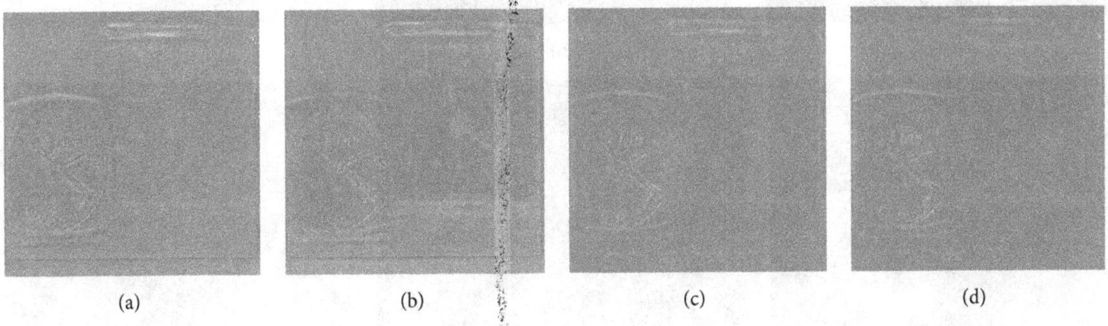

FIGURE 7: (a) Difference image between Figures 5(c) and 5(a); (b) difference image between Figures 5(d) and 5(a); (c) difference image between Figures 5(e) and 5(a); (d) difference image between Figures 5(f) and 5(a).

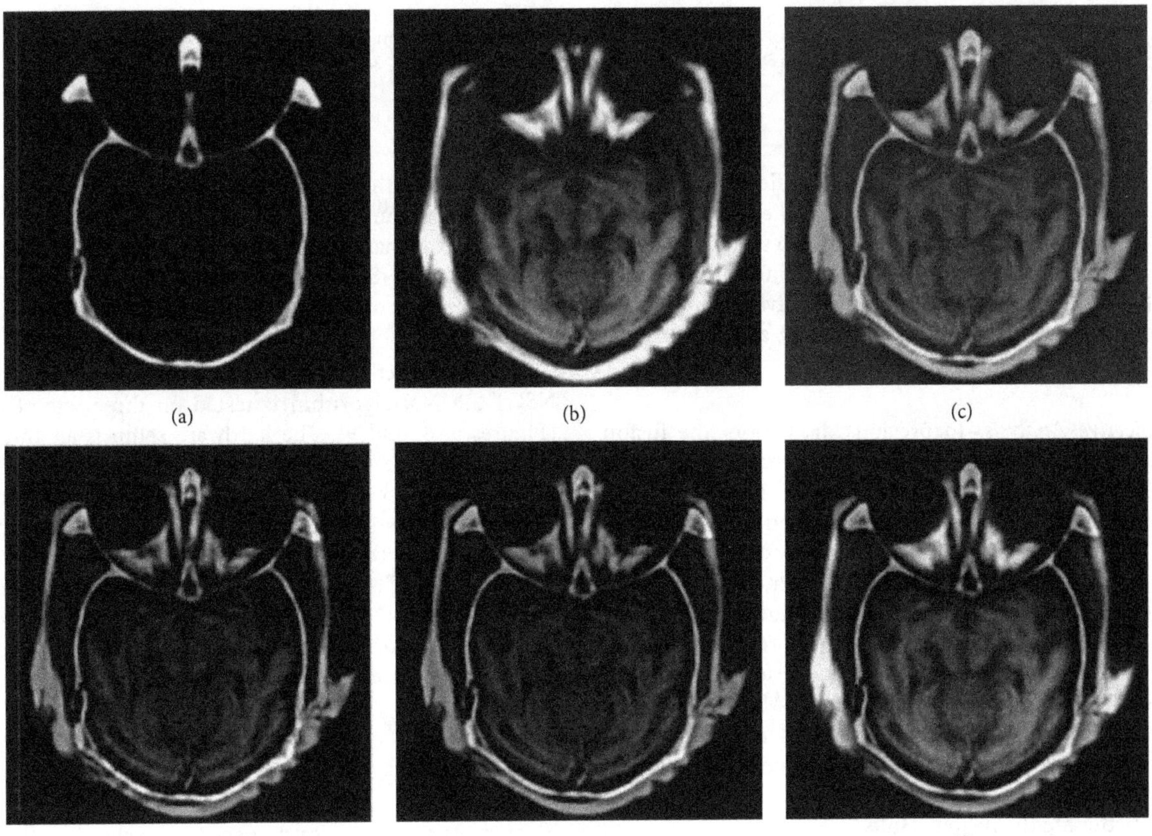

FIGURE 8: Medical source images (256 level, size of 256 × 256) and fused images: (a) and (b) source images and fused images using (c) LPT, (d) DWT, (e) NSCT-simple, and (f) NSCT-SPCNN methods.

(a) (b) (c)

(d) (e) (f)

FIGURE 9: Spatial source images (256 level, size of 512 × 512) and fused images: (a) focus on the right; (b) focus on the left and fused images using (c) LPT, (d) DWT, (e) NSCT-simple, and (f) NSCT-SPCNN methods.

which is obtained by artificial blurring different regions of the ground truth image using a Gaussian filter. Experimental results demonstrate the visual effects of two sets of images in Figures 8 and 9 coinciding with Figure 5. Thus, the proposed algorithm NSCT-SPCNN-based can effectively improve the quality of the fused image both multifocus, and multimodal images.

4.2. Objective Analysis. In previous discussion, the fusion results of different algorithms have been analyzed by visual aspect. However, the performance of fusion algorithms needs to be further evaluated using objective metric tools. A successful fusion technique has to satisfy many conditions, such as preserving important features of source images, enhancing contrast, and avoiding artifacts. Mutual information (MI) [25] and an objective image fusion performance measure ($Q_{AB/F}$) [26] are employed to evaluate the fusion performance of different fusion methods quantitatively. MI indicates how much of the input information the fused image contains. $Q_{AB/F}$ reflects the preservation of input edge information in the fused image. For the two metrics, the higher the values are, the better are the fusion results.

Figure 10 shows the quality measurement results for fused images in Figure 5 and Figures 8 and 9. Observing Figure 10, we can see that the LPT and DWT methods are the worst. This is consistent with the subjective visual analysis. Compared with other fusion algorithms, the NSCT-SPCNN yields the optimal performance. Experimental results demonstrate that the proposed NSCT-SPCNN algorithm can preserve the saliency regions of source images and improve the quality of the fused image.

Finally, the computational performance of the proposed NSCT-SPCNN algorithm is tested on three sets of images (Figures 5, 8, and 9). The hardware setup is an Intel Core i5-3479 PC with 4 GB RAMs. Our Matlab implementation takes about 16 seconds for Figures 5 and 8 and 73 seconds for Figure 9. Meanwhile, the NSCT-simple-based fusion method takes about 17 seconds for Figures 5 and 8 and 72 seconds for Figure 9. The LPT- and DWT-based fusion methods take less than 1 second. From the comparison, we can observe that the computational bottleneck lies in the NSCT transform. Therefore, a more efficient MST tool needs to be applied in the future.

5. Conclusion

The paper proposes a novel image fusion algorithm based on NSCT and saliency-motivated PCNN. In fusion for high-pass subbands, a saliency-motivated PCNN model is proposed. The key idea is that depending on the human visual attention

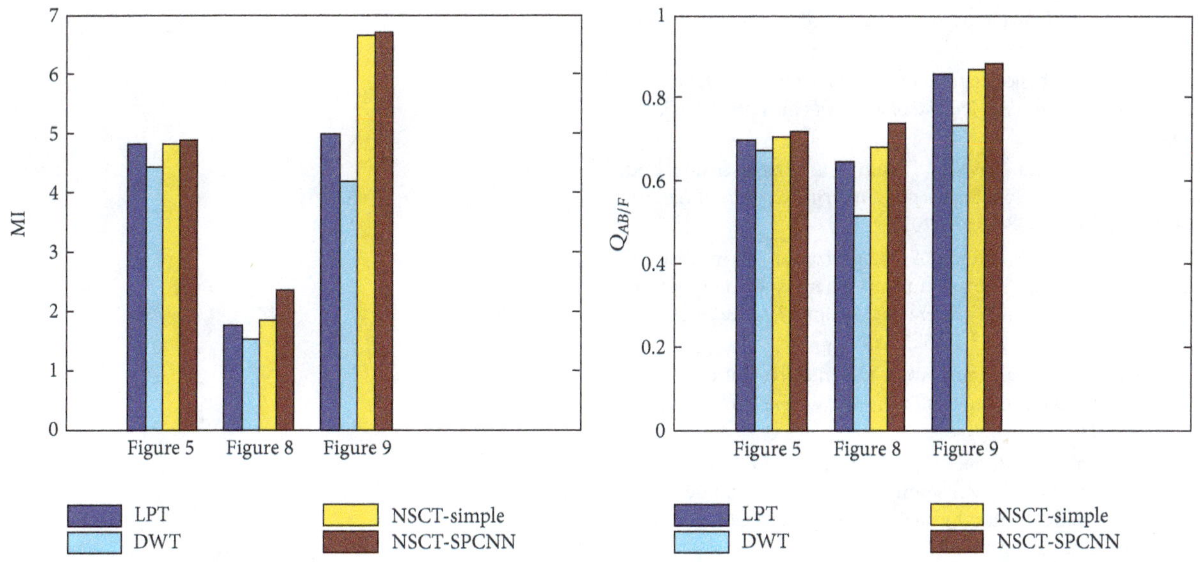

(a) MI metric

(b) $Q_{AB/F}$ metric

FIGURE 10: Quality metrics for the different fusion methods.

model, the visual saliency map is first built on high-pass subband coefficients of NSCT, and then the algorithm combines the visual saliency map with the coefficients of NSCT as input to motivate PCNN. Coefficients with large firing times are employed as the fused high-pass subband coefficients. Low-pass subband coefficients are merged to develop a weighted fusion rule based on firing times of PCNN. The algorithm can preserve the completeness and the sharpness of object regions. The fused image is more natural and can satisfy the requirement of HVS. Experiments illustrate that the proposed fusion algorithm improves greatly the quality of the fused images.

Acknowledgments

This work was supported by the National Basic Research Program of China (973 Program nos. 2012CB821200 2012CB821206), the National Natural Science Foundation of China (nos. 91024001, 61070142), and the Beijing Natural Science Foundation (no. 4111002).

References

[1] A. Leykin, Y. Ran, and R. Hammoud, "Thermal-visible video fusion for moving target tracking and pedestrian classification," in *Proceedings of the IEEE Computer Society Conference on Computer Vision and Pattern Recognition (CVPR' 07)*, June 2007.

[2] Y. F. Niu, S. T. Xu, L. Z. Wu, and W. D. Hu, "Airborne infrared and visible image fusion for target perception based on target region segmentation and discrete wavelet transform," *Mathematical Problems in Engineering*, vol. 2012, Article ID 275138, 10 pages, 2012.

[3] L. Yang, B. L. Guo, and W. Ni, "Multimodality medical image fusion based on multiscale geometric analysis of contourlet transform," *Neurocomputing*, vol. 72, no. 1–3, pp. 203–211, 2008.

[4] J. R. Raol, *Multi-Sensor Data Fusion with Matlab*, CRC Press Taylor and Francis Group, 2010.

[5] R. Blum, Z. Xue, and Z. Zhang, "Chapter 1: an overview of image fusion," in *Multi-Sensor Image Fusion and Its Applications*, pp. 1–36, CRC Press Taylor and Francis Group, 2006.

[6] P. J. Burt and E. H. Adelson, "The Laplacian pyramid as a compact image code," *IEEE Transactions on Communications*, vol. 31, no. 4, pp. 532–540, 1983.

[7] V. S. Petrović and C. S. Xydeas, "Gradient-based multiresolution image fusion," *IEEE Transactions on Image Processing*, vol. 13, no. 2, pp. 228–237, 2004.

[8] G. Pajares and J. M. de la Cruz, "A wavelet-based image fusion tutorial," *Pattern Recognition*, vol. 37, no. 9, pp. 1855–1872, 2004.

[9] X.-C. Yu, F. Ni, S.-L. Long, and W.-J. Pei, "Remote sensing image fusion based on integer wavelet transformation and ordered nonnegative independent component analysis," *GIScience and Remote Sensing*, vol. 49, no. 3, pp. 364–377, 2012.

[10] M. N. Do and M. Vetterli, "The finite ridgelet transform for image representation," *IEEE Transactions on Image Processing*, vol. 12, no. 1, pp. 16–28, 2003.

[11] M. N. Do and M. Vetterli, "The contourlet transform: an efficient directional multiresolution image representation," *IEEE Transactions on Image Processing*, vol. 14, no. 12, pp. 2091–2106, 2003.

[12] K. Liu, L. Guo, and J. Chen, "Contourlet transform for image fusion using cycle spinning," *Journal of Systems Engineering and Electronics*, vol. 22, no. 2, pp. 353–357, 2011.

[13] E. Candès, L. Demanet, D. Donoho, and L. Ying, "Fast discrete curvelet transforms," *Multiscale Modeling and Simulation*, vol. 5, no. 3, pp. 861–899, 2006.

[14] S. Ren, J. Cheng, and M. Li, "Multiresolution fusion of Pan and MS images based on the Curvelet transform," in *Proceedings of the 30th IEEE International Geoscience and Remote Sensing Symposium (IGARSS '10)*, pp. 472–475, July 2010.

[15] A. L. da Cunha, J. Zhou, and M. N. Do, "The nonsubsampled contourlet transform: theory, design, and applications," *IEEE*

Transactions on Image Processing, vol. 15, no. 10, pp. 3089–3101, 2006.

[16] G. Piella, "Image fusion for enhanced visualization: a variational approach," *International Journal of Computer Vision*, vol. 83, no. 1, pp. 1–11, 2009.

[17] C. Ludusan and O. Lavialle, "Multifocus image fusion and denoising: a variational approach," *Pattern Recognition Letters*, vol. 33, no. 10, pp. 1388–1396, 2012.

[18] Z. Zhang and R. S. Blum, "A categorization of multiscale-decomposition-based image fusion schemes with a performance study for a digital camera application," *Proceedings of the IEEE*, vol. 87, no. 8, pp. 1315–1326, 1999.

[19] G. Piella, "A general framework for multiresolution image fusion: from pixels to regions," *Information Fusion*, vol. 4, no. 4, pp. 259–280, 2003.

[20] R. Eckhorn, H. J. Reitboeck, M. Arndt, and P. Dicke, "Feature linking via synchronization among distributed assemblies: simulations of results from cat visual cortex," *Neural Computation*, vol. 2, no. 3, pp. 293–307, 1990.

[21] J. L. Johnson and M. L. Padgett, "PCNN models and applications," *IEEE Transactions on Neural Networks*, vol. 10, no. 3, pp. 480–498, 1999.

[22] R. P. Broussard, S. K. Rogers, M. E. Oxley, and G. L. Tarr, "Physiologically motivated image fusion for object detection using a pulse coupled neural network," *IEEE Transactions on Neural Networks*, vol. 10, no. 3, pp. 554–563, 1999.

[23] X.-B. Qu, J.-W. Yan, H.-Z. Xiao, and Z.-Q. Zhu, "Image fusion algorithm based on spatial frequency-motivated pulse coupled neural networks in nonsubsampled contourlet transform domain," *Acta Automatica Sinica*, vol. 34, no. 12, pp. 1508–1514, 2008.

[24] C. Guo, Q. Ma, and L. Zhang, "Spatio-temporal saliency detection using phase spectrum of quaternion fourier transform," in *Proceedings of the 26th IEEE Conference on Computer Vision and Pattern Recognition (CVPR '08)*, June 2008.

[25] G. Qu, D. Zhang, and P. Yan, "Information measure for performance of image fusion," *Electronics Letters*, vol. 38, no. 7, pp. 313–315, 2002.

[26] C. S. Xydeas and V. Petrović, "Objective image fusion performance measure," *Electronics Letters*, vol. 36, no. 4, pp. 308–309, 2000.

18

An Improved Generalized-Trend-Diffusion-Based Data Imputation for Steel Industry

Ying Liu, Zheng Lv, and Wei Wang

School of Control Sciences and Engineering, Dalian University of Technology, Dalian 116023, China

Correspondence should be addressed to Ying Liu; liu_ying@dlut.edu.cn

Academic Editor: Jun Zhao

Integrality and validity of industrial data are the fundamental factors in the domain of data-driven modeling. Aiming at the data missing problem of gas flow in steel industry, an improved Generalized-Trend-Diffusion (iGTD) algorithm is proposed in this study, where in particular it considers the sort of problem with data properties of consecutively missing and small samples. And, the imputation accuracy can be greatly increased by the proposed Gaussian membership-based GTD which expands the useful knowledge of data samples. In addition, the imputation order is further discussed to enhance the sequential forecasting accuracy of gas flow. To verify the effectiveness of the proposed method, a series of experiments that consists of three categories of data features in the gas system is presented, and the results indicate that this method is comprehensively better for the imputation of the periodical-like data and the time-series-like data.

1. Introduction

Data missing is one of the major obstacles to obtain valid data samples [1], which also might be common or even inevitable in some data-driven-based research fields, such as sample surveys, industrial productions, medical research, soft engineering, and wireless broadcast environment [2, 3]. The data missing problem might destroy the samples integrality since every cell in database may not be independent, and furthermore a single missing value might call for dropping the entire observed values or the useful information [4, 5]. As such, some useful information or knowledge could be lost from the data set. Moreover, the data missing will also lead to the nonresponse bias of samples which could be a serious concern for the data-driven-based studies [6–10]. In the literatures, most of the existing methods for such problem were mainly based on the statistical techniques. For instance, the multiple imputation (MI), a kind of popular technique, was used to resolve the missing data of gross domestic product (GDP) [11], cancer databases [12], and sample surveys [13]. And, a similar response pattern imputation (SRPI) was also implemented in [14]. In [15], the authors used the classic expectation-maximization (EM) algorithm, principal component analysis (PCA), and singular value decomposition, while [16, 17] utilized the maximum likelihood technique to carry out the missing data imputation. However, all the techniques mentioned previously might be hard to reflect the relationship among regression variables, since the imputed values were mere approximations of unknown values. Besides the statistical techniques, the machine learning was paying more and more attentions nowadays, as presented in [18, 19].

In industrial manufacturing process, the phenomenon of data missing often occurs due to the events such as data collector failures, transmission errors, or information storage errors, which directly result in some obstacles for establishing data-driven-models, such as scheduling models, data-driven based regression prediction models, and stochastic optimization models [20–22]. There were different types of approaches for industrial practitioners in the literatures to deal with these data missing problems. In [23, 24], the authors proposed a method called list-wise deletion that was easy to be implemented; however, it tended to reduce the sample data size. Considering that a lot of missing data in industry have the form of time series sampled in equal intervals in most cases, the integrality of sample data has to be broken by such deleting the missing points. Besides wasting a lot of costly collected data, this method also led to invalid results if the excluded group was a selective subsample from the entire sample [25]. Mean imputation presented in [26] was another widely employed method. However, the mean values of the sample might eliminate the samples diversity in time series whose amplitude dramatically changes, and the distortion of samples was usually unacceptable for industrial practitioners. With respect to the other statistical or machine learning techniques, the maximum likelihood estimation and the linear interpolator were, respectively, proposed in [27, 28], where the effective experiments were used to validate the time series imputation. Yet, all of these experiments showed high demands of samples, and as a result, their applications in real industrial process were rather limited. As for all of the above mentioned methods, few of them can bring satisfying imputation accuracy, once the consecutive missing happens, or the missing rate is high, and the sample size is too small.

The Generalized-Trend-Diffusion (GTD) is a method of sample construction aiming at small data sets. As the virtual examples presented in [29] and the functional virtual population in [30], the so-called shadow data and membership functions were employed to increase the knowledge of small data sets; see more details in [31]. And, the expanded samples were provided for Back Propagate-based (BP) neural networks to carry out the forecasting, resulting in the prediction accuracy higher than that without expanding. Thus, the most significant advantage of GTD was that it could bring satisfactory forecasting accuracy with relatively small data sets. On the other hand, the original GTD described the membership degree to the mean value of observed sample via a triangular membership function. As such, each observed data point deviation from the mean value is proportional to the difference between the membership function values; that is, the observed data points linearly deviate from the mean point. However, such description of deviation cannot bring excellent accuracy in the imputation tasks for real industrial manufacturing process.

This paper aims at the missing data imputation of blast furnace gas flow in steel energy system. An improved GTD modeling algorithm based on Gaussian membership function is proposed considering the diversity of the gas flow data and the complex missing situations. The Gaussian membership shows that the observed data deviate from the mean value nonuniformly, and this deviation makes the close-to-mean values more likely to appear in the imputation. The samples, expanded by the membership function, make the predicted values by BP-based network lean to the mean. And, such predicted values do not make the samples single as those by the mean imputation do. In addition, the imputation order is essential to the accuracy of time series problem. A both-side-toward-middle (BSTM) order is proposed in this paper which is indicated to be more appropriate than the chronological order. And the tests are implemented to verify the effectiveness of the proposed method, in which the sample data comes from the practice of Shanghai Baosteel Co. Ltd. The results demonstrate that the improved GTD method is much better than the original version and other methods in several cases.

This study is organized as follows. In Section 2, the practical conditions of blast furnace gas in Baosteel is described. And then, the original GTD and its improved version are established in Section 3, where the details of how to use the improved GTD for the industrial missing data imputation are discussed. In Section 4, the validity of the improved GTD is verified by a series of comparative experiments. Finally, this study is summarized in Section 5.

2. Problem Description

Blast furnace gas (BFG) is a kind of byproduct gas generated in the process of iron making [32]. As an important secondary energy for blast furnaces, coke ovens, power stations, and other units, its proper utility can not only reduce the energy consumption of steel enterprises but also improve their economic profits. Figure 1 shows the BFG system structure of a steel plant, where four blast furnaces supply the gas to consumers. However, BFG could be diffused if the flow prediction and the scheduling are inappropriately carried out, which will seriously pollute the environment. In this case, the supervision of BFG's generation and consumption becomes a crucial task for the steel enterprises.

Currently the on-site technicians perform the balance scheduling by estimating the BFG generation amount which comes from the observed data. However, the observed data often miss due to the collector failure, transmission errors, information storage errors, and so forth. Furthermore, the generating process of BFG is rather complex, and the output fluctuates irregularly, therefore the data missing makes the workers work hard to perceive the dynamics of gas flow via generic model. In practice, the gas engineers in Shanghai Baosteel employ the personal experience-based estimation as the current wide using method when encountering single point missing. However, there are more consecutive missing points in real manufacturing process, which make such method relatively weak. In addition, if the missing rate is high, the whole time series can be treated as a combination of several small size series. In this case, the existing methodologies like the recursive neural networks presented in [33, 34] cannot be utilized because they need a large amount of sample data to train the regression model.

Aiming at the various features of a large number of gas units of BFG system, we summarize the flow tendencies of the generation and consumption units as three categories, which involve (1) the periodicity-like flow data (the gas consumption amount of hot blast stove, see Figure 2), (2) the concussive flow data (the gas consumption of coke oven, see Figure 3), and (3) the ordinary time series flow (the generation amount of blast furnace, see Figure 4).

3. Improved Generalized-Trend-Diffusion

3.1. Generalized-Trend-Diffusion. The GTD is a method of sample construction aiming at small data sets which generates shadow data using the real data and the occurrence order of the observed data. The importance degree of those shadow data and observed data is quantified by the membership function values based on fuzzy theories. Both the

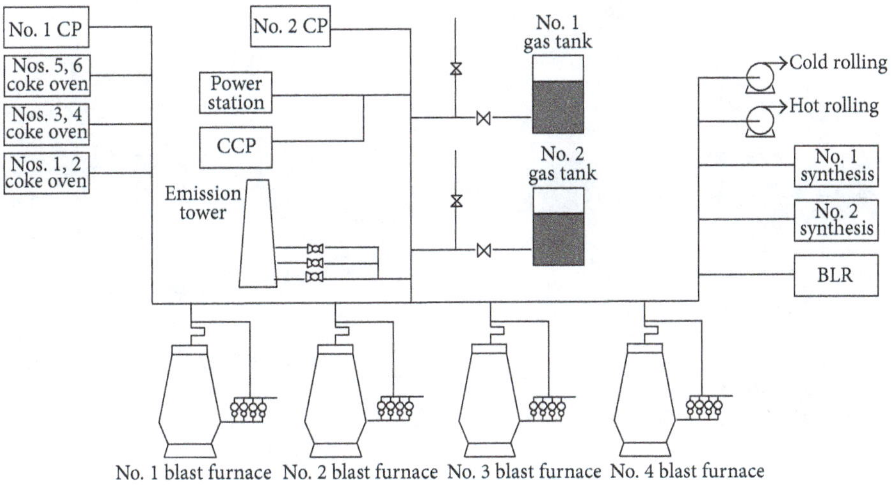

FIGURE 1: BFG network of Baosteel.

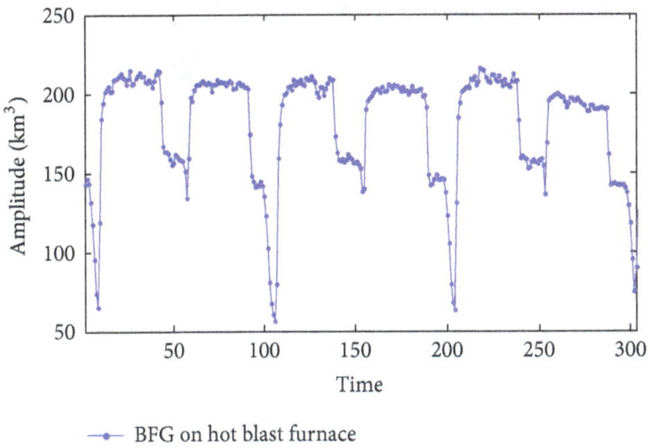

FIGURE 2: The consumption flow of hot blast stove.

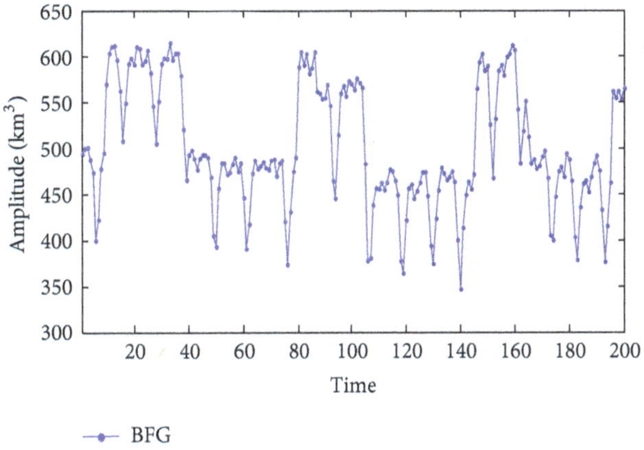

FIGURE 4: The generation flow of blast furnace.

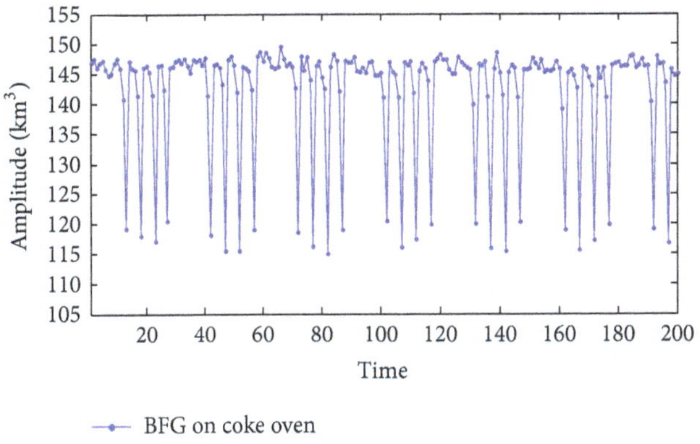

FIGURE 3: The consumption flow of coke oven.

membership values and the shadow data can be treated as the additional hidden data-related information, which helps to improve the imputation accuracy. All the previous features above make the GTD fit for the missing data imputation of time series because of their lack of more information except time [31]. One can start by considering that observations are collected with an empty set, where each point occurs with each observation (Figure 5). As the data increases, the central location, symbolized "C" in Figure 5, of the data for each observation moves from one location to another. If each point deviation from the central location can be obtained, then the detailed distribution of the whole sample is clear. As such, the GTD with membership function can be used to describe such deviation. Let the membership function value at "C" be 1, and let those of some missing data be MF_x. When these values get closer to 1, the missing data approach "C" and vice versa.

In the original GTD model, one can let MF_t be the membership function for the data collected at Step t. For example, MF_1 at Step 1 refers to Y_1 only, MF_2 at Step 2 refers to $\{Y_1, Y_2\}$, MF_3 at Step 3 refers to $\{Y_1, Y_2, Y_3\}$, and so forth. The data like Y_1 at Step 2 and $\{Y_1, Y_2\}$ at Step 3 are called the shadow data. They were called as such name because each of them was used repeatedly in each step when forming the corresponding membership functions, while it occurred actually once.

Then the imputation can be done by the shadow data. One can suppose that a sequence of data denoted as

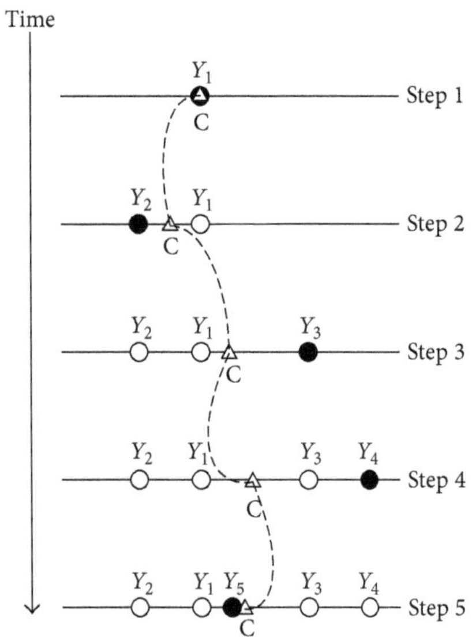

FIGURE 5: Time series and central location.

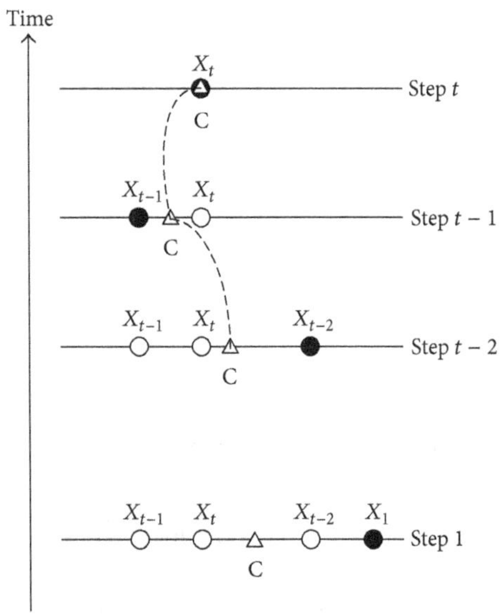

FIGURE 6: Backward tracking process.

$\{X_1, X_2, \ldots, X_{t-1}, X_t\}$ with X_{t+1} missing has been obtained. The shadow data can be built by unevenly repeating the more recent data which bring more important contemporary information of system variation than that provided by the previous data. As shown in Figure 6, the most recent point X_t is repeated t times, X_{t-1} is repeated $(t-1)$ times X_{t-2} is repeated $(t-2)$ times, and so forth. And, such repeating was called as the backward tracking progress, since it is done in the backward tracking progress. Then, these repeated data (shadow data) with their membership function values help to enlarge the sample knowledge.

3.2. Improved Generalized-Trend-Diffusion. A triangular membership function (Figure 7) was used to describe each point's deviation from the mean value in the original GTD.

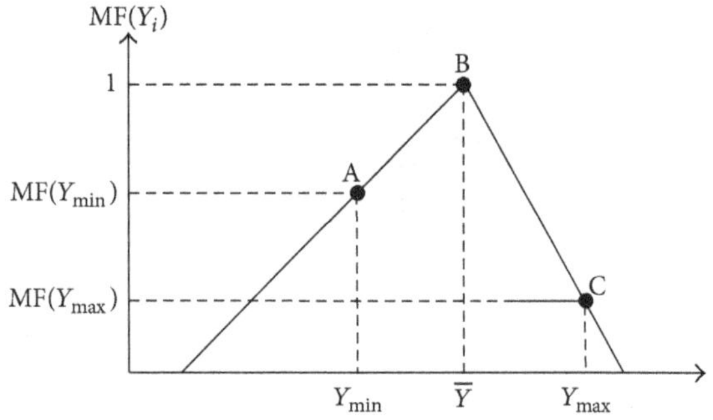

FIGURE 7: Triangular membership function.

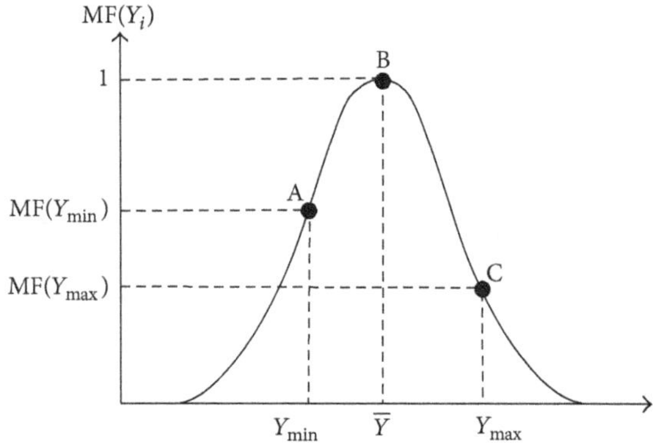

FIGURE 8: Gaussian membership function.

However, such description was somewhat unreasonable, since it restricted the deviation form as a linear one; that is, the deviation from the central location was proportional to the difference between the memberships. Under such condition, the possibility of the data value to appear in the imputation is equal. However, the mean-like data have actually a higher possibility of appearance in the industrial manufacturing process. In this study, we can call such data vividly as high frequency cloud. If a membership function can describe the high frequency cloud more like the mean value, then the data in the cloud (mean like) will reappear in the imputation with higher possibility. Considering such motivation, the Gaussian membership function (Figure 8) could be more competent to accomplish this job.

The information diffusion principle [35] is another reason for choosing the Gaussian membership function. Information diffusion has a function of filling in the blanks like the molecular diffusion, and its cause lies on that some data acquire little information from the sample knowledge, while molecular diffusion is caused by the heterogeneity in the space distribution. As for the molecular diffusion, it had been proved that current molecular density is proportional to the concentration gradient. If this principle is linked with the law of conservation of mass, the molecular diffusion can be described in the same form as the probability density of Gaussian distribution. As a kind of incremental learning method, the GTD is a representation of information diffusion. Since the causes of information diffusion and molecular diffusion are similar, we here get an inspiration to employ the Gaussian membership function in the improved GTD. The form of Gaussian function is as follows:

$$f(x) = ae^{-(x-b)^2/c^2}, \quad (1)$$

where a, b, and c are real constants and $a > 0$. In order to make the function adaptive to the sample construction in this study, we use (2) as the general form of Gaussian function instead of (1) as follows:

$$f(x) = e^{-(x-\mu)^2/\sigma^2}, \quad (2)$$

where μ is the mean value of sample and σ is the standard deviation. Here, we make a as 1, since the membership value at the mean value should be 1. After its form confirmed, the Gaussian membership is capable to enlarge the sample knowledge instead of the triangular one.

3.3. Data Imputation. The BP algorithm is a supervised learning method in a network, which is effected by altering the weights to minimize the difference between the output value and the desired output value [36]. The enlarged knowledge then can be utilized by BP neural networks to finish the prediction.

Missing data points need to be imputed one by one, so that the order of imputation should be another concern in this study. If the imputation is real time, the chronological order has to be taken because one cannot currently acquire the future data points. However, the study in this paper is a data mining job which does not need real-time imputations. Furthermore, if the imputation is not real time, the BSTM order is superior to the chronological one. For instance, let there be five consecutive missing data, as Figure 9 shows. If the imputation order is chronological, the forecast error of point number 1 will be amplified so as to affect the forecast accuracy of point number 2, and the error of point number 2 will be again propagated to that of point number 3. In such way, the errors will be cumulative.

Besides, data points in time series are always fluctuating, as Figure 9 shows. If the missing happens on the hillside, the chronological imputation result is very likely to be the same as △, since it continues the peak. However, the result may be like ◻ if we use the BSTM order. That is, points number 1 and number 2 are on the peak, while points number 4 and number 5 are on the plain which both continue the trends. As for point number 3, we impute that it sings the mean of points number 2 and number 4. Obviously, △ deviates from the real values more than ◻ which shows that the BSTM order is superior to the chronological one. This summary is consistent if analyzing the missing points on the peak or on the plain.

Let $(X_1, X_2, \ldots, X_{t-1}, X_t)$ be a time series, the index of the first missing point denotes as n, the number of the consecutive

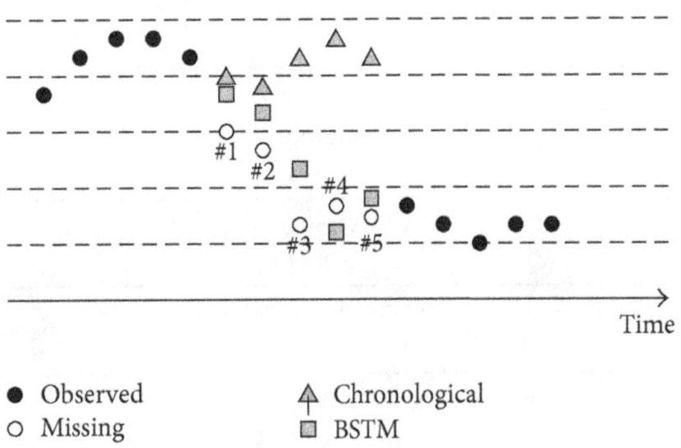

FIGURE 9: Imputation by BSTM and chronological manner.

missing points denotes as m, and the variable d represents the embedding dimension, then we have

$$\begin{aligned} \widehat{X}_i^{\text{former}} &= \text{net}\Big((X_{i-d}, \text{MF}_{i-d}), \\ &\qquad (X_{i-d+1}, \text{MF}_{i-d+1}), \ldots, (\widehat{X}_{i-1}, \text{MF}_{i-1}) \Big), \\ \widehat{X}_i^{\text{latter}} &= \text{net}\Big((\widehat{X}_{i+1}, \text{MF}_{i+1}), \\ &\qquad (\widehat{X}_{i+2}, \text{MF}_{i+2}), \ldots, (X_{i+d}, \text{MF}_{i+d}) \Big), \end{aligned} \quad (3)$$

where $\widehat{X}_i^{\text{former}}$ is the imputation of the former half, while $\widehat{X}_i^{\text{latter}}$ is in the latter half. Then all the imputations $\widehat{X}_i$ can be expressed as

$$\widehat{X}_i = \begin{cases} \widehat{X}_i^{\text{former}} & n \le i \le n + \dfrac{m}{2} - 1, \\ \widehat{X}_i^{\text{latter}} & n + \dfrac{m}{2} \le i \le n + m - 1, \end{cases} \quad \text{if } m = 2k,\ k \in N^*,$$

$$\widehat{X}_i = \begin{cases} \widehat{X}_i^{\text{former}} & n \le i \le n + \dfrac{m-1}{2} - 1, \\ \dfrac{1}{2}\left(\widehat{X}_i^{\text{former}} + \widehat{X}_i^{\text{latter}}\right) & i = n + \dfrac{m-1}{2}, \\ \widehat{X}_i^{\text{latter}} & n + \dfrac{m-1}{2} + 1 \le i \le n + m - 1, \end{cases} \quad \text{if } m = 2k - 1,\ k \in N^*. \quad (4)$$

4. Experimental Results and Analysis

The imputation tests of missing data in BFG flow are carried out with the proposed Gaussian membership function-based method, called iGTD here. First of all, the superiority of the BSTM order to the chronological one is tested and verified. A series of consecutive 800 data is picked from number 1 blast furnace in Baosteel dating from 14:34:00/13/8/2010 to 3:54:00/14/8/2010. Considering that it is difficult to guarantee quantities of consecutive valid data in real, industrial databases, and the small set of samples is our concerns in this study, the embedding dimension is empirically chosen

TABLE 1: Imputation accuracies with two kinds of orders.

Group	Order	RMSE	NRMSE	MAPE (%)
1	Chronological order	169.30	0.1486	45.65
	BSTM	110.54	0.0970	25.97
2	Chronological order	92.74	0.0669	18.80
	BSTM	82.40	0.0594	14.56
3	Chronological order	107.70	0.0842	23.95
	BSTM	57.32	0.0448	11.41
4	Chronological order	120.47	0.1022	29.75
	BSTM	95.64	0.0812	21.44

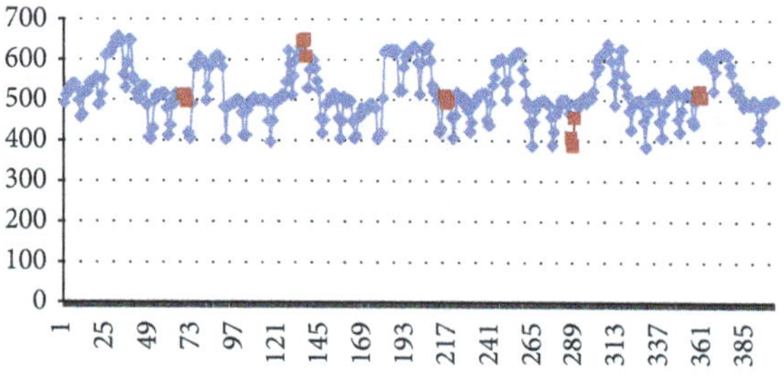

FIGURE 10: Overview of missing data.

as 15, and the hidden neuron size is chosen as 10 in the same manner. We divide the sample data into 4 groups. For each group, we randomly remove 3 consecutive points (A, B, and C) in 3 places. The tests are, respectively, implemented in the chronological order (A-B-C) and the BSTM order (A-C-B). Here, we use three indexes as the evaluation criterion of the imputation accuracy, which are root mean square error (RMSE), normalized root mean square error (NRMSE), and mean absolute percentage error (NRMSE) as follows:

$$\mathrm{RMSE} = \sqrt{\frac{1}{n}\sum_{i=1}^{n}\left(\widehat{Y}_i - Y_i\right)^2},$$

$$\mathrm{NRMSE} = \sqrt{\frac{1}{n\|Y\|^2}\sum_{i=1}^{n}\left(\widehat{Y}_i - Y_i\right)^2}, \qquad (5)$$

$$\mathrm{MAPE} = \frac{100}{n}\sum_{i=1}^{n}\frac{\left|\widehat{Y}_i - Y_i\right|}{Y_i},$$

where n is the total number of imputation, $\widehat{Y}$ is the imputation value, and Y is the real value. As for the separated 4 groups of data, the imputation accuracies for the different order are shown in Table 1. It is apparent that the effectiveness of BSTM is superior to that of the chronological order-based imputation method.

To further verify the effectiveness of the proposed Gaussian-based membership function, we comprehensively take the three categories of gas flow data mentioned in previous section, which include the periodicity-like flow data, shown like the BFG consumption amount by hot blast stove; the concussive flow data, shown like the gas consumption by coke oven; and the ordinary time series flow, shown like the generation amount by blast furnace. The comparative experiments are carried out by using the EM method, regression, spline, and the original GTD. EM algorithm is an iterative method for finding maximum likelihood or maximum a posteriori estimates of parameters in statistical models [37], which was widely used in dealing with missing data, since the maximum likelihood estimate of the unknown parameters can be determined by the incomplete data set. The regression method employed multiple linear regressions to estimate the missing values.

We still apply the real industrial data in Baosteel to complete the comparative experiment, where the collected data are divided into several groups, and some consecutive 3 points, 4 points, and 5 points are removed from the time series. In order to cover the all of the possible situations, the removed data involves the time series areas on peak, trough, and plain, as Figure 10 shows, in which the points in red are removed.

(1) BFG Consumption by Hot Blast Stove (Periodicity Like). The experimental data are from number 2 hot blast stove in Shanghai Baosteel randomly selected from 14:28:00/13/08/2010 to 21:55:00/14/08/2010. These data are divided into 3 groups, each of which is then divided into 3 subgroups, and each subgroup contains 200 points. The accuracies of the imputation result are presented in Table 2.

It can be found that the results by both the original GTD and the proposed iGTD are much more excellent in terms of the accuracy when the consecutive data missing occurs. Furthermore, the effectiveness of the iGTD is generally better than that of GTD. Then a conclusion can be drawn that the iGTD employed the Gaussian membership function can obtain the better data imputation results compared to the triangle-based membership of GTD.

(2) BFG Consumption by Coke Oven (Concussive). The experimental data are from number 1 coke oven in Shanghai Baosteel randomly selected from 07:14:00/14/08/2010 to 14:35:00/15/08/2010. The data-grouped measure is similar to that in the validation for periodicity-like data missing. And, the corresponding imputation accuracies are listed in Table 3. From the experiments results, all the five methods are almost same imputation accuracies, and in particular EM should be the best solution method of the five. However, it is

TABLE 2: Imputation accuracies with different methods (hot blast stove's BFG consumption).

Missing number	Methods	Group 1			Group 2			Group 3		
		RMSE	NRMSE	MAPE	RMSE	NRMSE	MAPE	RMSE	NRMSE	MAPE
3 points	Regression	57.16	0.0713	22.21	55.96	0.0777	28.56	35.56	0.0489	16.96
	EM	43.56	0.0543	17.88	45.82	0.0636	25.76	34.46	0.0474	17.58
	SPLINE	36.72	0.0458	13.80	60.62	0.0842	31.69	22.10	0.0304	10.68
	GTD	9.13	0.0114	2.88	14.90	0.0207	5.85	14.50	0.0199	3.42
	iGTD	7.19	0.0090	2.05	14.42	0.0200	5.71	6.23	0.0086	2.87
4 points	Regression	47.17	0.0578	23.07	29.10	0.0333	11.52	30.12	0.0356	11.56
	EM	26.15	0.0320	12.65	21.59	0.0247	10.96	26.44	0.0313	13.92
	SPLINE	25.18	0.0308	7.80	21.06	0.0241	6.51	14.99	0.0177	5.25
	GTD	15.73	0.0193	4.74	11.78	0.0135	3.82	17.42	0.0206	5.85
	iGTD	15.09	0.0185	4.25	9.80	0.0112	3.37	16.63	0.0197	5.13
5 points	Regression	28.84	0.0307	11.90	55.71	0.0612	30.84	39.92	0.0429	20.31
	EM	25.54	0.0272	13.35	39.99	0.0439	24.38	40.80	0.0439	25.78
	SPLINE	47.33	0.0505	21.45	15.21	0.0167	7.00	33.42	0.0359	19.18
	GTD	20.12	0.0215	6.70	7.89	0.0087	3.93	6.67	0.0072	2.70
	iGTD	15.03	0.0160	5.97	7.09	0.0078	3.55	5.15	0.0055	2.16

TABLE 3: Imputation accuracies with different methods (coke oven's BFG consumption).

Missing number	Methods	Group 1			Group 2			Group 3		
		RMSE	NRMSE	MAPE	RMSE	NRMSE	MAPE	RMSE	NRMSE	MAPE
3 points	Regression	16.73	0.0300	7.90	10.33	0.0185	3.56	9.37	0.0167	3.24
	EM	9.95	0.0178	5.43	7.25	0.0130	4.05	3.72	0.0066	2.40
	SPLINE	18.77	0.0336	7.73	16.81	0.0301	7.79	22.64	0.0402	9.49
	GTD	15.83	0.0283	8.30	15.82	0.0283	7.47	15.67	0.0278	7.49
	iGTD	13.94	0.0250	6.76	13.76	0.0246	6.55	13.74	0.0244	7.30
4 points	Regression	16.66	0.0260	7.81	11.10	0.0171	4.18	10.82	0.0168	4.49
	EM	10.32	0.0161	5.47	6.72	0.0104	3.80	6.71	0.0104	3.16
	SPLINE	15.77	0.0246	6.50	38.26	0.0591	9.33	14.97	0.0231	6.37
	GTD	15.93	0.0249	7.44	19.75	0.0305	10.55	29.51	0.0459	13.58
	iGTD	13.81	0.0216	6.07	19.36	0.0299	9.39	13.02	0.0203	6.51
5 points	Regression	13.24	0.0186	5.73	13.24	0.0184	5.55	15.98	0.0227	7.78
	EM	10.34	0.0145	5.31	7.52	0.0104	4.20	10.78	0.0153	5.27
	SPLINE	16.90	0.0237	9.06	21.57	0.0300	8.51	13.85	0.0196	6.20
	GTD	15.49	0.0217	8.33	22.83	0.0317	11.00	20.23	0.0287	11.48
	iGTD	11.81	0.0166	5.45	19.84	0.0276	9.02	19.27	0.0273	9.45

mentionable that the effectiveness of the proposed iGTD still does better than GTD in this test.

(3) BFG Generation Amount (Normal Time Series). The experimental data are from number 1 blast furnace in Shanghai Baosteel randomly selected from 02:28:00/27/03/2010 to 18:33:00/01/04/2010. And, the comparative accuracies are listed in Table 4.

From Table 3, we can discover that the regression method presents the worst performance, while iGTD obtains the best one. For the data with normal property of time series, iGTD is better than GTD, while GTD wins all the other three methods.

A conclusion can be drawn from Tables 2–4 that the proposed iGTD and the GTD are superior to regression, EM, and spline for the periodicity-like data and the normal time-series-like data. As for the data with concussive amplitude, both iGTD and GTD do not have an advantage, and yet iGTD still beats GTD which means the proposed Gaussian membership function is superior to the triangular one in the real industrial manufacturing process. And, for the visual imputation results of the BFG generation and consumption, the comparative imputation curves are randomly chosen as Figures 11, 12, and 13 show, where the advantage of the method proposed in this study can be easily presented.

5. Conclusion

This study aims at the imputation of missing data of gas flow in steel industry. In order to improve the imputation accuracy,

TABLE 4: Imputation accuracies with different methods (BFG output).

Missing number	Methods	Group 1			Group 2			Group 3		
		RMSE	NRMSE	MAPE	RMSE	NRMSE	MAPE	RMSE	NRMSE	MAPE
3 points	Regression	108.17	0.0598	21.94	97.04	0.0476	17.11	109.20	0.0537	16.60
	EM	87.43	0.0483	18.04	83.43	0.0409	15.38	71.65	0.0352	10.72
	SPLINE	92.99	0.0514	17.55	86.74	0.0437	15.41	129.67	0.0638	17.59
	GTD	75.21	0.0416	16.06	70.73	0.0347	12.50	56.69	0.0279	9.18
	iGTD	69.70	0.0385	14.74	68.64	0.0336	12.26	55.48	0.0273	9.32
4 points	Regression	98.25	0.0457	17.99	93.09	0.0394	15.45	95.38	0.0408	15.49
	EM	77.94	0.0363	14.70	75.53	0.0319	13.09	73.74	0.0315	11.57
	SPLINE	128.22	0.0597	20.51	85.83	0.0363	14.29	66.95	0.0286	10.86
	GTD	75.70	0.0352	14.21	63.36	0.0268	10.47	62.31	0.0266	9.44
	iGTD	70.91	0.0330	13.15	63.25	0.0267	10.32	41.51	0.0178	6.74
5 points	Regression	88.76	0.0362	17.09	79.13	0.0296	11.22	76.24	0.0289	11.47
	EM	72.61	0.0296	12.98	71.74	0.0268	11.98	71.39	0.0270	10.97
	SPLINE	97.83	0.0399	16.72	105.18	0.0393	16.78	102.59	0.0389	15.60
	GTD	67.44	0.0275	11.90	61.00	0.0228	9.63	67.23	0.0255	10.47
	iGTD	67.19	0.0274	11.95	57.75	0.0216	9.29	62.95	0.0238	10.14

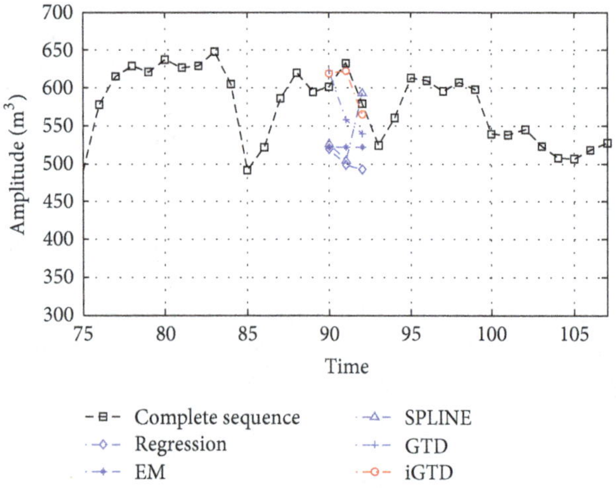

FIGURE 11: Comparison of different methods (3 points).

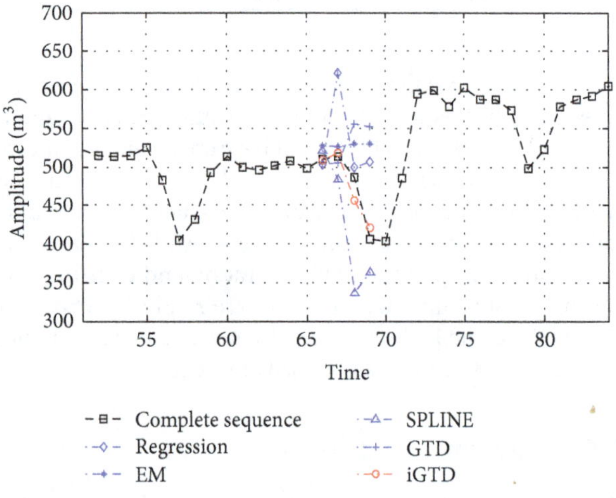

FIGURE 12: Comparison of different methods (4 points).

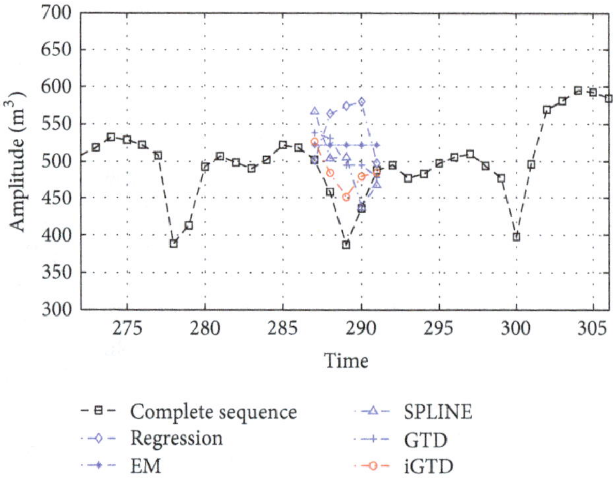

FIGURE 13: Comparison of different methods (5 points).

the proposed iGTD replaces the triangular membership function with the Gaussian one. Furthermore, the order of imputation is further discussed. The verification experiments show that the BSTM order brings less error than the chronological one does, since more observed data are utilized. As for the different data imputation method, compared to the original GTD, EM, regression, and spline, the proposed iGTD has some advantages in the problem with data properties of consecutively missing and small samples. And, the satisfying imputation accuracy provides the powerful support for the gas resources scheduling later. On the other hand, although the approach developed in this study can handle some types of missing in real industry, some theoretical analyses and the expanded application, for example, the type of concussive flow data, need to be given a further consideration in the future.

Acknowledgments

This work is supported by the National Natural Sciences Foundation of China (no. 61034003, no. 61104157, and no. 61273037) and the Fundamental Research Funds for the Central Universities of China (no. DUT11RC(3)07). The cooperation of energy center of Shanghai Baosteel Co. Ltd, China, in this work is greatly appreciated.

References

[1] J. Han and M. Kamber, *Data Mining: Concepts and Techniques*, Elsevier, 2006.

[2] P. C. Austin and M. D. Escobar, "Bayesian modeling of missing data in clinical research," *Computational Statistics and Data Analysis*, vol. 49, no. 3, pp. 821–836, 2005.

[3] S. Y. Yi, S. Jung, and J. Suh, "Better mobile client's cache reusability and data access time in a wireless broadcast environment," *Data and Knowledge Engineering*, vol. 63, no. 2, pp. 293–314, 2007.

[4] K. Lakshminarayan, S. A. Harp, and T. Samad, "Imputation of missing data in industrial databases," *Applied Intelligence*, vol. 11, no. 3, pp. 259–275, 1999.

[5] R. Lenz and M. Reichert, "IT support for healthcare processes—premises, challenges, perspectives," *Data and Knowledge Engineering*, vol. 61, no. 1, pp. 39–58, 2007.

[6] A. Mercatanti, "Analyzing a randomized experiment with imperfect compliance and ignorable conditions for missing data: theoretical and computational issues," *Computational Statistics and Data Analysis*, vol. 46, no. 3, pp. 493–509, 2004.

[7] T. Özacar, Ö. Öztürk, and M. O. Ünalir, "ANEMONE: An environment for modular ontology development," *Data and Knowledge Engineering*, vol. 70, no. 6, pp. 504–526, 2011.

[8] A. C. Favre, A. Matei, and Y. Tillé, "A variant of the Cox algorithm for the imputation of non-response of qualitative data," *Computational Statistics and Data Analysis*, vol. 45, no. 4, pp. 709–719, 2004.

[9] A. F. de Winter, A. J. Oldehinkel, R. Veenstra, J. A. Brunnekreef, F. C. Verhulst, and J. Ormel, "Evaluation of non-response bias in mental health determinants and outcomes in a large sample of pre-adolescents," *European Journal of Epidemiology*, vol. 20, no. 2, pp. 173–181, 2005.

[10] P. A. Patrician, "Multiple imputation for missing data," *Research in Nursing and Health*, vol. 25, no. 1, pp. 76–84, 2002.

[11] J. Honaker and G. King, "What to do about missing values in time-series cross-section data," *American Journal of Political Science*, vol. 54, no. 2, pp. 561–581, 2010.

[12] N. Sartori, A. Salvan, and K. Thomaseth, "Multiple imputation of missing values in a cancer mortality analysis with estimated exposure dose," *Computational Statistics and Data Analysis*, vol. 49, no. 3, pp. 937–953, 2005.

[13] J. M. Lepkowski, W. D. Mosher, K. E. Davis, R. M. Groves, and J. Van Hoewyk, "The 2006–2010 National Survey of Family Growth: sample design and analysis of a continuous survey," *Vital and Health Statistics*, no. 150, pp. 1–36, 2010.

[14] I. Myrtveit, E. Stensrud, and U. H. Olsson, "Analyzing data sets with missing data: an empirical evaluation of imputation methods and likelihood-based methods," *IEEE Transactions on Software Engineering*, vol. 27, no. 11, pp. 999–1013, 2001.

[15] A. L. Bello, "Imputation techniques in regression analysis: looking closely at their implementation," *Computational Statistics and Data Analysis*, vol. 20, no. 1, pp. 45–57, 1995.

[16] D. A. Newman, "Longitudinal modeling with randomly and systematically missing data: a simulation of Ad Hoc, maximum likelihood, and multiple imputation techniques," *Organizational Research Methods*, vol. 6, no. 3, pp. 328–362, 2003.

[17] R. H. Jones, "Maximum likelihood fitting of ARMA models to time series with missing observations," *Technometrics*, vol. 22, no. 3, pp. 389–395, 1980.

[18] G. Jagannathan and R. N. Wright, "Privacy-preserving imputation of missing data," *Data and Knowledge Engineering*, vol. 65, no. 1, pp. 40–56, 2008.

[19] J. Van Hulse and T. Khoshgoftaar, "Knowledge discovery from imbalanced and noisy data," *Data and Knowledge Engineering*, vol. 68, no. 12, pp. 1513–1542, 2009.

[20] A. Bagheri and M. Zandieh, "Bi-criteria flexible job-shop scheduling with sequence-dependent setup times—variable neighborhood search approach," *Journal of Manufacturing Systems*, vol. 30, no. 1, pp. 8–15, 2011.

[21] M. Zhang, J. Zhu, D. Djurdjanovic, and J. Ni, "A comparative study on the classification of engineering surfaces with dimension reduction and coefficient shrinkage methods," *Journal of Manufacturing Systems*, vol. 25, no. 3, pp. 209–220, 2006.

[22] M. S. Pishvaee, F. Jolai, and J. Razmi, "A stochastic optimization model for integrated forward/reverse logistics network design," *Journal of Manufacturing Systems*, vol. 28, no. 4, pp. 107–114, 2009.

[23] P. A. Ferrari, P. Annoni, A. Barbiero, and G. Manzi, "An imputation method for categorical variables with application to nonlinear principal component analysis," *Computational Statistics and Data Analysis*, vol. 55, no. 7, pp. 2410–2420, 2011.

[24] T. H. Lin, "A comparison of multiple imputation with EM algorithm and MCMC method for quality of life missing data," *Quality and Quantity*, vol. 44, no. 2, pp. 277–287, 2010.

[25] S. Van Buuren, H. C. Boshuizen, and D. L. Knook, "Multiple imputation of missing blood pressure covariates in survival analysis," *Statistics in Medicine*, vol. 18, pp. 681–694, 1999.

[26] K. Hron, M. Templ, and P. Filzmoser, "Imputation of missing values for compositional data using classical and robust methods," *Computational Statistics and Data Analysis*, vol. 54, no. 12, pp. 3095–3107, 2010.

[27] F. Dudbridge, "Likelihood-based association analysis for nuclear families and unrelated subjects with missing genotype data," *Human Heredity*, vol. 66, no. 2, pp. 87–98, 2008.

[28] Z. Lu and Y. V. Hui, "L_1 linear interpolator for missing values in time series," *Annals of the Institute of Statistical Mathematics*, vol. 55, no. 1, pp. 197–216, 2003.

[29] D. Decoste and B. Schölkopf, "Training invariant support vector machines," *Machine Learning*, vol. 46, no. 1–3, pp. 161–190, 2002.

[30] D. Li, L. Chen, and Y. Lin, "Using Functional Virtual Population as assistance to learn scheduling knowledge in dynamic manufacturing environments," *International Journal of Production Research*, vol. 41, no. 17, pp. 4011–4024, 2003.

[31] Y. S. Lin and D. C. Li, "The Generalized-Trend-Diffusion modeling algorithm for small data sets in the early stages of manufacturing systems," *European Journal of Operational Research*, vol. 207, no. 1, pp. 121–130, 2010.

[32] K. Goto, H. Okabe, F. A. Chowdhury, S. Shimizu, Y. Fujioka, and M. Onoda, "Development of novel absorbents for CO_2 capture from blast furnace gas," *International Journal of Greenhouse Gas Control*, vol. 5, no. 5, pp. 1214–1219, 2011.

[33] H. Jaeger, "A tutorial on training recurrent neural networks, covering BPTT, RTRL, EKF and "Echo State Network"

approach," GMD Report 159, German National Research Center for Information Technology, Berlin, German, 2002.

[34] H. Jaeger, "Adaptive nonlinear system identification with echo state networks," *Advances in Neural Information Processing Systems*, vol. 15, pp. 593–600, 2003.

[35] C. F. Huang, "Principle of information diffusion," *Fuzzy Sets and Systems*, vol. 91, no. 1, pp. 69–90, 1997.

[36] E. P. Zhou and D. K. Harrison, "Improving error compensation via a fuzzy-neural hybrid model," *Journal of Manufacturing Systems*, vol. 18, no. 5, pp. 335–344, 1999.

[37] G. Ward, T. Hastie, S. Barry, J. Elith, and J. R. Leathwick, "Presence-only data and the EM algorithm," *Biometrics*, vol. 65, no. 2, pp. 554–563, 2009.

19

Application of MM5-CAMx-PSAT Modeling Approach for Investigating Emission Source Contribution to Atmospheric SO_2 Pollution in Tangshan, Northern China

Li Li,[1,2] Shuiyuan Cheng,[1] Jianbing Li,[3] Jianlei Lang,[1] and Dongsheng Chen[1]

[1] *College of Environmental & Energy Engineering, Beijing University of Technology, Beijing 100124, China*
[2] *Beijing General Research Institute of Mining & Metallurgy, Beijing 100070, China*
[3] *Environmental Engineering Program, University of Northern British Columbia, Prince George, Canada V2N 4Z9*

Correspondence should be addressed to Shuiyuan Cheng; bjutpaper@gmail.com

Academic Editor: Guohe Huang

The MM5-CMAx-PSAT modeling approach was presented to identify the variation of emission contribution from each modeling grid to regional and urban air quality per unit emission rate change. The method was applied to a case study in Tangshan Municipality, a typical industrial region in northern China. The variation of emission contribution to the monthly atmospheric SO_2 concentrations in Tangshan from each modeling grid of 9 × 9 km per 1000 t/yr of emission rate change was simulated for four representative months in 2006. It was found that the northwestern part of Tangshan region had the maximum contribution variation ratio (i.e., greater than 0.36%) to regional air quality, while the lowest contribution variation ratio (i.e., less than 0.3%) occurred in the coastal areas. Principal component analysis (PCA), canonical correlation analysis (CCA), and Pearson correlation analysis indicated that there was an obvious negative correlation between the grid-based variation of emission contribution to regional air quality and planetary boundary layer height (PBLH) as well as wind speed, while terrain data presented insignificant impacts on emission contribution variation. The proposed method was also applied to analyze the variation of emission contribution to the urban air quality of Tangshan (i.e., a smaller scale).

1. Introduction

Air pollution is a serious environmental problem faced by many industrial cities in China as a consequence of many years' rapid economic expansion and insufficient environmental protection measures. It not only poses threats to human health, but also directly affects local economic development [1]. A variety of factors, such as emission sources, land surface characteristics, and meteorological conditions, could affect air pollution simulation. Thus, effective air quality management is usually a challenging task. To tackle such difficulties, it is of crucial importance to quantify the impacts of pollutant emission sources on the air quality of a planning region and understand the corresponding response of atmospheric pollutant concentration to perturbations in pollutant emission rate [2].

Previously, the method of wind rose based on wind speed and direction has been used for qualitatively investigating the impacts of emission sources on regional air quality [3]. Nowadays, computer modeling tools have been recognized as useful means to investigate such impacts [4]. Particularly, there has been a growing interest of applying advanced 3-D chemistry-transport models coupled with meteorological models for air quality studies, such as the Model-3 Community Multiscale Air Quality (Model-3/CMAQ) [5, 6], the Comprehensive Air quality Model with extensions

(CAMx) [7, 8], the PSU/NCAR mesoscale meteorological model MM5 [9], and the Weather Research and Forecast (WRF) model coupled with Chemistry (WRF-Chem) [10]. For example, Cheng et al. [11] used the coupled MM5-ARPS-CMAQ to examine contributions of various emission sources to ambient PM_{10} concentrations in Beijing, China; Titov et al. [12] applied a MM5-CAMx for predicting PM_{10} concentrations over the city of Christchurch in New Zealand during critical pollution episodes; Lee et al. [13] employed MM5-CAMx to simulate atmospheric pollutant transport and recirculation in the Santa Claria valley, USA; Shimadera et al. [14] applied MM5-CMAQ to estimate the contribution of transboundary transport of air pollutants from other Asian countries to Japan; Borrego et al. [15] applied MM5-CAMx to simulate surface concentrations of ozone and its precursors over the metropolitan area of Porto Alegre, Brazil, and identified the main emission sources of photochemical pollution.

In terms of examining the response of atmospheric pollutant concentration to perturbations in pollutant emission rate, a number of approaches have been proposed in the past years by using various models [4, 11, 16, 17]. Particularly, a technique named particulate matter source apportionment technology (PSAT) [18] has been implemented in CAMx to provide source apportionment for primary and secondary particulate matter (PM) species according to emission source categories and their geographic locations [19]. This technique is useful for identifying emission sources that significantly contribute to gaseous or PM pollution. For example, Wagstrom et al. [20] used PSAT to investigate the contribution of power plant SO_2 emissions to particulate sulfate concentrations in the Eastern United States, and the results illustrated that PSAT could provide a computationally efficient particulate matter apportionment algorithm to investigate pollutant transport and emission source contributions on regional scales; Koo et al. [21] compared two different methods of investigating relationships between PM concentrations and emission sources and found that PSAT was best at apportioning sulfate, nitrate, and ammonium to sources emitting SO_2, NO_x, and NH_3, respectively. In addition, there are some other methods to examine emission source apportionment [22–24], which used principal component analysis (PCA) and multilinear regression analysis (MLRA) to identify possible sources of particulate matter (PM) and to determine their contribution to air pollution.

In general, many of the previous source apportionment and emission contribution analysis works focused on examining relationship between the total emission amount of a source from a large-scale planning region and its air pollutant concentration [25]. In fact, for air pollution control strategy development, the more practical question is how pollutant concentrations would respond to emission changes within different small-scale areas of a large planning region [2]. The contribution of emission sources within different small-scale areas to regional and urban air quality could be quite different due to different land surface and meteorological conditions. Thus, it is of critical importance to identify the variation of atmospheric concentration to perturbations in emission rates of small-scale areas within a large planning region. The priority regulation of emissions with high contribution variation could result in significantly environmental and cost effectiveness. As an extension of our previous efforts, this study was focused on the establishment and application of the MM5-CAMx-PAST modeling approach for examining air quality variation due to perturbation in emission rates from small-scale areas within a large planning region, and it analyzed the possible affected factors. More accurate results can be obtained with the development of the advanced model simulation. The approach and results can provide sound decision making basis for effective air quality management. A case study for Tangshan, a typical industrial region in China, was presented to illustrate the proposed methodology. The MM5-CAMx was used to provide meteorological inputs and to simulate atmospheric SO_2 concentrations, and PSAT was applied to investigate the emission contribution variations. An air quality modeling domain with a spatial resolution of 9 by 9 km was adopted, and the regional and urban air quality variations due to SO_2 emission rate perturbation of 1000 t/yr within each modeling grid were simulated. Principal component analysis (PCA), canonical correlation analysis (CCA), and Pearson correlation analysis methods were then used to analyze the impacts of meteorological variables and terrain data on the emission source contribution variations.

2. Overview of the Study Area

Tangshan Municipality, located at about 300 km east of Beijing, is the biggest industrial center within Hebei province in northern China. It has a total population of 6.9 million in 2000 and a total area of 13,472 km^2 including 12 districts as shown in Figure 1. The municipality is situated on the alluvial plain formed by the diluvial sediments from the Yan Mountains in the north. Its mean sea level tends to decrease gradually from its northwest to southeast towards the Bohai Bay. It has a temperate continental climate influenced by wet monsoon, and there is an apparent distinction among four seasons, that is, windy and dry spring, hot and wet summer, mild and clear autumn, and cold and dry winter. The annual average temperature is 10–11.20°C, and annual average precipitation is about 600 mm. As one of the biggest industrial centers in northern China, Tangshan Municipality has experienced considerable changes through rapid industrialization and urbanization processes in the past decades. However, its growth has also been associated with a number of environmental concerns. Among them, the deteriorated air quality due to a combination of circumstances (i.e., increased energy consumption, population growth, increased industrial emissions, infrastructure construction/expansion, growth of passenger vehicles, and ineffective pollution control measures) posed significant challenges to the public, governments, and industries. Particularly, SO_2 pollution has been recognized as an important environmental issue.

3. Methodology

3.1. MM5-CAMx-PSAT Modeling. The fifth-generation NCAR/Penn State mesoscale meteorological model (MM5)

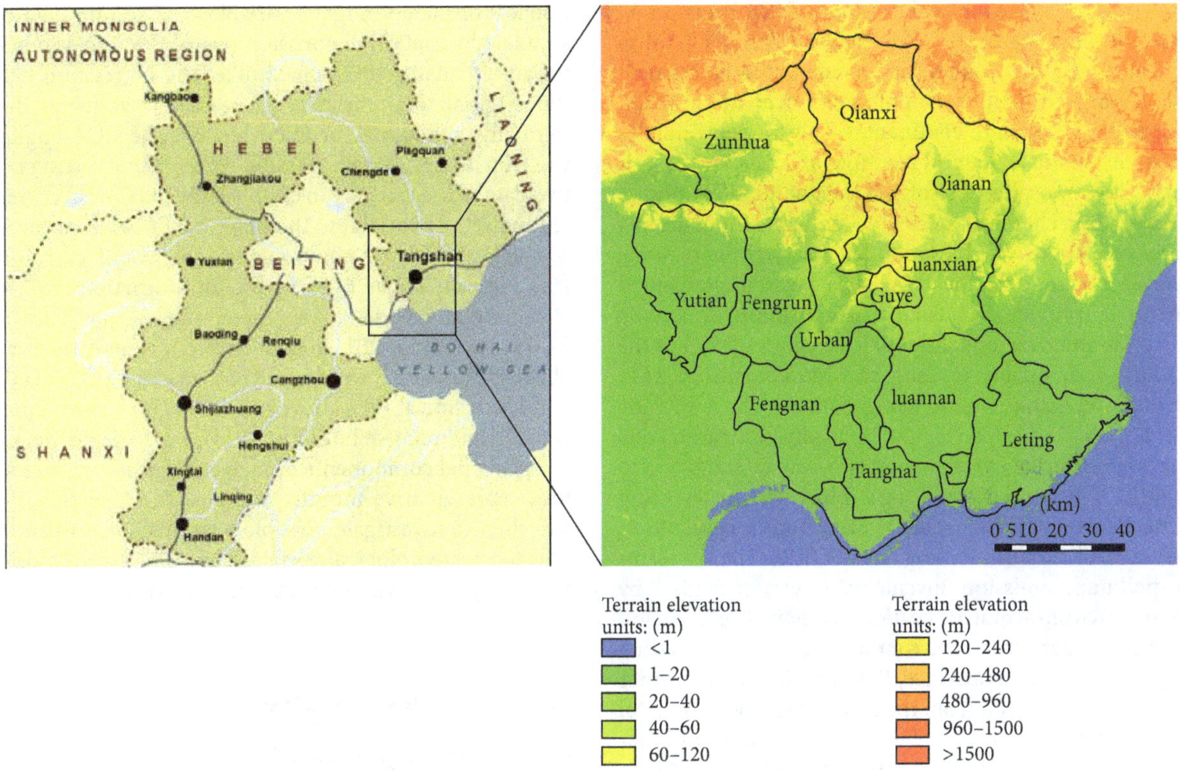

FIGURE 1: Tangshan Municipality and its surrounding cities.

is a limited-area, nonhydrostatic, terrain-following sigma-coordinate model designed to predict meso- and regional-scale atmospheric circulations [9]. It has been frequently used to provide meteorological inputs for many air quality modeling systems [26]. In this study, MM5 (version 3.7) model was applied and configured using two-level nested modeling domains (112–120°E, 37–43°N) as shown in Figure 2, where domain 1 has a spatial resolution of 27 km by 27 km and has been established with a dimension of 60 × 60 grid cells, and domain 2 has a spatial resolution of 9 by 9 km and has been established with a dimension of 94 × 82 grid cells. Twenty-four full σ levels extending from the ground surface to the top of modeling domain (i.e., 200 hpa) were applied. The 3-D first-guess meteorological fields for modeling were obtained from the Global Tropospheric Analyses datasets provided by the US National Center for Environmental Prediction (NCEP FNL data) and were available with six-hour resolution on a grid of 1°× 1°. The four-dimensional data assimilation (FDDA) was implemented using the meteorological observations from surface (eight times a day) and upper air (two times a day) monitoring stations of the Chinese Meteorological Information Comprehensive Analysis and Process System (MICAPS). The following physical parameters schemes in MM5 were selected, including (a) land-use scheme using five-layer LSM, (b) PBL scheme using medium-range forecasts (MRF), (c) cloud microphysics selecting mixed-phase, (d) cumulus parameterization schemes selecting Grell, and (e) radiation schemes selecting the highly accurate and efficient method (RRTM). The terrain and land-use data were obtained from USGS with a spatial resolution of 30 s.

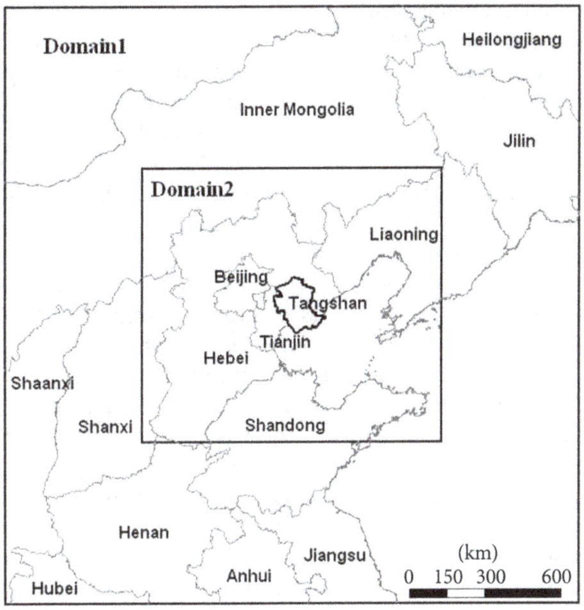

FIGURE 2: Two-level nested modeling domain for MM5.

The comprehensive air quality model CAMx version 5.1 was used in this study. It is an Eulerian photochemical dispersion model that allows for an integrated assessment of gaseous and particulate air pollutants over many scales ranging from suburban to continental. This model simulates emission, dispersion, chemical reaction, and removal of pollutants in the troposphere by solving the pollutant continuity equation for each chemical species. Its modeling input file formats are compatible with MM5 model. To study the regional emission contributions, the PSAT has been implemented in CAMx to provide SO_2 source apportionment among specific geographic regions and source categories [19]. For the simulation of air quality in Tangshan Municipality, CAMx was configured using one modeling domain which was the same as domain 2 of MM5 (Figure 2). Its physical parameters schemes were selected as follows: (a) two-way interactive grid nesting, (b) 12 vertical layers, (c) gas-phase chemistry using CB05 mechanism which includes 156 reactions formulations, and (d) aerosol chemistry using M4/ISORROPIA. In terms of air pollutant emission inventory, it was provided by Tangshan Environmental Protection Agency. The emission inventories of Tangshan's surrounding regions, including Hebei province, Shanxi province, Beijing, Tianjin, and Inner Mongolia, were obtained from the respective environmental protection administrations. The emission inventory of other regions was obtained from Zhang and Streets [27].

The MM5-CAMx was then used to simulate SO_2 concentrations in Tangshan for four representative months in 2006, including January, April, July, and October. Two scenarios were selected, including simulating the variation of emission contribution from emission rate perturbation in each modeling grid (9 × 9 km scale) to both regional and urban air quality, represented by the monthly average SO_2 concentration of the entire Tangshan region (i.e., large-scale receptor 1) and only its urban area (i.e., small-scale receptor 2) (Figure 3), respectively. The modeling procedures include (1) using MM5-CMAx to predict the temporal and spatial distributions of SO_2 concentrations within Tangshan Municipality based on its actual emissions in 2006 (i.e., base emission inventory), and the monthly average SO_2 concentrations within receptors 1 and 2 were then calculated based on the simulated hourly concentrations, respectively; (2) identifying the contribution variation of each modeling grid to the monthly average SO_2 concentrations through adding 1000 t/yr of SO_2 emission (i.e., an arbitrarily selected number) to each grid in addition to the base emission inventory, and the MM5-CMAx was used to predict the temporal and spatial distributions of SO_2 concentrations within Tangshan Municipality based on the new emission inventory (base emission inventory plus 1000 t/yr in a certain grid), and then the monthly average SO_2 concentrations within receptors 1 and 2 were calculated, respectively. The difference between the monthly average SO_2 concentrations calculated using base inventory and new inventory is regarded as the emission source contribution variation of that grid.

3.2. Multivariate Analysis. The multivariate analysis methods, including PCA and CCA, were used to analyze the impacts of meteorological variables and terrain data on the simulated variation of emission contribution to regional and urban air quality. PCA maximizes the correlation between the original total variance to form new variables that are mutually orthogonal, or uncorrelated. The CCA application was run to investigate possible relationship between these two data sets, especially to establish the maximum correlation among sets of variables. The objective of PCA was to obtain a small number of components that would explain most (i.e., typically above 60%) of the total variation [28]. In this study, the hourly data of six meteorological variables within MM5, including PBL height (PBLH), temperature at 2 m above ground (T2), wind speed at 10 m above ground (WS10), wind direction at 10 m above ground (WD), sea level pressure (PSLV), and relative humidity (RH), were selected to analyze the principal components of meteorological variables within four representative months in 2006. The objective of CCA was then to investigate possible relationship between the six selected meteorological variables as well as terrain data and the contribution variation of emission within each modeling grid [29].

4. Results and Discussions

4.1. Modeling Performance. The performance of the MM5-CAMx was evaluated using scatter plots [30]. The ground-based SO_2 observation results from three air quality monitoring stations located within Tangshan urban ("Urban" is showed in Figure 1) were averaged and were then compared with the predicted daily SO_2 concentration of the Tangshan urban area in the four selected months in 2006. Figure 4 displays the comparison results. The $y = x$ line on the scatter plots represents perfect agreement between the two data sets. A pair value above the $y = x$ line indicates a situation of overprediction, while the pair value below the line indicates underprediction. In general, Figure 4 shows that most of the scatter plots are adjacently distributed on both sides of $y = x$ line, which does highlight a consistent over- and underprediction for SO_2 concentration using the modeling system. Considering the inherent uncertain nature associated with meteorological parameters and air quality prediction, this fluctuation still indicates that the accuracy of model prediction is reasonable. In fact, the correlation coefficients between simulated and observed data were calculated as 0.781, 0.621, 0.690, and 0.801 for January, April, July, and October, respectively. Thus the performance of the coupled modeling system is satisfactory and acceptable [4].

4.2. Simulated SO_2 Concentration Distribution Using Base Emission Inventory. In the year of 2006, Tangshan Municipality had a total of 598 industrial establishments, including electrical, metallurgical, mining, chemical, construction materials, and textile industries. Spatial distributions and emission rates of SO_2 from these sources were investigated and shown in Figure 5(a). The hourly SO_2 concentrations in January 2006 in the entire Tangshan region were simulated using MM5-CAMx, and their corresponding monthly averages were then calculated. Figure 5(b) displays the simulated

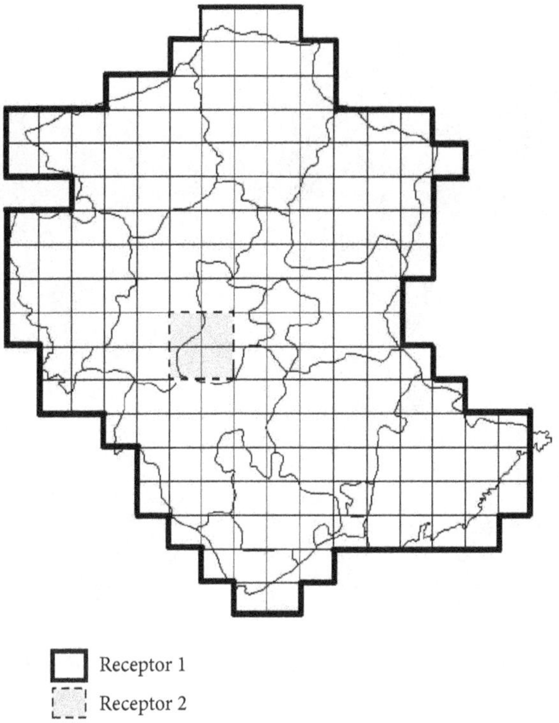

Figure 3: Schematic of modeling scenarios showing receptors 1 and 2.

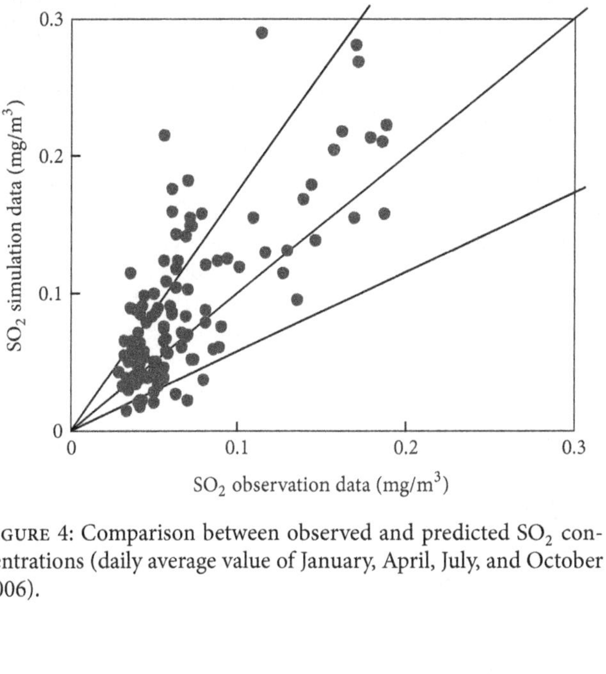

Figure 4: Comparison between observed and predicted SO_2 concentrations (daily average value of January, April, July, and October 2006).

SO_2 concentration in Tangshan region in January 2006, and it illustrates that SO_2 pollutions occurred in most areas of Tangshan region due to pollutant emissions and unfavorable meteorological conditions. In general, the air quality in Tangshan region was not satisfactory, and the municipal government and industries need to take actions to improve such situation. For cost-effective air quality management in Tangshan, the identification of the variation of emission source contribution to the regional and urban air quality due to emission perturbation in each small-scale emission area is of fundamental importance. Such information could provide sound basis for identifying emission areas requiring priority regulation.

4.3. Variation of Seasonal Emission Source Contribution to Regional Air Quality. The variation of emission contribution to the receptor 1 (shown in Figure 3) air quality due to SO_2 emission perturbation of 1000 t/yr of each modeling grid was calculated at first for the four representative months in 2006 using the MM5-CAMx-PSAT. Then, the corresponding monthly arithmetic averages were plotted using Geographic Information System (GIS) interpolation method. Figure 6 presents the spatial distribution of the variation of emission contribution to regional monthly SO_2 concentration in Tangshan. It is observed from Figure 6(a) that emissions in the junction of Yutian and Zunhua Counties had the maximum variation of emission contribution to air pollution in receptor 1, with more than 52 μg/m^3 of variation per 1000 t/yr of SO_2 emission per 9×9 km modeling grid in January, while emissions from the coastal areas of the southeast of Tangshan had the minimum variation of emission contribution (i.e., less than 35 μg/m^3 of variation per 1000 t/yr of SO_2 emission per modeling grid of 9×9 km). It can also be found that the grid-based variation of emission contribution to SO_2 pollution tended to gradually decrease from the northwest to southeast of Tangshan Municipality in January. In April, as shown in Figure 6(b), emissions from the southeast of Qianan County and the southern coastal areas of Tangshan displayed the maximum variations of contribution to SO_2 pollution in receptor 1, with more than 13.9 μg/m^3 of contribution variation per 1000 t/yr of SO_2 emission per grid. Meanwhile, emissions from the eastern area of Leting and northern part of Qianxi County showed minimum contribution variations. In terms of July, as shown in Figure 6(c), emissions from the coastal areas of Tangshan made the largest contribution variations (i.e., greater than 14.6 μg/m^3 of variation per 1000 t/yr of SO_2 emission per grid), while the spatial distribution of emission contribution variations showed several local high-value points, and the minimum contribution variation occurred in Qianxi County. It is shown in Figure 6(d) that the variation of emission contributions to the average SO_2 concentration in receptor 1 in October displayed a relatively even distribution, tending to gradually decrease from the high-value area of Yutian County (i.e., with contribution variation of greater than 26.0 μg/m^3) to the east and southeast of Tangshan. Consequently, the simulation results indicate that the largest variations of emission contribution to air pollution occurred in January, and the contribution variation distribution displayed an apparent seasonal difference. This is due to the fact that Tangshan has the temperate continental climate, and different meteorological conditions among four seasons would cause such seasonal differences.

4.4. Impacts of Meteorological Factors on Emission Source Contribution Variation. PCA was used to identify the principal

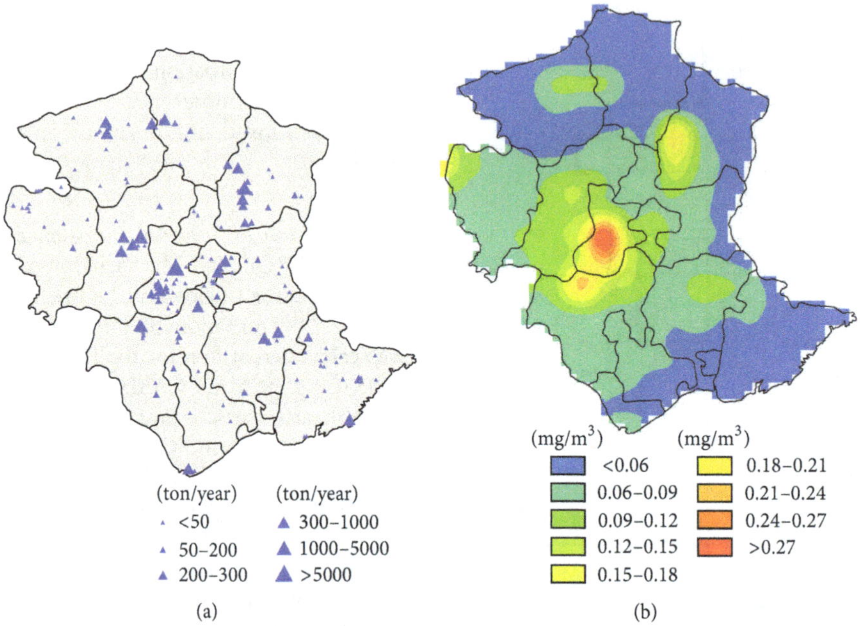

FIGURE 5: Annual emission rate of SO_2 and simulated monthly SO_2 concentration in January 2006: (a) emission rate; (b) concentration distribution.

components from six meteorological variables in Tangshan. Table 1 lists the PCA results for January, April, July, and October 2006, respectively, and the eigenvalues of PCA for the meteorological variables are also presented. It is found from Table 1 that examination of 30-day data for each modeling grid in January led to three principal components accounting for 81.8% of the total variance. Using the values of the respective principal component loadings presented in Table 1, there is a reasonable interpretation for these components. Only loadings with absolute values greater than 50% were selected for PC interpretation [31]. The first PC was PBL height (with component loadings of −0.503), and the second PCs showed that a main source of variation was wind speed (with component loadings of 0.529) and temperature (with component loadings of 0.598), while the third PC was wind direction (with component loadings of 0.801). Thus, the PCA results for January indicated low PBL height and prevalent northwest winds as well as inversion weather. These meteorological conditions could result in higher atmospheric stability in surface layer in Tangshan which then facilitated the accumulation of pollutants near the ground, leading to the highest variation of emission contributions to regional air quality from the modeling grids as compared to other months (Figure 6(a)). In terms of meteorological conditions in April, Table 1 illustrates that the first PCs were PBL height (with component loadings of 0.574) and relative humidity (with component loadings of −0.515). The second PCs were temperature (with component loadings of 0.715) and sea level pressure (with component loadings of −0.603), while the third PC was wind speed (with component loadings of 0.723). The PCA results for April indicate a dry spring with high PBLH, high temperature, low sea level pressure, and strong wind, and such meteorological conditions were conducive for dispersion of pollutants, leading to relatively low variation of emission contribution to regional air quality from modeling grids (Figure 6(b)). For meteorological conditions in July, the PCA results illustrate that the first PCs were PBL height (with component loadings of 0.603) and relative humidity (with component loadings of 0.549), and the second PC was wind direction (with component loadings of −0.698), while the third PC was wind speed (with component loadings of −0.832). The PCA results indicate a wet and rainy summer with high PBL height and prevalent southeast winds influenced by the maritime climate. Such meteorological conditions would help disperse and reduce pollutant concentrations, leading to minimum variation of emission contribution to regional air quality in July as compared to other months (Figure 6(c)). For October, the PCA results showed a mild and clear autumn, with first PCs being the temperature (with component loading of 0.52) and sea level pressure (with component loading of −0.576), the second PCs being PBLH (with component loading of 0.684) and wind speed (with component loading of 0.649), and the third PC being relative humidity (with component loading of −0.696). These values illustrate that the temperature in autumn was slightly higher than that in spring, wind was not stronger than that in spring, and the prevalently northwest wind was influenced by the invasion of cold air. Due to the impact of such meteorological conditions, the variation of emission contribution to regional air quality in October from modeling grids was between the minimum and maximum (Figure 6(d)).

Results of CCA between grid-based variation of emission contribution to regional air quality and meteorology-terrain data in Tangshan are presented in Table 2. In this study, there was only one canonical variable (CV). The correlations of

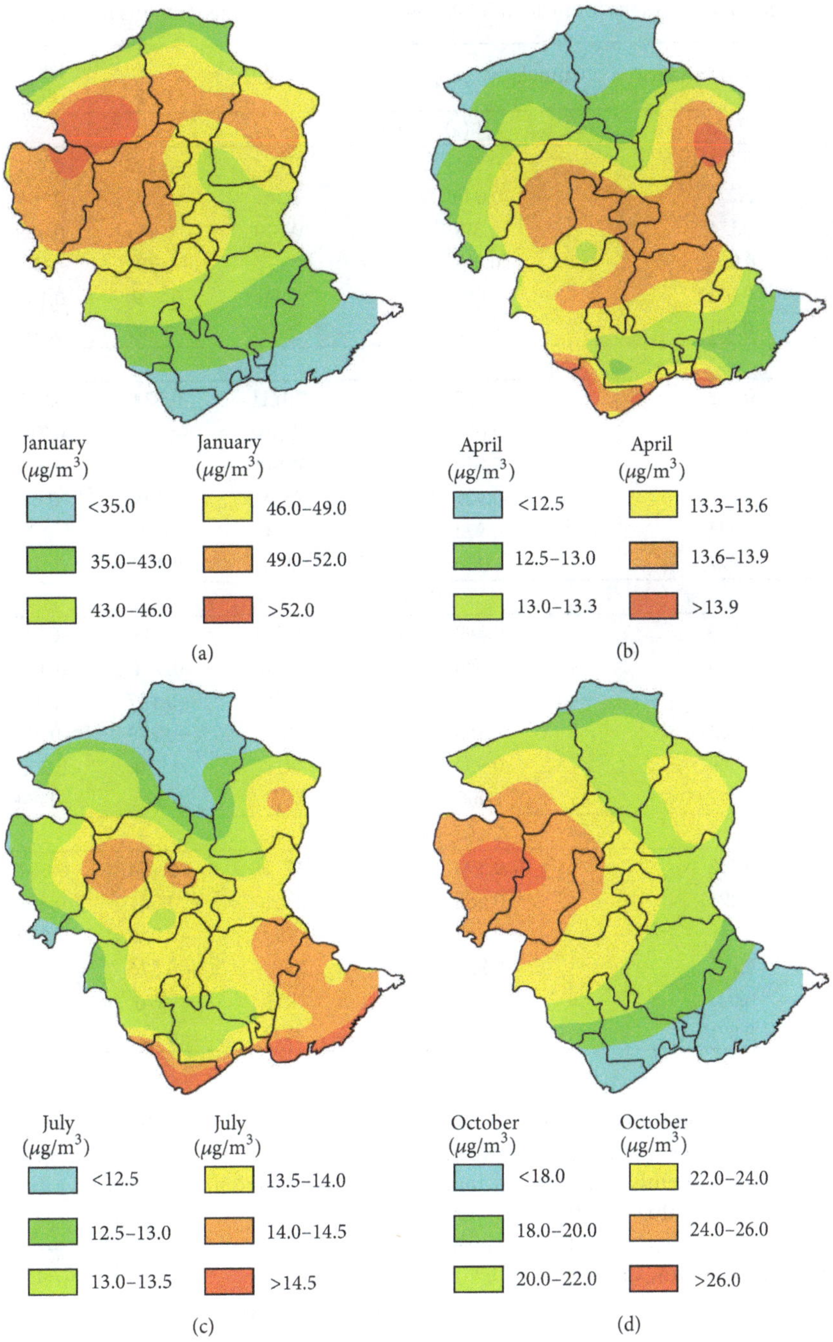

FIGURE 6: Simulated monthly average emission contribution response to regional average SO_2 concentration (i.e., receptor 1) due to SO_2 emission perturbation of 1000 t/yr.

CV1 were 0.781, 0.748, 0.725, and 0.807 for the four selected months, respectively, and all CCAs passed the statistical test of significance. According to the variable loading values shown in Table 2, the main meteorological variables were PBL height and wind speed in January which showed a negative correlation with grid-based variation of emission contribution to air quality in receptor 1. Pearson correlation analysis also gave the same results as CCA. Figure 7 presents the monthly average PBLH and WS10 in January, and the contours exhibit negative correlation with Figure 6(a). This indicates that high variation of emission contribution was related to low PBL height and low wind speed conditions. It is found from Table 2 that the variable loading values for April and October gave similar results for January. However, CCA and Pearson correlation analysis gave different results for July. The CCA results showed that relative humidity was associated

TABLE 1: PCA results for meteorological variables in four selected months in 2006.

PC	Eigenvalue	Proportion variance	Cumulative proportion	Variable	Principal component loadings			
					PC1	PC2	PC3	PC4
			(a) January					
PC1	1.492	0.371	0.371	PBLH	−0.503	0.421	−0.209	0.000
PC2	1.279	0.273	0.644	T2m	0.326	0.598	0.000	−0.26
PC3	1.022	0.174	0.818	WS10	−0.411	0.529	−0.132	0.28
PC4	0.772	0.099	0.918	PSLV	−0.391	−0.405	−0.488	0.21
PC5	0.554	0.051	0.969	RH	0.479	0.147	−0.244	0.799
PC6	0.433	0.031	1.000	WD	−0.297	0.000	0.801	0.412
			(b) April					
PC1	1.516	0.383	0.383	PBLH	0.574	0.207	−0.143	0.000
PC2	1.280	0.273	0.656	T2m	−0.106	0.715	−0.215	0.000
PC3	0.982	0.161	0.817	WS10	0.318	0.171	0.723	−0.56
PC4	0.813	0.110	0.927	PSLV	0.294	−0.603	−0.305	−0.23
PC5	0.523	0.046	0.973	RH	−0.515	−0.22	0.453	0.227
PC6	0.405	0.027	1.000	WD	0.455	0.000	0.336	0.758
			(c) July					
PC1	1.435	0.343	0.343	PBLH	0.603	−0.242	0.000	−0.141
PC2	1.182	0.233	0.576	T2m	0.484	0.337	−0.207	−0.465
PC3	1.056	0.186	0.762	WS10	−0.137	−0.171	−0.832	−0.33
PC4	0.785	0.103	0.865	PSLV	−0.279	−0.497	0.404	−0.71
PC5	0.729	0.089	0.953	RH	0.549	0.255	−0.188	0.000
PC6	0.529	0.047	1.000	WD	0.000	−0.698	−0.257	0.381
			(d) October					
PC1	1.535	0.392	0.392	PBLH	−0.134	0.684	−0.127	0.000
PC2	1.324	0.292	0.685	T2m	0.52	0.301	0.397	0.000
PC3	0.951	0.151	0.835	WS10	−0.226	0.649	−0.132	0.000
PC4	0.785	0.103	0.938	PSLV	−0.576	−0.138	−0.352	−0.13
PC5	0.498	0.041	0.979	RH	0.376	0.000	−0.696	0.607
PC6	0.351	0.021	1.000	WD	−0.433	0.000	0.448	0.781

with the second highest absolute loading value (i.e., −0.715) which indicated an obvious negative correlation between humidity and grid-based variation of emission contribution. However, the Pearson correlation value for RH was just −0.066. Since it is widely recognized that wet deposition has the function of removing pollutant, the results of CCA seemed more reasonable to find the relationship between more than two variables. The terrain data did not show obvious correlation with grid-based emission contribution variation through CCA and Pearson correlation analysis. This can be explained by the fact that most areas of Tangshan are flat although it is located in the alluvial plains of the Yanshan Mountains, with higher elevation in the northwestern part and lower elevation in the southeastern region.

4.5. Variation of Annual Emission Contribution to Regional Air Quality. The modeling results (Figure 6) indicated significant seasonal change of emission contribution variation for each modeling grid due to the impacts of many meteorological factors such as PBL height and wind speed. Thus, a parameter of emission contribution variation ratio was introduced in this study for investigating the variation of annual average emission contribution to regional air quality for the convenience of air quality management. The calculation of emission contribution variation ratio is as follows:

$$R_i = \frac{1}{4}\left|\sum \frac{C_{i,j}}{\sum_{i=1}^{n} C_{i,j}}\right|, \quad (1)$$

where $C_{i,j}$ is the variation of emission contribution to the monthly average SO_2 concentration of the receptor area in month j (i.e., January, April, July, and October) per 1000 t/yr of emission rate change in grid i ($\mu g/m^3$); R_i is the annual average emission contribution variation ratio of grid i due to 1000 t/yr of emission rate change; n is the total number of modeling grids. Figure 8(a) presents the annual emission contribution variation ratio of each grid to average SO_2 concentration in receptor 1 in 2006. It is found that the northwestern part of Tangshan such as the junction area of Yutian and Fengrun Counties had the maximum emission contribution variation ratio (i.e., greater than 0.36%) to the air quality of receptor 1, indicating that the regional air quality was more sensitive to the emissions from the northwestern part of Tangshan. The contribution variation ratio tended to decrease towards the north and southeast of Tangshan, while

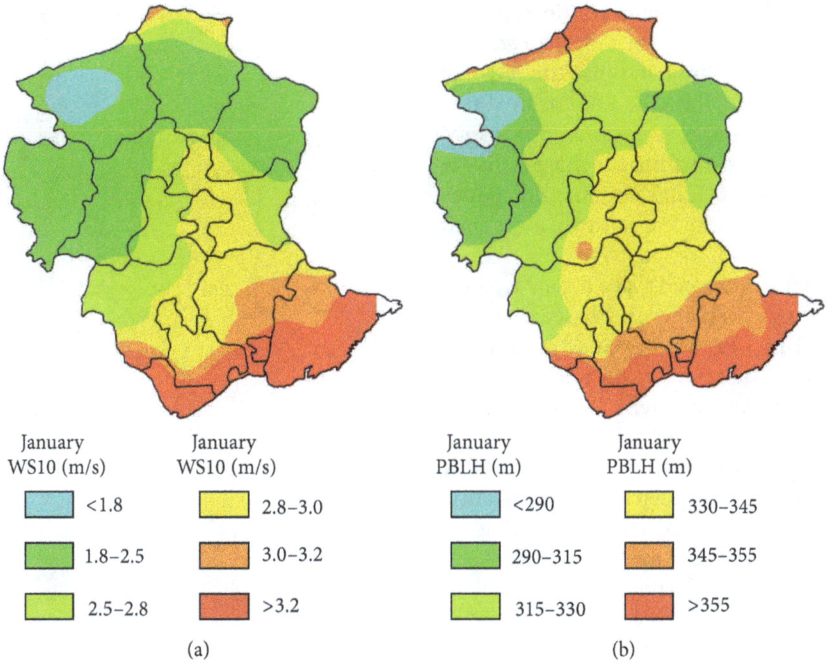

FIGURE 7: Monthly average wind speed at 10 m above ground (WS10) (a) and planetary boundary layer height (PBLH) (b) in January 2006.

TABLE 2: Results of CCA between grid-based variation of emission contribution to air quality in receptor 1 and meteorological/terrain data in Tangshan, 2006.

	(a) January			(b) April	
CV	Correlation	Pearson	CV	Correlation	Pearson
CV1	0.781	Correlation	CV1	0.748	Correlation
Variable	Loadings		Variable	Loadings	
Contribution	0.987	1.000	Contribution	0.987	1.000
PBLH	−0.439	−0.677	PBLH	−0.639	−0.483
T2m	−0.062	−0.163	T2m	−0.387	−0.188
WS10	−0.626	−0.746	WS10	−0.672	−0.588
TERRAIN	0.009	0.041	TERRAIN	−0.039	0.242
PSLV	0.061	−0.002	PSLV	−0.038	0.205
RH	−0.160	0.094	RH	−0.479	0.221
WD	−0.069	−0.150	WD	−0.011	−0.243
	(c) July			(d) October	
CV	Correlation	Pearson	CV	Correlation	Pearson
CV1	0.725	Correlation	CV1	0.807	Correlation
Variable	Loadings		Variable	Loadings	
Contribution	0.987	1.000	Contribution	0.987	1.000
PBLH	−0.989	−0.416	PBLH	−0.434	−0.724
T2m	−0.061	−0.167	T2m	−0.088	−0.113
WS10	−0.502	−0.347	WS10	−0.591	−0.729
TERRAIN	−0.005	0.346	TERRAIN	−0.003	0.045
PSLV	−0.050	0.029	PSLV	0.010	−0.001
RH	−0.715	−0.066	RH	−0.120	0.029
WD	0.091	−0.086	WD	0.010	−0.035

some local higher values occurred within Qianan County. The lowest emission contribution variation ratio (i.e., less than 0.3%) occurred in the coastal areas of Tangshan and the northern part of Qianxi County, implying that the regional air quality was less sensitive to the emissions from these areas. As a result, in order to improve regional air quality, the industries (Figure 5(a)) located within the more sensitive areas (i.e., northwestern part of Tangshan) should reduce their emissions or be relocated to the less sensitive areas such as the coastal area of Tangshan Municipality.

As described earlier, CCA indicated that a negative correlation existed between wind speed and grid-based emission contribution variation. This can be proved from another perspective. The data from three state-controlled weather stations located in Zunhua (northwestern area), Tangshan urban (center area), and Leting (southeastern area) (Figure 1) were used for meteorological factor analysis. The monthly average wind speed and calm frequency were previously identified as the main meteorological factors affecting air pollution [25] and thus were used for analysis in this study. Table 3 lists the average wind speed and calm frequency of the four representative months in 2006. Previous studies suggested that the greater the wind speed and the smaller the calm frequency, the more beneficial for pollutant dispersion. It can be observed from Table 3 that the ranking of monthly and yearly average wind speed from large to small is Leting, Tangshan urban, and Zunhua. This would indicate that the dispersion capability of pollutants gradually decreases from the coast (i.e., Leting) to inland area (i.e., Zunhua), leading to gradually increased emission contribution variation from the coastal area to inland area as shown in Figure 8(a). However, the order of calm frequency for the three selected areas does not hold the same as that of average wind speed. Although the calm frequency in Zunhua area was higher than that in other two areas in all seasons which was less conducive to the dispersion of pollutants, the calm frequencies in Tanghan urban area in April and July were significantly lower than those in Leting, which could give a good explanation for the local low emission contribution variation values shown in the center area of Tangshan in Figures 6(b) and 6(c). In addition, Figure 8(a) not only displays the annual average emission contribution variation ratios of the modeling grids to air quality in the entire Tangshan region, but also gives a visual representation of the dominant wind direction. It is found from Figure 8(a) that the east-west direction modeling grids had higher contribution variation ratios than north-south direction girds, implying that east-west was the dominant wind direction in Tangshan Municipality.

TABLE 3: Wind speed and calm wind frequency in Tangshan in 2006.

Area	Month	Wind speed (m/s)	Calm frequency (%)
Zunhua	January	1.49	5.71
	April	2.61	3.00
	July	1.66	4.12
	October	1.43	11.07
	Annual	1.80	6.01
Tangshan urban	January	1.91	1.22
	April	2.63	0.43
	July	1.91	0.82
	October	1.77	7.79
	Annual	2.06	2.59
Leting	January	2.09	3.67
	April	3.03	2.14
	July	2.08	1.23
	October	1.86	4.51
	Annual	2.27	2.90

4.6. Variation of Annual Emission Contribution to Urban Air Quality. Air quality control within a smaller area than regional scale is usually important and more practical in urban environmental management. In this study, the urban area of Tangshan was selected as a control area (i.e., receptor 2), and the grid-based variation of emission contribution to the average air quality of receptor 2 was then simulated using MM5-CAMx-PSAT. Figure 8(b) shows the distribution of grid-based annual emission contribution variation ratio. It is found that the grids with largest contribution variation ratios were receptor 2 itself (with contribution variation ratio of greater than 10.0%), and the second were the grids mainly surrounding receptor 2. Figure 8(b) also reveals that emission contribution variation ratio had correlation with the distance between emission grids and the receptor area. The east-west modeling grids around receptor 2 had slightly higher emission contribution variation ratio than the north-south grids. This could be explained by the fact that east and west winds were the main wind directions in the study area as observed from the monitoring data in 2006. The contribution variation ratios of the remaining parts of Tangshan Municipality were very small, with minimum contribution variation ratios occurring in the coastal areas and northern parts of Tangshan (i.e., less than 0.2%). The results indicated that the receptor itself as emission grids had significant contribution to the urban air quality. The obtained emission contribution variation analysis results are of practical importance for air quality management. For example, to improve the urban air quality in Tangshan, the industries (Figure 5(a)) within the more sensitive areas (i.e., Tangshan urban, Fengrun, and Fengnan) should be relocated to the less sensitive areas (i.e., coastal area of Tangshan), and the new industrial projects with SO_2 emissions such as power plants should also be located within the less sensitive coastal areas.

5. Conclusions

A modeling grid-based emission contribution analysis approach was proposed to identify emission areas with higher response of regional and urban air quality change due to emission rate perturbation. This approach relied on a coupled MM5-CAMx where MM5 was used to provide meteorological inputs for the air quality model CAMx, while CAMx was used to predict air pollutant concentration distributions. The particulate matter source apportioning

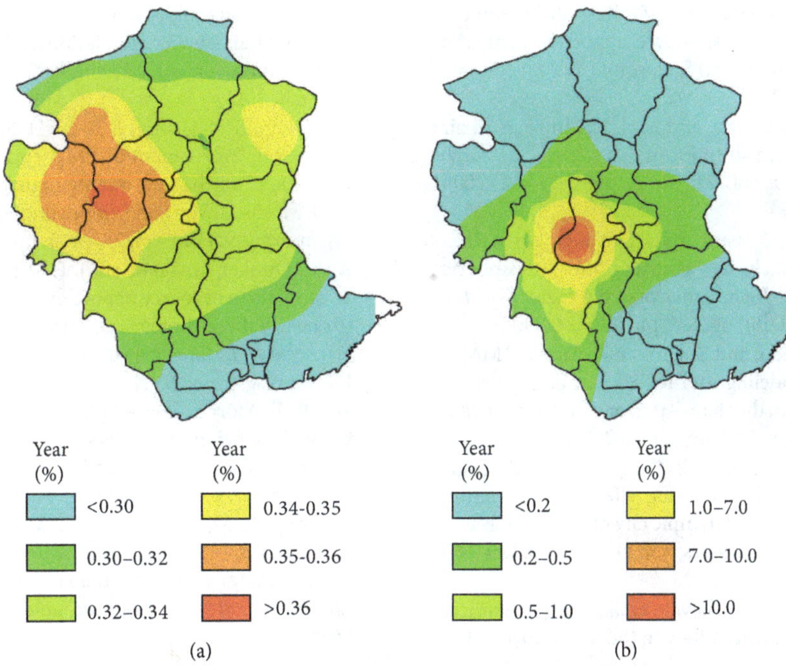

FIGURE 8: Distribution of annual average emission contribution variation ratio of modeling grids to air quality in (a) receptor 1 and (b) receptor 2.

technology (PSAT) within CAMx was used to calculate the variation of emission contribution to air quality from emission rate perturbation within each modeling grid. The method was applied to a case study in Tangshan Municipality in northern China. The MM5-CMAx was implemented to predict hourly SO_2 concentrations based on the base emission inventory of SO_2 in 2006 with modeling grid scale of 9×9 km, and the impact of emission perturbation in each modeling grid to atmospheric SO_2 concentrations was calculated by using PSAT technology through adding 1000 t/yr of SO_2 emission to the grid in addition to the base emission inventory. The variation of emission contribution to regional air quality from each modeling grid per 1000 t/yr of emission rate change was obtained for four representative months (January, April, July, and October) in 2006. PCA and CCA were conducted to examine the impacts of meteorological factors on the variation of emission source contribution, and the results indicated that there was an obvious negative correlation between emission contribution variation and planetary boundary layer height (PBLH) as well as wind speed. The analysis of the variation of emission contribution to annual regional SO_2 concentration (i.e., larger scale) indicated that the northwestern part of Tangshan was the most sensitive area with emission contribution variation ratio of more than 0.36%, while the southern coastal area had the lowest contribution variation ratio of less than 0.30%. The proposed method was also applied to analyze the variation of emission contribution to the SO_2 pollution in the urban area in Tangshan (i.e., at a smaller scale), and it was found that the largest contribution grids were the urban area itself (with contribution variation ratio of greater than 10.0%), and the minimum contribution variation ratios (i.e., less than 0.2%) occurred in the coastal areas and northern parts of Tangshan. Based on the modeling results, the emission sources within the areas with higher contribution variation ratios should be regulated with priority or relocated to other areas with lower contribution variation ratios such as the coastal areas in Tangshan. In summary, the proposed methodology can be applied to address many other regional and urban air pollution problems, and the results would provide sound scientific basis for effective air quality management.

Acknowledgments

This research was supported by the Natural Sciences Foundation of China (no. 51038001) and the Ministry of Environmental Protection Special Funds for Scientific Research on Public Causes (no. 201209003). The authors would like to thank Natural Science Foundation of Beijing (no. 8092004), Beijing NOVA Program of China (no. 2009B07), Innovation Team Project of Beijing Municipal Education Commission (PHR201007105), and the Cultivation Fund of the Key Scientific and Technical Innovation Project, Ministry of Education of China (708017), for supporting this work.

References

[1] D. S. Chen, S. Y. Cheng, L. Liu, T. Chen, and X. R. Guo, "An integrated MM5—CMAQ modeling approach for assessing trans-boundary PM_{10} contribution to the host city of 2008 Olympic summer games—Beijing, China," *Atmospheric Environment*, vol. 41, no. 6, pp. 1237–1250, 2007.

[2] D. S. Cohan, A. Hakami, Y. Hu, and A. G. Russell, "Nonlinear response of ozone to emissions: source apportionment and sensitivity analysis," *Environmental Science and Technology*, vol. 39, no. 17, pp. 6739–6748, 2005.

[3] F. Wang, D. S. Chen, S. Y. Cheng, and M. J. Li, "Impacts of air pollutant transport based on air trajectory clustering," *Research of Environmental Sciences*, vol. 22, no. 6, pp. 637–642, 2009 (Chinese).

[4] S. Cheng, D. Chen, J. Li, X. Guo, and H. Wang, "An ARPS-CMAQ modeling approach for assessing the atmospheric assimilative capacity of the Beijing metropolitan region," *Water, Air, and Soil Pollution*, vol. 181, no. 1–4, pp. 211–224, 2007.

[5] S. M. Lee, H. J. S. Fernando, and S. Grossman-Clarke, "MM5-SMOKE-CMAQ as a modeling tool for 8-h ozone regulatory enforcement: application to the state of Arizona," *Environmental Modeling and Assessment*, vol. 12, no. 1, pp. 63–74, 2007.

[6] Y. Zhou, S. Y. Cheng, L. Liu, and D. S. Chen, "A Coupled MM5-CMAQ modeling system for assessing effects of restriction measures on PM_{10} pollution in Olympic city of Beijing, China," *Journal of Environmental Informatics*, vol. 19, no. 2, pp. 120–127, 2012.

[7] E. Angelino, M. Bedogni, C. Carnevale et al., "PM_{10} chemical model simulations over Northern Italy in the framework of the citydelta exercise," *Environmental Modeling and Assessment*, vol. 13, no. 3, pp. 401–413, 2008.

[8] U. Nopmongcol, B. Koo, E. Tai et al., "Modeling Europe with CAMx for the air quality model evaluation international initiative (AQMEII)," *Atmospheric Environment*, vol. 53, no. 7, pp. 177–185, 2012.

[9] J. Dudhia, D. Gill, K. Manning, W. Wang, and C. Bruyere, *PSU/NCAR Mesoscale Modeling System Tutorial Class Notes and User's Guide: MM5 Modeling System Version 3*, National Center for Atmospheric Research, 2004.

[10] X. Tie, S. Madronich, G. Li et al., "Characterizations of chemical oxidants in Mexico City: a regional chemical dynamical model (WRF-Chem) study," *Atmospheric Environment*, vol. 41, no. 9, pp. 1989–2008, 2007.

[11] S. Cheng, D. Chen, J. Li, H. Wang, and X. Guo, "The assessment of emission-source contributions to air quality by using a coupled MM5-ARPS-CMAQ modeling system: a case study in the Beijing metropolitan region, China," *Environmental Modelling and Software*, vol. 22, no. 11, pp. 1601–1616, 2007.

[12] M. Titov, A. P. Sturman, and P. Zawar-Reza, "Application of MM5 and CAMx4 to local scale dispersion of particulate matter for the city of Christchurch, New Zealand," *Atmospheric Environment*, vol. 41, no. 2, pp. 327–338, 2007.

[13] S. M. Lee, M. Princevac, S. Mitsutomi, and J. Cassmassi, "MM5 simulations for air quality modeling: an application to a coastal area with complex terrain," *Atmospheric Environment*, vol. 43, no. 2, pp. 447–457, 2009.

[14] H. Shimadera, A. Kondo, A. Kaga, K. L. Shrestha, and Y. Inoue, "Contribution of transboundary air pollution to ionic concentrations in fog in the Kinki Region of Japan," *Atmospheric Environment*, vol. 43, no. 37, pp. 5894–5907, 2009.

[15] C. Borrego, A. Monteiro, J. Ferreira et al., "Modelling the photochemical pollution over the metropolitan area of Porto Alegre, Brazil," *Atmospheric Environment*, vol. 44, no. 3, pp. 370–380, 2010.

[16] D. G. Streets, J. S. Fu, C. J. Jang et al., "Air quality during the 2008 Beijing Olympic Games," *Atmospheric Environment*, vol. 41, no. 3, pp. 480–492, 2007.

[17] Y. Zhou, S. Y. Cheng, J. B. Li, J. L. Lang, L. Li, and D. S. Chen, "A new statistical modeling and optimization framework for establishing high-resolution PM_{10} emission inventory—II. Integrated air quality simulation and optimization for performance improvement," *Atmospheric Environment*, vol. 60, pp. 623–631, 2012.

[18] Q. Huang, S. Y. Cheng, J. B. Li, D. S. Chen, H. Y. Wang, and X. R. Guo, "Assessment of PM_{10} emission sources for priority regulation in urban air quality management using a new coupled MM5-CAMx-PSAT modeling approach," *Environmental Engineering Science*, vol. 29, no. 5, pp. 343–349, 2012.

[19] A. M. Dunker, G. Yarwood, J. P. Ortmann, and G. M. Wilson, "Comparison of source apportionment and source sensitivity of ozone in a three-dimensional air quality model," *Environmental Science and Technology*, vol. 36, no. 13, pp. 2953–2964, 2002.

[20] K. M. Wagstrom, S. N. Pandis, G. Yarwood, G. M. Wilson, and R. E. Morris, "Development and application of a computationally efficient particulate matter apportionment algorithm in a three-dimensional chemical transport model," *Atmospheric Environment*, vol. 42, no. 22, pp. 5650–5659, 2008.

[21] B. Koo, G. M. Wilson, R. E. Morris, A. M. Dunker, and G. Yarwood, "Comparison of source apportionment and sensitivity analysis in a particulate matter air quality model," *Environmental Science and Technology*, vol. 43, no. 17, pp. 6669–6675, 2009.

[22] S. M. Almeida, C. A. Pio, M. C. Freitas, M. A. Reis, and M. A. Trancoso, "Source apportionment of fine and coarse particulate matter in a sub-urban area at the Western European Coast," *Atmospheric Environment*, vol. 39, no. 17, pp. 3127–3138, 2005.

[23] S. S. Park and Y. J. Kim, "Source contributions to fine particulate matter in an urban atmosphere," *Chemosphere*, vol. 59, no. 2, pp. 217–226, 2005.

[24] A. Srivastava, S. Gupta, and V. K. Jain, "Source apportionment of total suspended particulate matter in coarse and fine size ranges over Delhi," *Aerosol and Air Quality Research*, vol. 8, no. 2, pp. 188–200, 2008.

[25] Z. H. Chen, S. Y. Cheng, J. B. Li, X. R. Guo, W. H. Wang, and D. S. Chen, "Relationship between atmospheric pollution processes and synoptic pressure patterns in northern China," *Atmospheric Environment*, vol. 42, no. 24, pp. 6078–6087, 2008.

[26] F. Wang, D. S. Chen, S. Y. Cheng, J. B. Li, M. J. Li, and Z. H. Ren, "Identification of regional atmospheric PM_{10} transport pathways using HYSPLIT, MM5-CMAQ and synoptic pressure pattern analysis," *Environmental Modelling and Software*, vol. 25, no. 8, pp. 927–934, 2010.

[27] Q. Zhang and D. G. Streets, "2006 Asia Emissions for INTEX-B," December 2009, http://www.cgrer.uiowa.edu/EMISSION_DATA_new/index_16.html.

[28] M. Viana, X. Querol, A. Alastuey, J. I. Gil, and M. Menéndez, "Identification of PM sources by principal component analysis (PCA) coupled with wind direction data," *Chemosphere*, vol. 65, no. 11, pp. 2411–2418, 2006.

[29] M. Statheropoulos, N. Vassiliadis, and A. Pappa, "Principal component and canonical correlation analysis for examining air pollution and meteorological data," *Atmospheric Environment*, vol. 32, no. 6, pp. 1087–1095, 1998.

[30] V. Isakov, A. Venkatram, J. S. Touma, D. Koračin, and T. L. Otte, "Evaluating the use of outputs from comprehensive meteorological models in air quality modeling applications," *Atmospheric Environment*, vol. 41, no. 8, pp. 1689–1705, 2007.

[31] S. A. Abdul-Wahab, C. S. Bakheit, and S. M. Al-Alawi, "Principal component and multiple regression analysis in modelling of ground-level ozone and factors affecting its concentrations," *Environmental Modelling and Software*, vol. 20, no. 10, pp. 1263–1271, 2005.

20

Contact Problem for an Elastic Layer on an Elastic Half Plane Loaded by Means of Three Rigid Flat Punches

T. S. Ozsahin and O. Taskıner

Civil Engineering Department, Karadeniz Technical University, 61080 Trabzon, Turkey

Correspondence should be addressed to T. S. Ozsahin; talat@ktu.edu.tr

Academic Editor: Safa Bozkurt Coskun

The frictionless contact problem for an elastic layer resting on an elastic half plane is considered. The problem is solved by using the theory of elasticity and integral transformation technique. The compressive loads P and Q (per unit thickness in z direction) are applied to the layer through three rigid flat punches. The elastic layer is also subjected to uniform vertical body force due to effect of gravity. The contact along the interface between elastic layer and half plane is continuous, if the value of the load factor, λ, is less than a critical value, λ_{cr}. In this case, initial separation loads, λ_{cr} and initial separation points, x_{cr} are determined. Also the required distance between the punches to avoid any separation between the punches and the elastic layer is studied and the limit distance between punches that ends interaction of punches is investigated for various dimensionless quantities. However, if tensile tractions are not allowed on the interface, for $\lambda > \lambda_{cr}$ the layer separates from the interface along a certain finite region. Numerical results for distance determining the separation area, vertical displacement in the separation zone, contact stress distribution along the interface between elastic layer and half plane are given for this discontinuous contact case.

1. Introduction

Contact between deformable bodies abounds in industry and everyday life. Because of the industrial importance of the physical processes that take place during contact, a considerable effort has been made in their modeling, analysis, and numerical simulations.

The range of application in contact mechanics starts with problems like foundations in civil engineering, where the lift off the foundation from soil due to eccentric forces acting on a building, railway ballasts, foundation grillages, continuous foundation beams, runaways, liquid tanks resting on the ground, and grain silos is considered. Furthermore, foundation including piles as supporting members or the driving of piles into the soil is of interest. Also classical bearing problem of steel constructions and the connection of structural members by bolts or screws are areas in which contact analysis enters the design process in civil engineering [1].

A complete analysis of the interaction problem for elastic bodies generally requires the determination of stresses and strain within the individual bodies in contact, together with information regarding the distribution of displacements and stresses at the contact regions.

The contact problem for an elastic layer has attracted considerable attention in the past because of its possible application to a variety of structures of practical interest. The layer usually rests on a foundation which may be either elastic or rigid and the body force due to gravity may be neglected [2–27] or effect of gravity may be taken into account [28–35]. Studies regarding the frictionless contact along the interface may be found in [2–15]. Also contact region may exhibit frictional characteristics, giving rise to normal and shear traction at the contact surface [16–27].

While initial contact is determined by the geometric features of the bodies, the extent of the contact generally changes not only by the particular loads applied to the bodies but also with the elastic constants of the materials. Due to the bending of the layer under local compressive loads, in the absence of gravity effects, the contact area would decrease to a finite size which is independent of the magnitude of the applied load [2–7]. If the effect of gravity was taken into account, the normal stress along the layer subspace interface will be compressive and the contact is maintained through

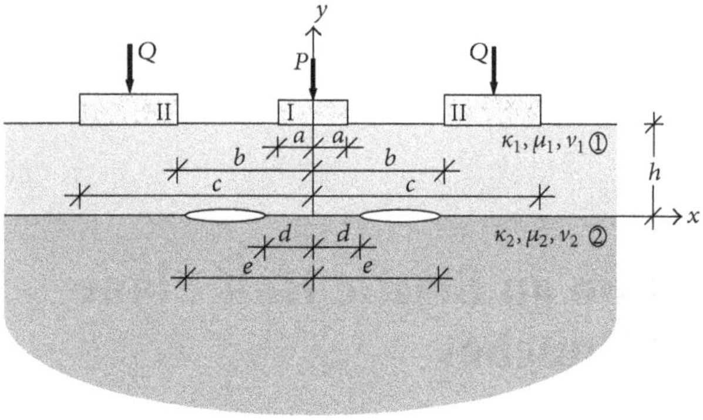

FIGURE 1: Elastic layer resting on an elastic half plane loaded by means of three rigid flat punches.

the frictionless interface unless the compressive applied load exceeds a certain critical value. When the magnitude of the compressive external load exceeds a certain value, a separation will take place between the layer and the foundation. The length of separation region along the interface is unknown, and the problem becomes a discontinuous contact problem [28–35].

Interaction between an elastic medium and a rigid punch forms another group of contact problems. Rigid punches may be structural elements such as foundations, beams, and plates of finite or infinite extent resting on idealized linearly deformable elastic media. Here, the shape of the contact region may be known a priori and remains constant, or contact region may be changed due to the shape of punch profile [8–27]. The problem of flat-ended rigid punch has important applications in soil mechanics, particularly in estimating the safety of foundations. The application of the three punches for an elastic layer resting on an elastic half plane in soil mechanics is obvious; for example, the punches can be taken as foundations placed on layered soil. When the foundations are placed on soil, there is a possibility of pressure isobars of adjacent foundations overlapping each other. The soil is highly stressed in the zones of overlapping, or the difference of settlement between two adjacent foundations, commonly referred to as differential settlement may cause damage to the structure. It is possible to avoid overlapping of pressures or differential settlement by installing the foundations at considerable distance apart from each other.

In this study, contact problem of the three punches for an elastic layer resting on an elastic half plane is considered according to the theory of elasticity with integral transformation technique. The compressive loads P and Q (per unit thickness in z direction) are applied to the layer through three rigid flat punches. The width of midmost punch can be different from the other two punches and thickness of the layer is constant, h. The layer is subjected to homogeneous vertical body force due to gravity, $\rho_1 g$. All surfaces are frictionless. The layer remains in contact with the elastic half plane where the magnitude of the load factor λ is less than a critical value, λ_{cr} ($\lambda = P/\rho_1 g h^2$). If $\lambda > \lambda_{cr}$, the contact is discontinuous and a separation takes place between the layer and the half plane. A numerical integration procedure is performed for the solution of the problems, and different parameters are researched for various dimensionless quantities for both continuous and discontinuous contact cases. Finally, numerical results are analyzed and conclusions are drawn.

2. General Expressions for Stresses and Displacements

Consider a frictionless elastic layer of thickness h lying on an elastic half plane. The geometry and coordinate system are shown in Figure 1. The governing equations are

$$\mu_k \nabla^2 u_k + \frac{2\mu_k}{(\kappa_k - 1)} \frac{\partial}{\partial x} \left(\frac{\partial u_k}{\partial x} + \frac{\partial v_k}{\partial y} \right) = 0, \tag{1a}$$

$$\mu_k \nabla^2 v_k + \frac{2\mu_k}{(\kappa_k - 1)} \frac{\partial}{\partial y} \left(\frac{\partial u_k}{\partial x} + \frac{\partial v_k}{\partial y} \right) = \rho_k g, \quad (k = 1, 2) \tag{1b}$$

where $\rho_k g$ is the intensity of the body force acting vertically in which ρ_k and g are mass density and gravity acceleration. u_k and v_k are the x and y components of the displacement vector, μ_k and κ_k represent shear modulus and elastic constant of the layer and the half plane, respectively. $\kappa_k = (3 - \gamma_k)/(1 + \gamma_k)$ for plane stress and $\kappa_k = (3 - 4\gamma_k)$ for plane strain. γ_k is the Poisson ratio ($k = 1, 2$). Subscript 1 indicates the elastic layer and subscript 2 indicates the elastic half plane.

u_p and v_p represent the displacements for the case in which gravity forces are considered. u_h and v_h are the displacements when the gravity forces are ignored, and total field of displacements may be expressed as

$$u = u_p + u_h, \tag{2a}$$

$$v = v_p + v_h. \tag{2b}$$

Observing that $x = 0$ is a plane of symmetry, it is sufficient to consider the problem in the region $0 \leq x \leq \infty$ only. Using the symmetry consideration, the following expressions may be written:

$$u_1(x, y) = -u_1(-x, y), \tag{3a}$$

$$v_1(x, y) = v_1(-x, y), \tag{3b}$$

$$u_1(x, y) = \frac{2}{\pi} \int_0^\infty \phi_1(\alpha, y) \sin(\alpha x)\, d\alpha, \tag{3c}$$

$$v_1(x, y) = \frac{2}{\pi} \int_0^\infty \Psi_1(\alpha, y) \cos(\alpha x)\, d\alpha, \tag{3d}$$

where ϕ_1 and Ψ_1 functions are inverse Fourier transforms of u_1 and v_1 respectively. Taking necessary derivatives of (3c) and (3d), substituting them into (1a) and (1b), and solving the second-order differential equations, the following equations may be obtained for displacements:

$$u_{1_h}(x,y) = \frac{2}{\pi}\int_0^{\infty}[(A+By)e^{-\alpha y} + (C+Dy)e^{\alpha y}]\sin(\alpha x)\,d\alpha, \tag{4a}$$

$$v_{1_h}(x,y) = \frac{2}{\pi}\int_0^{\infty}\left[\left[A+\left(\frac{\kappa_1}{\alpha}\kappa+y\right)B\right]e^{-\alpha y} + \left[-C+\left(\frac{\kappa_1}{\alpha}-y\right)D\right]e^{\alpha y}\right]\cos(\alpha x)\,d\alpha. \tag{4b}$$

Using Hooke's law and (4a) and (4b), stress components which do not include the gravity force may be expressed as follows:

$$\sigma_{x_{1_h}}(x,y) = \frac{4\mu_1}{\pi}\int_0^{\infty}\left[\left[\alpha(A+By)-\left(\frac{3-\kappa_1}{2}\right)B\right]e^{-\alpha y} + \left[\alpha(C+Dy)+\left(\frac{3-\kappa_i}{2}\right)D\right]e^{\alpha y}\right] \times\cos(\alpha x)\,d\alpha, \tag{5a}$$

$$\sigma_{y_{1_h}}(x,y) = \frac{4\mu_1}{\pi}\int_0^{\infty}\left[-\left[\alpha(A+By)+\left(\frac{\kappa_1+1}{2}\right)B\right]e^{-\alpha y} + \left[-\alpha(C+Dy)+\left(\frac{\kappa_1+1}{2}\right)D\right]e^{\alpha y}\right] \times\cos(\alpha x)\,d\alpha, \tag{5b}$$

$$\tau_{xy_{1_h}}(x,y) = \frac{4\mu_1}{\pi}\int_0^{\infty}\left[-\left[\alpha(A+By)+\left(\frac{\kappa_i-1}{2}\right)B\right]e^{-\alpha y} + \left[\alpha(C+Dy)-\left(\frac{\kappa_i-1}{2}\right)D\right]e^{\alpha y}\right] \times\sin(\alpha x)\,d\alpha. \tag{5c}$$

For the case in which gravity force exist, particular part of the displacement components corresponding to $\rho_1 g$, the following expressions are obtained, that is, special solution of the Navier equations for a layer with a height h:

$$u_{1_p} = \frac{3-\kappa_1}{8\mu_1}\frac{\rho_1 gh}{2}x, \tag{6a}$$

$$v_{1_p} = \frac{\rho_1 gy}{2\mu_1}\left[\frac{(\kappa_1-1)}{(\kappa_1+1)}(y-h)-\frac{(\kappa_1+1)}{8}h\right], \tag{6b}$$

$$\sigma_{y_{1_p}} = \rho_1 g(y-h), \tag{6c}$$

$$\sigma_{x_{1_p}} = \rho_1 g\left(y-\frac{h}{2}\right)\frac{(1+\kappa_2)}{(1+\kappa_2)+2\mu_2(1+\kappa_1)}, \tag{6d}$$

$$\tau_{xy_{1_p}} = 0. \tag{6e}$$

Considering the orthogonal axes shown in Figure 1, displacements will be zero for $y=-\infty$, and if μ_2, ν_2 are the elastic constants of the half plane, then the homogenous field of displacements and stresses of the elastic half plane may be obtained as

$$u_{2_h}(x,y) = \frac{2}{\pi}\int_0^{\infty}[(E+Fy)e^{\alpha y}]\sin(\alpha x)\,d\alpha, \tag{7a}$$

$$v_{2_h}(x,y) = \frac{2}{\pi}\int_0^{\infty}\left[\left[-E+\left(\frac{\kappa_2}{\alpha}-y\right)F\right]e^{\alpha y}\right]\cos(\alpha x)\,d\alpha, \tag{7b}$$

$$\sigma_{x_{2_h}}(x,y) = \frac{4\mu_2}{\pi}\int_0^{\infty}\left[\left[\alpha(E+Fy)+\left(\frac{3-\kappa_2}{2}\right)F\right]e^{\alpha y}\right] \times\cos(\alpha x)\,d\alpha, \tag{7c}$$

$$\sigma_{y_{2_h}}(x,y) = \frac{4\mu_2}{\pi}\int_0^{\infty}\left[\left[-\alpha(E+Fy) + \left(\frac{\kappa_2+1}{2}\right)F\right]e^{\alpha y}\right]\cos(\alpha x)\,d\alpha, \tag{7d}$$

$$\tau_{xy_{2_h}}(x,y) = \frac{4\mu_2}{\pi}\int_0^{\infty}\left[\left[\alpha(E+Fy)-\left(\frac{\kappa_2-1}{2}\right)F\right]e^{\alpha y}\right] \times\sin(\alpha x)\,d\alpha, \tag{7e}$$

subscript 2 indicates the elastic half plane. Note that the body force acting in the foundation is neglected since it does not disturb the contact pressure distribution. A, B, C, D, E, and F are the unknown constants, which will be determined from the boundary and continuity conditions at $y=0$ and $y=h$.

3. Case of Continuous Contact

An elastic layer with a height of h resting on an elastic half plane, shown in Figure 1, is analyzed for unit thickness in z direction. Widths of punches at both sides are similar, $(c-b)/h$, and each of these punches transmits a concentrated load of Q to the elastic layer. The width of the midmost punch is different, $2a/h$ and it is subjected to a concentrated load, P. All surfaces are frictionless. Particularly, the initial separation load (λ_{cr}) and point (x_{cr}) where the layer separated from the elastic half plane and the variation of the stress distribution between elastic layer and elastic half plane is examined depending on material properties, width of punches, and magnitude of the external loads, P and Q. Due to the different settlement of punches, a separation takes place between punch I and elastic layer, if punches are close enough. Therefore, the critical distance between the punches indicating the initiation of separation between the punch I and the elastic layer is researched and also limit distance between the punches where the interaction of punches ends is investigated.

If load factor (λ) is sufficiently small, then the contact along the layer-subspace, $y = 0, 0 < x < \infty$, will be continuous, and A, B, C, D, E, and F must be determined from the following boundary and continuity conditions:

$$\sigma_{y_1}(x,h) = \begin{cases} -P(x), & 0<x<a \\ -Q(x), & b<x<c \\ 0, & a<x<b,\ c<x<\infty, \end{cases} \tag{8a}$$

$$\tau_{xy_1}(x,h) = 0, \quad 0<x<\infty, \tag{8b}$$

$$\tau_{xy_1}(x,0) = 0, \quad 0<x<\infty, \tag{8c}$$

$$\tau_{xy_2}(x,0) = 0, \quad 0<x<\infty, \tag{8d}$$

$$\sigma_{y_1}(x,0) = \sigma_{y_2}(x,0), \quad 0<x<\infty, \tag{8e}$$

$$\frac{\partial}{\partial x}[v_2(x,0) - v_1(x,0)] = 0, \quad 0<x<\infty, \tag{8f}$$

$$\frac{\partial}{\partial x}[v_1(x,h)] = 0, \quad 0<x<a, \tag{8g}$$

$$\frac{\partial}{\partial x}[v_1(x,h)] = 0, \quad b<x<c, \tag{8h}$$

in which subscripts 1 and 2 indicate relation to the elastic layer and the elastic half plane, respectively. $P(x)$ is the unknown contact pressure under punch I and $Q(x)$ is the unknown contact pressure under punch II, which have not been determined yet. If a separation occurs between the elastic layer and elastic half plane, this will give rise to a discontinuous contact position and the following results for former solution will no longer be valid and new solution will be attained for the latter case.

Equilibrium conditions of the problem may be expressed as

$$\int_0^a P(x)\,dx = \frac{P}{2}, \tag{9a}$$

$$\int_b^c Q(x)\,dx = Q. \tag{9b}$$

Displacement and stress expressions (4a), (4b), (5a)–(5c), (6a)–(6e), and (7a)–(7e) are substituted into boundary conditions (8a)–(8f), and unknown constants A, B, C, D, E, and F are determined in terms of unknown functions $P(x)$ and $Q(x)$. By making use of (8g) and (8h), after some simple manipulations, one may obtain the following singular integral equations for $P(x)$ and $Q(x)$ [36, 37]:

$$\begin{aligned} &-\frac{1}{\pi\mu_1}\int_0^a \left[k_1(x,t) + \frac{(1+\kappa_1)}{4}\left(\frac{1}{t+x}+\frac{1}{t-x}\right)\right] P(t)\,dt \\ &-\frac{1}{\pi\mu_1}\int_b^c \left[k_1(x,t) + \frac{(1+\kappa_1)}{4}\left(\frac{1}{t+x}+\frac{1}{t-x}\right)\right] \\ &\times Q(t)\,dt = 0, \quad 0<x<a, \end{aligned} \tag{10a}$$

$$\begin{aligned} &-\frac{1}{\pi\mu_1}\int_0^a \left[k_1(x,t) + \frac{(1+\kappa_1)}{4}\left(\frac{1}{t+x}+\frac{1}{t-x}\right)\right] P(t)\,dt \\ &-\frac{1}{\pi\mu_1}\int_b^c \left[k_1(x,t) + \frac{(1+\kappa_1)}{4}\left(\frac{1}{t+x}+\frac{1}{t-x}\right)\right] \\ &\times Q(t)\,dt = 0, \quad b<x<c, \end{aligned} \tag{10b}$$

where

$$\begin{aligned} k_1(x,t) = \int_0^\infty \Bigg\{ \Bigg\{ &\frac{\alpha^3}{8(1+\kappa_1)} \\ &\times \left[\left[(1+\kappa_2) + \frac{\mu_2}{\mu_1(1+\kappa_1)}\right]\right. \\ &+ e^{-2\alpha h}\left[4\alpha h(1+\kappa_2) - \frac{2\mu_2}{\mu_1(1+\kappa_1)}\right] \\ &\left.+ e^{-4\alpha h}\left[-1-\kappa_2+\frac{\mu_2}{\mu_1(1+\kappa_1)}\right]\right]\Bigg\}(\Delta)^{-1} \\ &- \frac{(1+\kappa_1)}{4}\Bigg\} \\ &\times \{\sin\alpha(t+x) - \sin\alpha(t-x)\}\,d\alpha, \end{aligned} \tag{11}$$

in which

$$\begin{aligned} \Delta = \alpha^3\Bigg(2\Bigg\{&\left(-1-\kappa_2-\frac{\mu_2}{\mu_1(1+\kappa_1)}\right) \\ &+ e^{-4\alpha h}\left[-1-\kappa_2+\frac{\mu_2}{\mu_1(1+\kappa_1)}\right] \\ &+ e^{-2\alpha h}\Bigg[(1+\kappa_2)\left(2+4\alpha^2h^2\right) \\ &+\frac{4\alpha h\mu_2}{\mu_1(-1-\kappa_1)}\Bigg]\Bigg\}\Bigg)^{-1}. \end{aligned} \tag{12}$$

If evaluated values of A, B, C, and D in terms of $P(x)$ and $Q(x)$ are substituted into (5b), the expression of the contact stress between elastic layer and half plane may be obtained as

$$\begin{aligned} \sigma_{y_1}(x,0) = &-\rho_1 gh - \frac{1}{\pi}\int_0^a k_2(x,t)P(t)\,dt \\ &-\frac{1}{\pi}\int_b^c k_2(x,t)Q(t)\,dt, \quad 0<x<\infty, \end{aligned} \tag{13}$$

in which ρ_1 and g are mass density and gravity acceleration, respectively, where

$$k_2(x,t) = \int_0^\infty \left\{ \frac{\alpha^3 \mu_2}{\mu_1 (1+\kappa_1)} \times \left[e^{-3\alpha h}(-1+\alpha h) + e^{-\alpha h}(1+\alpha h) \right] \right\} (\Delta)^{-1} \times \{\cos\alpha(t+x) + \cos\alpha(t-x)\} d\alpha. \tag{14}$$

To simplify the numerical analysis, the following dimensionless quantities are introduced:

$$x_1 = ar_1, \tag{15a}$$

$$t_1 = as_1, \tag{15b}$$

$$x_2 = \frac{c-b}{2} r_2 + \frac{c+b}{2}, \tag{15c}$$

$$t_2 = \frac{c-b}{2} s_2 + \frac{c+b}{2}, \tag{15d}$$

$$g_1(s_1) = \frac{P(as_1)}{P/h}, \tag{15e}$$

$$g_2(s_2) = \frac{Q(((c-b)/2)s_2 + (c+b)/2)}{P/h}, \tag{15f}$$

$$\alpha = wh, \tag{15g}$$

$$\lambda = \frac{P}{\rho_1 g h^2}. \tag{15h}$$

Substituting from (15a)–(15h), (9a), (9b) and (10a), (10b) may be expressed as

$$\int_0^1 g_1(s_1) \frac{a}{h} ds_1 = \frac{1}{2}, \tag{16a}$$

$$\int_{-1}^1 g_2(s_2) \frac{c-b}{2h} ds_2 = \frac{Q}{P}, \tag{16b}$$

$$\begin{aligned} &-\frac{1}{\pi}\int_0^1 g_1(s_1)\frac{a}{h}ds_1 \left[m_1(r_1,s_1) + \frac{(1+\kappa_1)}{4} \times \left(\frac{1}{a(s_1+r_1)} - \frac{1}{a(s_1-r_1)} \right) \right] \\ &-\frac{1}{\pi}\int_{-1}^1 g_2(s_2)\frac{c-b}{2h}ds_2 \times \left[m_2(r_1,s_2) + \frac{(1+\kappa_1)}{4} \times \left(\frac{1}{(((c-b)/2)s_2 + (c+b)/2) + ar_1} - \frac{1}{(((c-b)/2)s_2 + (c+b)/2) - ar_1} \right) \right] = 0, \\ &0 < r_1 < 1, \end{aligned} \tag{16c}$$

$$\begin{aligned} &-\frac{1}{\pi}\int_0^1 g_1(s_1)\frac{a}{h}ds_1 \times \left[m_3(r_2,s_1) + \frac{(1+\kappa_1)}{4} \times \left(\frac{1}{as_1 + (((c-b)/2)r_2 + (c+b)/2)} - \frac{1}{as_1 - (((c-b)/2)r_2 + (c+b)/2)} \right) \right] \\ &-\frac{1}{\pi}\int_{-1}^1 g_2(s_2)\frac{c-b}{2}ds_2 \times \left[m_4(r_2,s_2) + \frac{(1+\kappa_1)}{4} \times \left(\frac{1}{((c-b)/2)(s_2+r_2) + (c+b)} - \frac{1}{((c-b)/2)(s_2-r_2)} \right) \right] = 0, \\ &-1 < r_2 < 1, \end{aligned} \tag{16d}$$

$$\frac{\sigma_{y_1}(x,0)}{P/h} = -\frac{1}{\lambda} - \frac{1}{\pi}\int_0^1 m_5(r_1,s_1) g_1(s_1) \frac{a}{h} ds_1 - \frac{1}{\pi}\int_{-1}^1 m_6(r_2,s_2) g_2(s_2) \frac{c-b}{2h} ds_2, \tag{16e}$$

where

$$m_1(r_1,s_1) = k_1(x_1,t_1), \tag{17a}$$

$$m_2(r_1,s_2) = k_1(x_1,t_2), \tag{17b}$$

$$m_3(r_2,s_1) = k_1(x_2,t_1), \tag{17c}$$

$$m_4(r_2,s_2) = k_1(x_2,t_2), \tag{17d}$$

$$m_5(r_1,s_1) = k_2(x_1,t_1), \tag{17e}$$

$$m_6(r_2,s_2) = k_2(x_2,t_2). \tag{17f}$$

The index of the integral equations in (16a)–(16e) is +1; so the functions $g_1(s_1)$ and $g_2(s_2)$ may be expressed in the following forms:

$$g_1(s_1) = G_1(s_1)\left(1 - s_1^2\right)^{-1/2}, \quad (0 < s_1 < 1), \tag{18a}$$

$$g_2(s_2) = G_2(s_2)\left(1 - s_2^2\right)^{-1/2}, \quad (-1 < s_2 < 1), \tag{18b}$$

where $G_1(s_1)$ is bounded in $[0,1]$ and $G_2(s_2)$ is bounded in $[-1,1]$. Then using appropriate Gauss-Chebyshev integration formula given in [36], (16a)–(16d) are replaced by the following algebraic equations:

$$\sum_{i=1}^{n} \pi W_i \frac{a}{h} G_1(s_{1_i}) = \frac{1}{2}, \tag{19a}$$

$$\sum_{i=1}^{n} \pi W_i \frac{c-b}{2h} G_2\left(s_{2_i}\right) = \frac{Q}{P}, \tag{19b}$$

$$\begin{aligned}
&-\sum_{i=1}^{n} W_i G_1\left(s_{1_i}\right) \frac{a}{h}\left[m_1\left(r_{1_j}, s_{1_i}\right) + \frac{(1+\kappa_1)}{4} \times \left(\frac{1}{a\left(s_{1_i}+r_{1_j}\right)} - \frac{1}{a\left(s_{1_i}-r_{1_j}\right)}\right)\right] \\
&-\sum_{i=1}^{n} W_i G_2\left(s_{2_i}\right) \frac{c-b}{2h} \times \left[m_2\left(r_{1_j}, s_{2_i}\right) + \frac{(1+\kappa_1)}{4} \times \left(\frac{1}{\left(((c-b)/2)\, s_{2_i} + (c+b)/2\right) + ar_{1_j}} - \frac{1}{\left(((c-b)/2)\, s_{2_i} + (c+b)/2\right) - ar_{1_j}}\right)\right] = 0, \\
&(j = 1, \ldots, n-1),
\end{aligned} \tag{19c}$$

$$\begin{aligned}
&-\sum_{i=1}^{n} W_i G_1\left(s_{1_i}\right) \frac{a}{h} \times \left[m_3\left(r_{2_j}, s_{1_i}\right) + \frac{(1+\kappa_1)}{4} \times \left(\frac{1}{as_{1_i} + \left(((c-b)/2)\, r_{2_j} + (c+b)/2\right)} - \frac{1}{as_{1_i} - \left(((c-b)/2)\, r_{2_j} + (c+b)/2\right)}\right)\right] \\
&-\sum_{i=1}^{n} W_i G_2\left(s_{2_i}\right) \frac{c-b}{2h} \times \left[m_4\left(r_{2_j}, s_{2_i}\right) + \frac{(1+\kappa_1)}{4} \times \left(\frac{1}{((c-b)/2)\left(s_{2_i}+r_{2_j}\right) + (c+b)} - \frac{1}{((c-b)/2)\left(s_{2_i}-r_{2_j}\right)}\right)\right] = 0, \\
&(j = 1, \ldots, n-1),
\end{aligned} \tag{19d}$$

$$W_1 = W_n = \frac{1}{2n-2}, \qquad W_i = \frac{1}{n-1}, \qquad (i = 2, \ldots, n-1), \tag{19e}$$

$$s_{1_i} = \cos\left(\frac{i-1}{2n-1}\pi\right), \qquad s_{2_i} = \cos\left(\frac{i-1}{n-1}\pi\right), \qquad (i = 1, \ldots, n), \tag{19f}$$

$$r_{1_j} = \cos\left(\frac{2j-1}{4n-2}\pi\right), \qquad r_{2_j} = \cos\left(\frac{2j-1}{2n-2}\pi\right), \qquad (j = 1, \ldots, n-1). \tag{19g}$$

The unknowns $G_1(s_{1_i})$ and $G_2(s_{2_i})$ $(i = 1, \ldots, n)$ are determined from the system (19a)–(19d). By using (18a), (18b), substituting the results into (16e), and using Gauss-Chebyshev integration formula, the contact stress $\sigma_{y_1}(x,0)h/P$ is evaluated. It should be observed that the integral equations (16c) and (16d) are valid provided the contact stress $\sigma_{y_1}(x,0)h/P$ is compressive everywhere; that is, $0 < x < \infty$. The critical load factor, λ_{cr} and the corresponding location of interface separation, x_{cr} can be determined through the use of the following condition for various dimensionless quantities:

$$\frac{\sigma_{y_1}(x,0)}{P/h} = 0. \tag{20}$$

4. Case of Discontinuous Contact

Since the interface cannot carry tensile tractions for $\lambda > \lambda_{cr}$, there will be separation between the elastic layer and the elastic half plane in the neighborhood of $x = x_{cr}$ on the contact plane $y = 0$, as shown in Figure 1. Assuming that the separation region is described by $d < x < e$, $y = 0$, where d and e are unknowns and functions of λ, boundary and continuity conditions for the discontinuous contact case are defined as follows:

$$\sigma_{y_1}(x,h) = \begin{cases} -P(x), & 0 < x < a \\ -Q(x), & b < x < c \\ 0, & a < x < b,\ c < x < \infty, \end{cases} \tag{21a}$$

$$\tau_{xy_1}(x,h) = 0, \quad 0 < x < \infty, \tag{21b}$$

$$\tau_{xy_1}(x,0) = 0, \quad 0 < x < \infty, \tag{21c}$$

$$\tau_{xy_2}(x,0) = 0, \quad 0 < x < \infty, \tag{21d}$$

$$\sigma_{y_1}(x,0) = \sigma_{y_2}(x,0), \quad 0 < x < \infty, \tag{21e}$$

$$\frac{\partial}{\partial x}\left[v_2(x,0) - v_1(x,0)\right] = \begin{cases} \varphi(x), & d < x < e \\ 0, & 0 < x < d,\ e < x < \infty, \end{cases} \tag{21f}$$

$$\sigma_{y_1}(x,0) = \sigma_{y_2}(x,0) = 0, \quad d < x < e, \tag{21g}$$

$$\frac{\partial}{\partial x}\left[v_1(x,h)\right]=0,\quad 0<x<a,\tag{21h}$$

$$\frac{\partial}{\partial x}\left[v_1(x,h)\right]=0,\quad c<x<b.\tag{21i}$$

After utilizing the boundary and continuity conditions defined in (21a)–(21f), new values for the constants A, B, C, D, E, and F which appear in (4a), (4b), (5a)–(5c), and (7a)–(7e) may be obtained in terms of new unknown functions $P(x)$, $Q(x)$, and $\varphi(x)$. Unknown functions are then determined from the conditions (21h)-(21i) which have not yet been satisfied. These conditions give the following system of singular integral equations:

$$-\frac{1}{\pi\mu_1}\int_0^a\left[k_1(x,t)+\frac{(1+\kappa_1)}{4}\left(\frac{1}{t+x}-\frac{1}{t-x}\right)\right]P(t)\,dt$$
$$-\frac{1}{\pi\mu_1}\int_b^c\left[k_1(x,t)+\frac{(1+\kappa_1)}{4}\left(\frac{1}{t+x}-\frac{1}{t-x}\right)\right]Q(t)\,dt$$
$$+\frac{1}{\pi}\int_d^e k_2(x,t)\,\varphi(t)\,dt=0,\quad 0<x<a,\tag{22a}$$

$$-\frac{1}{\pi\mu_1}\int_0^a\left[k_1(x,t)+\frac{(1+\kappa_1)}{4}\left(\frac{1}{t+x}-\frac{1}{t-x}\right)\right]P(t)\,dt$$
$$-\frac{1}{\pi\mu_1}\int_b^c\left[k_1(x,t)+\frac{(1+\kappa_1)}{4}\left(\frac{1}{t+x}-\frac{1}{t-x}\right)\right]Q(t)\,dt$$
$$+\frac{1}{\pi}\int_d^e k_2(x,t)\,\varphi(t)\,dt=0,\quad b<x<c,\tag{22b}$$

$$\frac{1}{\pi}\int_0^a k_2(x,t)\,P(t)\,dt+\frac{1}{\pi}\int_b^c k_2(x,t)\,Q(t)\,dt$$
$$-\frac{\mu_1}{\pi}\int_d^e\left[k_3(x,t)-\frac{4\mu_2/\mu_1}{(1+\kappa_2)+\mu_2/\mu_1(1+\kappa_1)}\right.$$
$$\left.\times\left(\frac{1}{t+x}+\frac{1}{t-x}\right)\right]\varphi(t)\,dt-\rho_1 gh=0,$$
$$d<x<e,\tag{22c}$$

where kernels $k_1(x,t)$ and $k_2(x,t)$ are given by (11) and (14) and

$$k_3(x,t)=\int_0^\infty\left\{\left\{-\frac{2\alpha^3\mu_2}{\mu_1}\left\{e^{-2\alpha h}\left(2+4\alpha^2h^2\right)-e^{-4\alpha h}-1\right\}\right.\right.$$
$$\left.\times(\Delta)^{-1}+\frac{4\mu_2/\mu_1}{(1+\kappa_2)+\mu_2/\mu_1(1+\kappa_1)}\right\}$$
$$\times\{\sin\alpha(t+x)+\sin\alpha(t-x)\}\,d\alpha,\tag{23}$$

in which Δ is given by (12).

The index of integral equations (22a) and (22b) is +1. On the other hand, the index of the singular integral equation (22c) is −1 due to the physical requirement of smooth contact at the end points d and e [37]. Thus, in solving the problem the two conditions which would account for the unknowns d and e are the consistency condition of integral equation (22c) and the single-valuedness condition

$$\int_d^e\varphi(x)\,dx=0.\tag{24}$$

Defining the following dimensionless quantities

$$x_3=\frac{e-d}{2}r_3+\frac{e+d}{2},\qquad t_3=\frac{e-d}{2}s_3+\frac{e+d}{2},\tag{25a}$$

$$g_3(s_3)=\frac{\mu_1\varphi\left(((e-d)/2)\,s_3+(e+d)/2\right)}{P/h};\tag{25b}$$

by making use of (15a)–(15h), the integral equations (22a)–(22c) may be expressed as follows:

$$-\frac{1}{\pi}\int_0^1 g_1(s_1)\frac{a}{h}ds_1\left[m_1^*(r_1,s_1)+\frac{(1+\kappa_1)}{4}\right.$$
$$\left.\times\left(\frac{1}{a(s_1+r_1)}-\frac{1}{a(s_1-r_1)}\right)\right]$$
$$-\frac{1}{\pi}\int_{-1}^1 g_2(s_2)\frac{c-b}{2h}ds_2$$
$$\times\left[m_2^*(r_1,s_2)+\frac{(1+\kappa_1)}{4}\right.$$
$$\times\left(\frac{1}{(((c-b)/2)\,s_2+(c+b)/2)+ar_1}\right.$$
$$\left.\left.-\frac{1}{(((c-b)/2)\,s_2+(c+b)/2)-ar_1}\right)\right]$$
$$+\frac{1}{\pi}\int_{-1}^1 g_3(s_3)\,m_3^*(r_1,s_3)\frac{e-d}{2h}ds_3=0,$$
$$0<r_1<1,\tag{26a}$$

$$-\frac{1}{\pi}\int_{-1}^1 g_1(s_1)\frac{a}{h}ds_1$$
$$\times\left[m_4^*(r_2,s_1)+\frac{(1+\kappa_1)}{4}\right.$$
$$\times\left(\frac{1}{as_1+(((c-b)/2)\,r_2+(c+b)/2)}\right.$$
$$\left.\left.-\frac{1}{as_1-(((c-b)/2)\,r_2+(c+b)/2)}\right)\right]$$
$$-\frac{1}{\pi}\int_{-1}^1 g_2(s_2)\frac{c-b}{2h}ds_2\left[m_5^*(r_2,s_2)+\frac{(1+\kappa_1)}{4}\right.$$
$$\times\left(\frac{1}{((c-b)/2)(s_2+r_2)+(c+b)}\right.$$
$$\left.\left.-\frac{1}{((c-b)/2)(s_2-r_2)}\right)\right]$$
$$+\frac{1}{\pi}\int_{-1}^1 g_3(s_3)\,m_6^*(r_2,s_3)\frac{e-d}{2h}ds_3=0,$$
$$-1<r_2<1,\tag{26b}$$

$$-\frac{1}{\pi}\int_{-1}^{1} g_1(s_1)\, m_7^*(r_3,s_1)\,\frac{a}{h}ds_1$$
$$-\frac{1}{\pi}\int_{-1}^{1} g_2(s_2)\, m_8^*(r_3,s_2)\,\frac{c-b}{2h}ds_2$$
$$-\frac{1}{\pi}\int_{-1}^{1} g_3(s_3)\,\frac{e-d}{2h}ds_3$$
$$\times\left[m_9^*(r_3,s_3)-\frac{4\mu_2/\mu_1}{(1+\kappa_2)+\mu_2/\mu_1(1+\kappa_1)}\right.$$
$$\times\left(\frac{1}{((e-d)/2)(s_3+r_3)+(e+d)}\right.$$
$$\left.\left.+\frac{1}{((e-d)/2)(s_3-r_3)}\right)\right]-\frac{1}{\lambda}=0,$$
$$-1<r_3<1, \qquad (26c)$$

where

$$m_1^*(r_1,s_1)=k_1(x_1,t_1), \qquad (27a)$$
$$m_2^*(r_1,s_2)=k_1(x_1,t_2), \qquad (27b)$$
$$m_3^*(r_1,s_3)=k_2(x_1,t_3), \qquad (27c)$$
$$m_4^*(r_2,s_1)=k_1(x_2,t_1), \qquad (27d)$$
$$m_5^*(r_2,s_2)=k_1(x_2,t_2), \qquad (27e)$$
$$m_6^*(r_2,s_3)=k_2(x_2,t_3), \qquad (27f)$$
$$m_7^*(r_3,s_1)=k_2(x_3,t_1), \qquad (27g)$$
$$m_8^*(r_3,s_2)=k_2(x_3,t_2), \qquad (27h)$$
$$m_9^*(r_3,s_3)=k_3(x_3,t_3). \qquad (27i)$$

Similar to (16a), (16b), additional condition (24) may be expressed as

$$\int_{-1}^{1} g_3(s_3)\,ds_3=0. \qquad (28)$$

To solve the system of integral equations, it is found to be more convenient to assume that (26c) as well as (26a) and (26b) has an index +1 [29]; consequently, the function $g_i(s_i)$ $(i=1,\ldots,3)$ may be expressed in the form

$$g_1(s_1)=G_1(s_1)\left(1-s_1^2\right)^{-1/2},\quad 0<s_1<1, \qquad (29a)$$

$$g_i(s_i)=G_i(s_i)\left(1-s_i^2\right)^{-1/2},$$
$$-1<s_i<1,\quad (i=2,3), \qquad (29b)$$

where $G_i(s_i)$ is a bounded function. In order to insure smooth contact at the end points of the separation region, we then impose the following conditions on $G_3(s_3)$:

$$G_3(-1)=0,\qquad G_3(1)=0. \qquad (30)$$

Equations (26a)–(26c), (16a), (16b), and (28) can easily be reduced to the following system of linear algebraic equations by employing the appropriate Gauss-Chebyshev integration formula [36]:

$$-\sum_{i=1}^{n} W_iG_1\left(s_{1_i}\right)\frac{a}{h}\left[m_1^*\left(r_{1_j},s_{1_i}\right)+\frac{(1+\kappa_1)}{4}\right.$$
$$\left.\times\frac{1}{a\left(s_{1_i}+r_{1_j}\right)}-\frac{1}{a\left(s_{1_i}-r_{1_j}\right)}\right]$$
$$-\sum_{i=1}^{n} W_iG_2\left(s_{2_i}\right)\frac{c-b}{2h}$$
$$\times\left[m_2^*\left(r_{1_j},s_{2_i}\right)+\frac{(1+\kappa_1)}{4}\right.$$
$$\times\left(\frac{1}{\left(((c-b)/2)s_{2_i}+(c+b)/2\right)+ar_{1_j}}\right.$$
$$\left.\left.-\frac{1}{\left(((c-b)/2)s_{2_i}+(c+b)/2\right)-ar_{1_j}}\right)\right]$$
$$+\sum_{i=2}^{n-1} W_iG_3\left(s_{3_i}\right)\frac{e-d}{2h}m_3^*\left(r_{1_j},s_{3_i}\right)=0,$$
$$(j=1,\ldots,n-1), \qquad (31a)$$

$$-\sum_{i=1}^{n} W_iG_1\left(s_{1_i}\right)\frac{a}{h}$$
$$\times\left[m_4^*\left(r_{2_j},s_{1_i}\right)+\frac{(1+\kappa_1)}{4}\right.$$
$$\times\left(\frac{1}{as_{1_i}+\left(((c-b)/2)r_{2_j}+(c+b)/2\right)}\right.$$
$$\left.\left.-\frac{1}{as_{1_i}-\left(((c-b)/2)r_{2_j}+(c+b)/2\right)}\right)\right]$$
$$-\sum_{i=1}^{n} W_iG_2\left(s_{2_i}\right)\frac{c-b}{2h}\left[m_5^*\left(r_{2_j},s_{2_i}\right)+\frac{(1+\kappa_1)}{4}\right.$$
$$\times\left(\frac{1}{((c-b)/2)\left(s_{2_i}-r_{2_j}\right)+(c+b)}\right.$$
$$\left.\left.\pm\frac{1}{((c-b)/2)\left(s_{2_i}-r_{2_j}\right)}\right)\right]$$
$$+\sum_{i=2}^{n-1} W_iG_3\left(s_{3_i}\right)\frac{e-d}{2h}m_6^*\left(r_{2_j},s_{3_i}\right)=0$$
$$(j=1,\ldots,n-1), \qquad (31b)$$

$$
\begin{aligned}
&-\sum_{i=1}^{n} W_i G_1\left(s_{1_i}\right)\frac{a}{h} m_7^*\left(r_{3_j}, s_{1_i}\right) \\
&-\sum_{i=1}^{n} W_i G_2\left(s_{2_i}\right)\frac{c-b}{2h} m_8^*\left(r_{3_j}, s_{2_i}\right) \\
&-\sum_{i=2}^{n-1} W_i G_3\left(s_{3_i}\right)\frac{e-d}{2h} \\
&\times\left[m_9^*\left(r_{3_j}, s_{3_i}\right) - \frac{4\mu_2/\mu_1}{(1+\kappa_2)+\mu_2/\mu_1(1+\kappa_1)}\right. \\
&\times\left(\frac{1}{((e-d)/2)\left(s_{3_i}-r_{3_j}\right)+(e+d)}\right. \\
&\left.\left.+\frac{1}{((e-d)/2)\left(s_{3_i}-r_{3_j}\right)}\right)\right] - \frac{1}{\lambda} = 0, \\
&(j = 1,\ldots,n-1),
\end{aligned}
\tag{31c}
$$

$$\sum_{i=1}^{n} \pi W_i \frac{a}{h} G_1\left(s_{1_i}\right) = \frac{1}{2}, \tag{31d}$$

$$\sum_{i=1}^{n} \pi W_i \frac{c-b}{2h} G_2\left(s_{2_i}\right) = \frac{Q}{P}, \tag{31e}$$

$$\sum_{i=2}^{n-1} \pi W_i G_3\left(s_{3_i}\right) = 0, \tag{31f}$$

where W_i, s_i, and r_j are given by (19e)–(19g) ($r_2 = r_3$, $s_2 = s_3$). It was shown in [36] that the consistency condition is automatically satisfied if the Gauss-Chebyshev integration formula is used for solving integral equations. Thus, (31a)–(31d) and (31e)-(31f) give $3n$ equation for $3n$ unknowns $G_1(s_i)$, $G_2(s_i)$, $G_3(s_j)$ ($i = 1,\ldots,n$), ($j = 2,\ldots,n-1$), d and e. The equation system is nonlinear in d and e; so an interpolation scheme is required for the solution. Selected values of d and e are substituted into (31a)–(31d), and $G_1(s)$, $G_2(s)$, and $G_3(s)$ are obtained which must satisfy (31e), (31f) at the same time for known $\lambda > \lambda_{cr}$. If (31e), (31f) are not satisfied, then solution must be repeated with new values of d and e until the (31e), (31f) are satisfied at the same time.

It should be noted that (22c) gives the $\sigma_{y_1}(x,0)h/P$ outside as well as inside the separation region (e, f). Thus, once the functions $G_1(s)$, $G_2(s)$, and $G_3(s)$ and the constants d and e are determined, contact stress $\sigma_{y_1}(x,0)h/P$ may be easily evaluated. The displacement component $v^*(x,0)$ in the separation region (d, e), referring to (21f) and (25b), may be obtained from

$$v^*(x,0) = v_2(x,0) - v_1(x,0) = \int_d^x g_3(t)\,dt, \qquad d < x < e \tag{32a}$$

TABLE 1: Variation of minimum value of distance between two punches $(b-a)/h$ to avoid separation under first punch with a/h and $(c-b)/h$ for various values of Q/P ($\mu_2/\mu_1 = 6.48$).

$(c-b)/h$	a/h	$(b-a)/h$ 2P	4P	6P	8P	10P	12P
0.25	0.5	0.2773	0.5465	0.7263	0.8564	0.9569	1.0383
	0.75	0.4645	0.7676	0.9485	1.075	1.1716	1.250
	1.0	0.603	0.9107	1.089	1.2138	1.3105	1.3893
0.5	0.5	0.185	0.4403	0.617	0.7466	0.8474	0.9291
	0.75	0.3617	0.6597	0.8405	0.9675	1.065	1.144
	1.0	0.4983	0.8035	0.983	1.1086	1.206	1.2856
0.75	0.5	0.108	0.3432	0.516	0.645	0.746	0.8285
	0.75	0.2715	0.5613	0.742	0.8699	0.9685	1.0485
	1.0	0.4045	0.7073	0.8875	1.015	1.1134	1.4445
1.0	0.5	0.0357	0.2504	0.4195	0.548	0.6497	0.7332
	0.75	0.191	0.4694	0.6502	0.7797	0.8798	0.9613
	1.0	0.320	0.6188	0.8005	0.930	1.030	1.1123

or

$$\frac{\mu_1}{P/h} v^*(x,0) = \frac{e-d}{2h}\int_{-1}^{r_3} G_3(s_3)\,ds_3, \qquad -1 < r_3 < 1, \tag{32b}$$

where

$$x = \frac{e-d}{2} r_3 + \frac{e+d}{2}. \tag{32c}$$

Also using appropriate Gauss-Chebyshev integration formula and taking +1 as the index of (32b), the following expression may be written for the vertical displacement in the separation region:

$$\frac{\mu_1}{P/h} v^*(x,0) = \frac{e-d}{2h}\sum_{i=2}^{k-1} W_i G_3\left(s_{3_i}\right), \qquad (k = 2,\ldots,n-1), \tag{33}$$

where W_i and s_i are given by (19e)–(19g).

5. Results and Discussion

Some of the calculated results obtained from the solution of the continuous and discontinuous contact problems for various dimensionless quantities such as μ_2/μ_1, a/h, $(c-b)/h$, Q/P, λ, and $(b-a)/h$ are presented in Figures 2, 3, 4, 5, and 6 and Tables 1, 2, 3, 4, and 5. First separation point between elastic layer and elastic half plane is determined and first separation region is investigated. Contact pressure $\sigma_{y_1}(x,0)h/P$ is presented. Depending on load factor λ, possibilities of other separation regions between elastic layer and elastic half plane are determined. Also possibility of separation between first punch and elastic layer is researched. Besides, the distance $(b-a)/h$, that ends the interaction of punches is examined. It is assumed that $Q/h \geq P/h$.

TABLE 2: Variation of distance between two punches $(b-a)/h$ that ends interaction of punches with Q/P ($\mu_2/\mu_1 = 2.75$, $a/h = 0.25$, and $(c-b)/h = 0.5$).

Q	$(b-a)/h$	Punch I		Punch II	
		$\lambda_{\mathrm{cr}_{\mathrm{right}}}$	$(x_{\mathrm{cr}_{\mathrm{right}}}-a)/h$	$\lambda_{\mathrm{cr}_{\mathrm{left}}}=\lambda_{\mathrm{cr}_{\mathrm{right}}}$	$(b-x_{\mathrm{cr}_{\mathrm{left}}})/h=(x_{\mathrm{cr}_{\mathrm{right}}}-c)/h$
P	10.5585	96.6748	2.091	96.6175	2.092
$2P$	9.5874	96.0544	2.091	48.3612	2.092
$4P$	8.6032	94.2026	2.094	24.1974	2.092
$6P$	8.1313	91.6959	2.100	16.1360	2.092
$8P$	7.8582	88.6769	2.111	12.1037	2.091
$10P$	7.6742	85.2495	2.128	9.6838	2.091
$12P$	7.5378	81.4924	2.153	8.0704	2.091
$14P$	7.4308	81.4924	2.192	6.9178	2.090

TABLE 3: Variation of distance between two punches $(b-a)/h$ that ends interaction of punches with μ_2/μ_1 ($Q = 6P$, $a/h = 0.25$, and $(c-b)/h = 0.5$).

μ_2/μ_1	$(b-a)/h$	Punch I		Punch II	
		$\lambda_{\mathrm{cr}_{\mathrm{right}}}$	$(x_{\mathrm{cr}_{\mathrm{right}}}-a)/h$	$\lambda_{\mathrm{cr}_{\mathrm{left}}}=\lambda_{\mathrm{cr}_{\mathrm{right}}}$	$(b-x_{\mathrm{cr}_{\mathrm{left}}})/h=(x_{\mathrm{cr}_{\mathrm{right}}}-c)/h$
0.15	14.8404	164.5971	5.309	36.6977	4.795
0.36	12.1858	156.6651	3.678	30.6651	3.596
0.61	10.8897	143.5372	3.066	26.8566	3.027
1.65	9.2688	114.5372	2.418	20.5185	2.402
2.75	8.1313	91.6959	2.100	16.1360	2.092
6.48	6.4981	65.5235	1.812	11.4136	1.793

Figure 2 shows limit distance between punches that initiates separation under first punch for various values of material constant, μ_2/μ_1 with Q/P. The distance $(b-a)/h$ increases, maintaining continuous contact between first punch and elastic layer with Q/P. Besides, for bigger values of μ_2/μ_1, limit distance that initiates separation under first punch decreases. In such a case, elastic half plane gets stiffer and it becomes easy to separate first punch from the elastic layer.

Variation of critical distance between punches with a/h and $(c-b)/h$ for various values of Q/P is presented in Table 1. For fixed values of second punch width, $(c-b)/h$, an increase in first punch width requires longer distance between punches to avoid separation under first punch. On the contrary, for fixed values of first punch width, a/h, an increase in second punch width decreases $(b-a)/h$ and punches can be placed closer to each other without separation.

Interaction between punches ends for a definite value of $(b-a)/h$. Tables 2–4 show the critical value of the distance that ends interaction between punches with dimensionless quantities μ_2/μ_1, a/h, $(c-b)/h$, and Q/P. In such a case, there is no need to consider punches together. Also these tables show the values of the load factor that cause separation between elastic layer and elastic half plane, λ_{cr}. For $\lambda = \lambda_{\mathrm{cr}}$, $\sigma_{y_1}(x,0)h/P$ is zero. Contact between punches and elastic layer is continuous.

Table 3 shows the critical distance between punches that ends interaction of punches with elastic constant, μ_2/μ_1. For small values of μ_2/μ_1; that is, it is easy to bend elastic layer, interaction between punches ends in a longer distance. Initial separation point x_{cr} between elastic layer and elastic half plane from the origin $x = 0$ decreases with an increase in μ_2/μ_1. Critical load factor also decreases in this situation.

Distance $(b-a)/h$ that ends interaction between punches increases with a decrease in second punch width while first punch width is fixed. If both first and second widths are increased, interaction of punches ends in a shorter distance. This situation is presented in Table 4. In this case, critical load factor, λ_{cr} increases but initial separation point, x_{cr} decreases with increment in punch widths. Separation also occurs at the right-hand side of the second punch.

For fixed values of $a/h = 0.5$, $(c-b)/h = 1$, $Q = 2P$, and $\mu_2/\mu_1 = 1.65$, variations of critical load factor, λ_{cr} and initial separation point, x_{cr} are given in Table 5. For small values of $(b-a)/h$, initial separation point between elastic layer and elastic half plane appears at the right-hand side of second punch. If $(b-a)/h$ increases, in this case separation takes place between two punches. Keeping on increasing, the distance $(b-a)/h$ ends the interaction of punches. For $(b-a)/h = 7.9446$, there is no need to consider punches together.

In Figures 3–5, the normalized contact stress distribution $\sigma_{y_1}(x,0)h/P$ at the interface of elastic layer and elastic half plane is given for the problems described in Section 3, and Section 4. Different scales have been used for continuous and discontinuous contact cases in order to include the entire pressure distribution and to give sufficient details in compact forms.

TABLE 4: Variation of distance between two punches $(b-a)/h$ that ends interaction of punches with a/h and $(c-b)/h$ ($Q = 4P$, $\mu_2/\mu_1 = 0.36$).

a/h	$(c-b)/h$	$(b-a)/h$	Punch I		Punch II	
			$\lambda_{cr_{right}}$	$(x_{cr_{right}} - a)/h$	$\lambda_{cr_{left}} = \lambda_{cr_{right}}$	$(b - x_{cr_{left}})/h = (x_{cr_{right}} - c)/h$
0.25	1.5	10.8345	150.2590	3.775	54.1240	3.3816
0.25	1.0	11.8775	164.4214	3.655	49.0259	3.4495
0.25	0.5	12.7008	169.3000	3.628	45.9626	3.5968
0.50	1.0	11.6754	173.4645	3.523	49.0235	3.4494
0.75	1.5	10.4960	176.6011	3.421	54.1289	3.3810

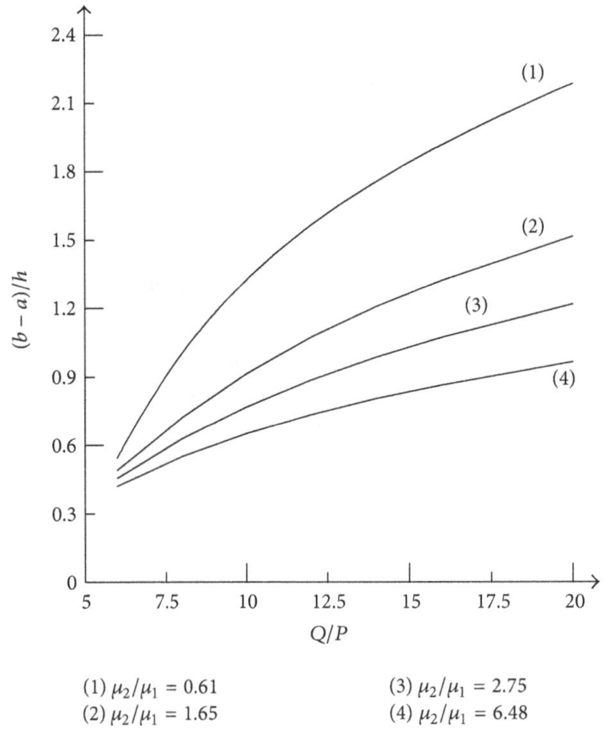

FIGURE 2: Variation of minimum value of distance between two punches $(b-a)/h$ to avoid separation under first punch with Q/P for various values of μ_2/μ_1 ($a/h = 0.5$, $(c-b)/h = 1$).

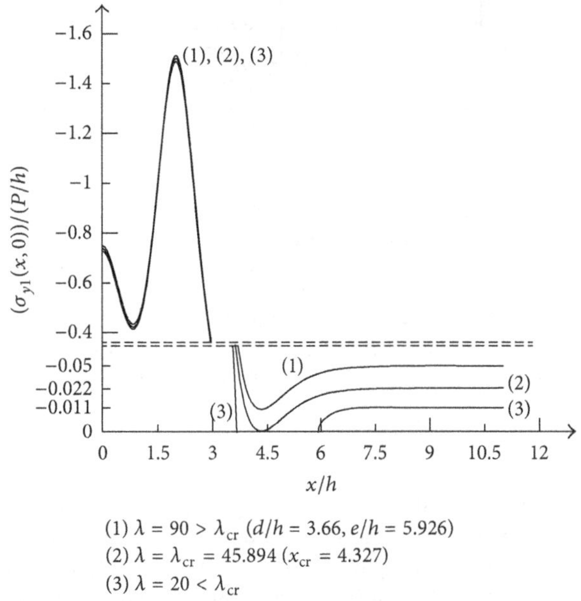

FIGURE 3: Contact stress distribution between elastic layer and elastic half plane for the cases of continuous ($\lambda < \lambda_{cr}$) and discontinuous contact ($\lambda > \lambda_{cr}$) ($Q = 2P$, $\mu_2/\mu_1 = 2.75$, $a/h = 0.25$, $(b-a)/h = 1.5$, and $(c-b)/h = 0.5$).

TABLE 5: Variation of load factor values with distance between two punches $(b-a)/h$ ($Q = 2P$, $\mu_2/\mu_1 = 1.65$, $a/h = 0.5$, and $(c-b)/h = 1$).

$(b-a)/h$	Punch I		Punch II			
	$\lambda_{cr_{right}}$	$x_{cr_{right}}$	$\lambda_{cr_{left}}$	$x_{cr_{left}}$	$\lambda_{cr_{right}}$	$x_{cr_{right}}$
0.5					**30.6814**	**4.294**
1					**32.4279**	**4.806**
3					**34.0783**	**6.808**
5	**28.4331**	**3.167**	**28.4331**	**3.167**	34.2756	8.803
6	**32.0727**	**4.184**	**32.0727**	**4.184**	34.3055	9.801
7	**33.8531**	**5.211**	**33.8531**	**5.211**	34.3175	10.800
7.9446	123.1714	2.835	**34.3182**	**6.148**	**34.3182**	**11.743**

Variation of the contact stress distribution $\sigma_{y_1}(x,0)h/P$ with load factor $\lambda = P/\rho_1 g h^2$ for fixed values of $Q = 2P$, $\mu_2/\mu_1 = 2.75$, $a/h = 0.25$, $(b-a)/h = 1.5$, and $(c-b)/h = 0.5$ is presented in Figure 3, $\lambda < \lambda_{cr}$ and $\lambda > \lambda_{cr}$ show contact stress distribution for the continuous and discontinuous contact cases, respectively. It can be seen from the figure that another separation region is possible if distance between punches $(b-a)/h$ or load factor λ is increased. Contact pressure has peaks around the edges of the rigid punches.

In Figure 4, the variation of the normalized contact stress $\sigma_{y_1}(x,0)h/P$ with Q/P is given for the discontinuous contact case ($\lambda = 75 > \lambda_{cr}$). Figure 4 shows that both contact stress and separation zone $(e-d)/h$ increase with an increase in Q/P. Variation of the normalized contact stress $\sigma_{y_1}(x,0)h/P$ for the discontinuous contact case shows three different regions between elastic layer and elastic half plane. These are continuous contact region, separation zone, and also continuous contact region, where the effects of external load Q/h and P/h decrease and disappear infinitely.

Variation of contact stress $\sigma_{y_1}(x,0)h/P$ with $\mu_2/\mu_1 = 1.65$ is shown in Figure 5. Separation zone increases as elastic half plane gets stiffer than the elastic layer. In this case, peak value of the contact stress also increases.

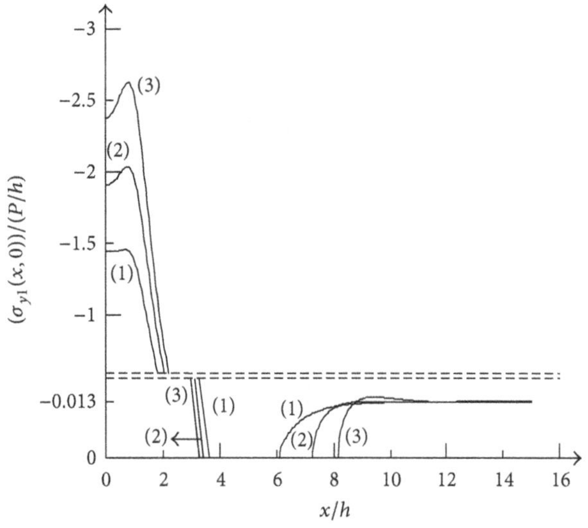

(1) $Q = 2P$ ($dh = 3.634$, $e/h = 6.085$)
(2) $Q = 3P$ ($dh = 3.396$, $e/h = 7.245$)
(3) $Q = 4P$ ($dh = 3.27$, $e/h = 8.153$)

FIGURE 4: Contact stress distribution between elastic layer and elastic half plane for the case of discontinuous contact ($\mu_2/\mu_1 = 0.36$, $a/h = 0.125$, $(b-a)/h = 0.5$, $(c-b)/h = 0.5$, and $\lambda = 75 > \lambda_{cr}$).

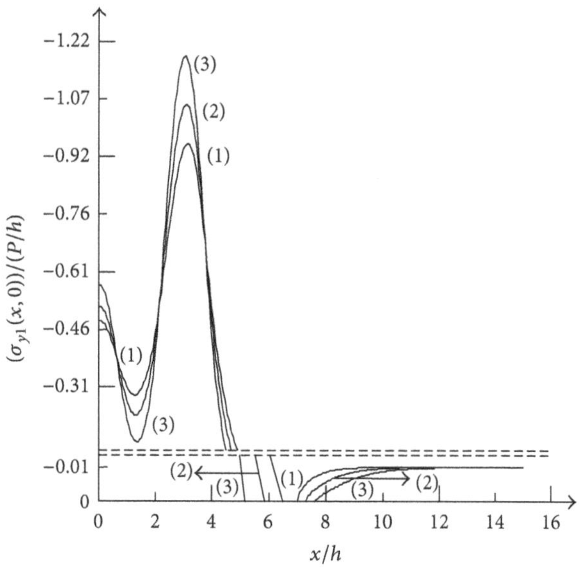

(1) $\mu_2/\mu_1 = 0.36$ ($dh = 6.498$, $e/h = 7.612$)
(2) $\mu_2/\mu_1 = 0.61$ ($dh = 5.869$, $e/h = 7.299$)
(3) $\mu_2/\mu_1 = 1.65$ ($dh = 5.175$, $e/h = 7.003$)

FIGURE 5: Contact stress distribution between elastic layer and elastic half plane for the case of discontinuous contact ($Q = 2P$, $a/h = 0.5$, $(b-a)/h = 2$, $(c-b)/h = 1$, and $\lambda = 100 > \lambda_{cr}$).

Results calculated from (33) giving the displacement $v^*(x,0)$ in the separation region $d < x < e$ as a function of x are shown in Figure 6. Also results are compared with those of [35]. It is seen that separation region $d < x < e$ and displacement values increase with an increase in Q/P ratio as this is the case in [35]. Separation zone is nearly the same, but displacement values are smaller than those of [35] for fixed values of $\mu_2/\mu_1 = 6.48$, $a/h = 0.75$, $(b-a)/h = 1$, and $(c-b)/h = 1.5$.

6. Conclusion

In this paper, continuous and discontinuous contact problems for an elastic layer resting on an elastic half plane loaded by means of three rigid flat punches are considered. Numerical procedures developed in this study can be used to find approximate solutions to problems of engineering interest.

Numerical results show that the punch widths, elastic constants, external loads, distance between punches play a very important role in the formation of the continuous and discontinuous contact areas, initial separation point and load, separation displacement, limit distance that ends the interaction of punches, critical distance that causes separation under first punch, and the contact pressure distribution.

From the study, the following conclusions may be drawn:

(i) it is easy to separate first punch from elastic layer if Q/P or a/h increases. This results also decrease in μ_2/μ_1 or $(c-b)/h$.

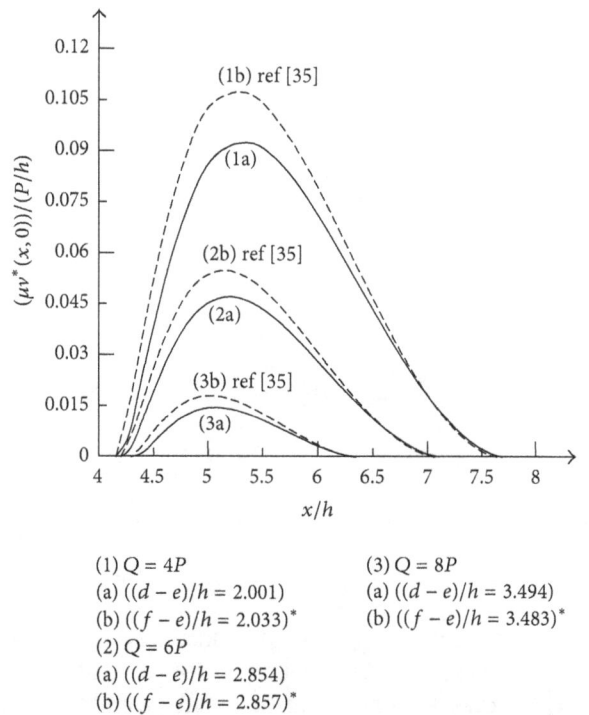

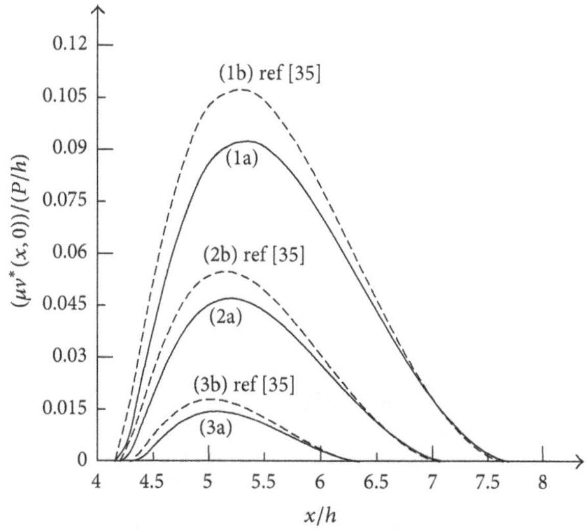

(1) $Q = 4P$
(a) $((d-e)/h = 2.001)$
(b) $((f-e)/h = 2.033)^*$
(2) $Q = 6P$
(a) $((d-e)/h = 2.854)$
(b) $((f-e)/h = 2.857)^*$
(3) $Q = 8P$
(a) $((d-e)/h = 3.494)$
(b) $((f-e)/h = 3.483)^*$

FIGURE 6: Separation displacement $v^*(x,0)$ between elastic layer and elastic half space as a function of x for various values of second punch load Q ($\mu_2/\mu_1 = 6.48$, $a/h = 0.75$, $(b-a)/h = 1$, $(c-b)/h = 1.5$, and $\lambda = 50 > \lambda_{cr}$), (*: [35]).

(ii) Increment in Q/P, a/h or $(c-b)/h$ ends interaction of punches in a shorter distance. Also an increase in μ_2/μ_1 causes the same result.

(iii) First separation between elastic layer and elastic half plane occurs between punches or in the region $c < x < \infty$ depending on λ. If $(b-a)/h$ is big enough, interaction of punches disappears.

(iv) Size of separation region is not affected much with Q/P, but separation displacement is increased with an increase in Q/P.

Nomenclature

P: Compressive load per unit thickness in z direction on punch I, (N/m)
Q: Compressive load per unit thickness in z direction on punch II and punch III, (N/m)
h: Thickness of the layer, (m)
λ: Load factor
u: x-component of the displacement, (m)
v: y-component of the displacement, (m)
ρ_k: Mass density, (kg/m^3)
g: Gravity acceleration, (m/sn2)
μ: Shear modulus, (Pa)
κ: Elastic constant
γ: Poisson's ratio
$(c-b)$: Width of punch II and punch III, (m)
a: Half width of punch I, (m)
$P(x)$: Contact pressure under punch I, (Pa)
$Q(x)$: Contact pressure under punch II and punch III, (Pa)
$(e-d)$: Separation length, (m).

References

[1] P. Wriggers, *Computational Contact Mechanics*, John Wiley & Sons, Chichester, UK, 2002.

[2] M. R. Gecit, "The axisymmetric double contact problem for a frictionless elastic layer indented by an elastic cylinder," *International Journal of Engineering Science*, vol. 24, no. 9, pp. 1571–1584, 1986.

[3] V. Kahya, T. S. Ozsahin, A. Birinci, and R. Erdol, "A receding contact problem for an anisotropic elastic medium consisting of a layer and a half plane," *International Journal of Solids and Structures*, vol. 44, no. 17, pp. 5695–5710, 2007.

[4] I. Comez, A. Birinci, and R. Erdol, "Double receding contact problem for a rigid stamp and two elastic layers," *European Journal of Mechanics, A/Solids*, vol. 23, no. 2, pp. 301–309, 2004.

[5] L. M. Keer, J. Dondurs, and K. C. Tsai, "Problems involving a receding contact between a layer and a half space," *Journal of Applied Mechanics, Transactions ASME*, vol. 39, no. 4, pp. 1115–1120, 1972.

[6] F. Erdogan and M. Ratwani, "The contact problem for an elastic layer supported by two elastic quarter planes," *Journal of Applied Mechanics, Transactions ASME*, vol. 41, no. 3, pp. 673–678, 1974.

[7] M. B. Civelek and F. Erdogan, "The axisymmetric double contact problem for a frictionless elastic layer," *International Journal of Solids and Structures*, vol. 10, no. 6, pp. 639–659, 1974.

[8] Q. Lan, G. A. C. Graham, and A. P. S. Selvadurai, "Certain two-punch problems for an elastic layer," *International Journal of Solids and Structures*, vol. 33, no. 19, pp. 2759–2774, 1996.

[9] M. J. Jaffar, "Frictionless contact between an elastic layer on a rigid base and a circular flat-ended punch with rounded edge or a conical punch with rounded tip," *International Journal of Mechanical Sciences*, vol. 44, no. 3, pp. 545–560, 2002.

[10] G. Fu and A. Chandra, "Normal indentation of elastic half-space with a rigid frictionless axisymmetric punch," *Journal of Applied Mechanics, Transactions ASME*, vol. 69, no. 2, pp. 142–147, 2002.

[11] Y. Pinyochotiwong, J. Rungamornrat, and T. Senjuntichai, "Analysis of rigid frictionless indentation on half-space with surface elasticity," in *Proceedings of the 12th East Asia-Pasific Conference on Structural Engineering and Construction*, vol. 14, pp. 2403–2410, 2011.

[12] G. M. L. Gladwell, "On some unbonded contact problems in plane elasticity theory," *Journal of Applied Mechanics, Transactions ASME*, vol. 43, no. 2, pp. 263–267, 1976.

[13] O. I. Zhupanska, "Contact problem for elastic spheres: applicability of the Hertz theory to non-small contact areas," *International Journal of Engineering Science*, vol. 49, no. 7, pp. 576–588, 2011.

[14] M. Porter and D. A. Hills, "A flat punch pressed against an elastic interlayer under conditions of slip and separation," *International Journal of Mechanical Sciences*, vol. 44, no. 3, pp. 465–474, 2002.

[15] M. R. . Geçit and S. Gökpınar, "Frictionless contact between an elastic layer and a rigid rounded support," *Arabian Journal of Science and Engineering*, vol. 10, no. 3, pp. 243–251, 1985.

[16] F. Yang, "Adhesive contact between an elliptical rigid flat-ended punch and an elastic half space," *Journal of Physics D*, vol. 38, no. 8, pp. 1211–1214, 2005.

[17] M. I. Porter and D. A. Hills, "Note on the complete contact between a flat rigid punch and an elastic layer attached to a dissimilar substrate," *International Journal of Mechanical Sciences*, vol. 44, no. 3, pp. 509–520, 2002.

[18] M. E. R. Shanahan, "Adhesion of a punch to a thin membrane," *Comptes Rendus de l'Academie des Sciences IV*, vol. 1, no. 4, pp. 517–522, 2000.

[19] A. Klarbring, A. Mikelić, and M. Shillor, "The rigid punch problem with friction," *International Journal of Engineering Science*, vol. 29, no. 6, pp. 751–768, 1991.

[20] V. I. Fabrikant and T. S. Sankar, "Concentrated force underneath a punch bonded to a transversely isotropic half-space," *International Journal of Engineering Science*, vol. 24, no. 1, pp. 111–117, 1986.

[21] S. Karuppanan, C. M. Churchman, D. A. Hills, and E. Giner, "Sliding frictional contact between a square block and an elastically similar half-plane," *European Journal of Mechanics, A/Solids*, vol. 27, no. 3, pp. 443–459, 2008.

[22] I. Çömez and R. Erdöl, "Frictional contact problem of a rigid stamp and an elastic layer bonded to a homogeneous substrate," *Archive of Applied Mechanics*, vol. 83, no. 1, pp. 15–24, 2013.

[23] A. P. S. Selvadurai, "Mechanics of a rigid circular disc bonded to a cracked elastic half-space," *International Journal of Solids and Structures*, vol. 39, no. 24, pp. 6035–6053, 2002.

[24] A. P. S. Selvadurai, "Boussinesq indentation of an isotropic elastic halfspace reinforced with an inextensible membrane," *International Journal of Engineering Science*, vol. 47, no. 11-12, pp. 1339–1345, 2009.

[25] R. Artan and M. Omurtag, "Two plane punches on a nonlocal elastic half plane," *International Journal of Engineering Science*, vol. 38, no. 4, pp. 395–403, 2000.

[26] R. S. Dhaliwal, "Punch problem for an elastic layer overlying an elastic foundation," *International Journal of Engineering Science*, vol. 8, no. 4, pp. 273–288, 1970.

[27] I. Çömez, "Frictional contact problem for a rigid cylindrical stamp and an elastic layer resting on a half plane," *International Journal of Solids and Structures*, vol. 47, no. 7-8, pp. 1090–1097, 2010.

[28] M. R. Gecit, "A tensionless contact without friction between an elastic layer and an elastic foundation," *International Journal of Solids and Structures*, vol. 16, no. 5, pp. 387–396, 1980.

[29] A. O. Cakiroglu and F. L. Cakiroglu, "Continuous and discontinuous contact problems for strips on an elastic semi-infinite plane," *International Journal of Engineering Science*, vol. 29, no. 1, pp. 99–111, 1991.

[30] M. R. Gecit and F. Erdogan, "Frictionless contact problem for an elastic layer under axisymmetric loading," *International Journal of Solids and Structures*, vol. 14, no. 9, pp. 771–785, 1978.

[31] A. Birinci and R. Erdöl, "Continuous and discontinuous contact problem for a layered composite resting on simple supports," *Structural Engineering and Mechanics*, vol. 12, no. 1, pp. 17–34, 2001.

[32] M. B. Civelek and F. Erdogan, "The frictionless contact problem for an elastic layer under gravity," *Journal of Applied Mechanics, Transactions ASME*, vol. 42, no. 97, pp. 136–140, 1975.

[33] D. Schmueser, M. Comninou, and J. Dundurs, "Separation and slip between a layer and a substrate caused by a tensile load," *International Journal of Engineering Science*, vol. 18, no. 9, pp. 1149–1155, 1980.

[34] M. B. Civelek, F. Erdogan, and A. O. Çakıroglu, "Interface separation for an elastic layer loaded by a rigid stamp," *International Journal of Engineering Science*, vol. 16, no. 9, pp. 669–679, 1978.

[35] T. S. Ozsahin, "Frictionless contact problem for a layer on an elastic half plane loaded by means of two dissimilar rigid punches," *Structural Engineering and Mechanics*, vol. 25, no. 4, pp. 383–403, 2007.

[36] F. Erdogan and G. D. Gupta, "On the numerical solution of singular integral equations," *Quarterly of Applied Mathematics*, vol. 29, pp. 525–534, 1972.

[37] N. I. Muskhelishvili, *Singular Integral Equations*, Noordhoff, Leyden, The Netherlands, 1958.

21

Prediction of Banking Systemic Risk Based on Support Vector Machine

Shouwei Li, Mingliang Wang, and Jianmin He

School of Economics and Management, Southeast University, Nanjing, Jiangsu 211189, China

Correspondence should be addressed to Shouwei Li; lswseu@126.com

Academic Editor: Wei-Chiang Hong

Banking systemic risk is a complex nonlinear phenomenon and has shed light on the importance of safeguarding financial stability by recent financial crisis. According to the complex nonlinear characteristics of banking systemic risk, in this paper we apply support vector machine (SVM) to the prediction of banking systemic risk in an attempt to suggest a new model with better explanatory power and stability. We conduct a case study of an SVM-based prediction model for Chinese banking systemic risk and find the experiment results showing that support vector machine is an efficient method in such case.

1. Introduction

Financial crises occurred over the past decade have shown that banking crises are usually in the core position of financial crises. Therefore, banking sectors' stability is a key to maintain financial stability. However, the major threat to banking sectors is systemic risk, because its contagion effect makes a single bank crisis evolve into the whole banking crisis, which may cause a financial crisis. The 2007–2009 financial crisis has threatened the stability of the international monetary market, shed light on the importance of systemic risk, and revealed the necessity of predicting one in order to safeguard financial stability.

The prediction of banking systemic crisis has been done since the mid-1990s. Frankel and Rose [1] use a probit model to evaluate the predictive power of several indicators for studying the determinants of currency crashes. Kaminsky and Reinhart [2] extend the analysis of Frankel and Rose [1] to a wider set of crises by introducing the so-called "signal" approach to evaluate the leading properties of indicators. Based on the study of Kaminsky and Reinhart [2], Demirgüç-Kunt and Detragiache [3] use a multivariate logit model for the prediction of banking crises. Other approaches are introduced to early warning models for systemic risk in the subsequent years, such as the regression model [4–6], the neurofuzzy approach [7], binary recursive trees [8, 9], and the alternative approach [10, 11].

In fact, banking systemic risk is a complex nonlinear phenomenon, which originates from diversity and uncertainty of risk sources, multiple contagion channels and their relationships as well as complexity and evolution of banking system structures [12–14]. The above prediction approaches either have certain limitations to analyze the complex nonlinear property of banking systemic risk or need a lot of samples. Recently, support vector machine (SVM) introduced by Vapnik [15] has become a promising tool for solving nonlinear regression problems and has been applied in prediction in many areas [16–18]. SVM has many advantages; for example, it combines the advantages of multivariate nonlinear regression in that only a small amount of data is required to produce a good generalization. In addition, the weakness of the transformational models in multivariate nonlinear regression can be overcome by mapping the data points to a sufficiently high-dimensional feature space [17]. Therefore, it is believed that SVM will provide a new approach for predicting banking systemic risk.

In this paper, we attempt to develop an SVM-based predictive model for banking systemic risk. Our paper is organized as follows. Section 2 provides the brief description of SVM theory. Section 3 elaborates a case study of an SVM-based prediction model for Chinese banking systemic risk. Section 4 draws conclusions.

2. Support Vector Machine

Based on statistical learning theory, Vapnik [15] put forward support vector machine. Its basic idea is to construct an optimal separating hyperplane with high classification accuracy [19]. A brief description of the SVM algorithm is provided as follows [16, 18, 19].

Consider the training dataset $D = \{x_i, y_i\}_{i=1}^{M}$, where $x_i \in R^m$ is an input vector and $y_i \in \{-1, 1\}$ denotes a target label. In the linear separable case, SVM algorithm is to find an optimal separating plane $w \cdot x + b = 0$. In the nonlinear separable case, a nonlinear function $\Phi(x)$ must be applied to map input space to a higher dimensional feature space, and the constrained optimization model of soft margin based on SVM is as follows [19]:

$$\text{Minimize} \quad \frac{1}{2} w^T w + C \sum_{i=1}^{M} \xi_i \tag{1}$$

subject to

$$\begin{aligned} y_i \left[w^T \Phi(x_i) + b \right] &\geq 1 - \xi_i \quad (i = 1, 2, \ldots, M), \\ \xi_i &\geq 0 \quad (i = 1, 2, \ldots, M), \end{aligned} \tag{2}$$

where ξ_i is a slack variable and C is a tuning parameter. By applying the Lagrangian technique to (2), we can obtain its dual problem as follows:

$$\text{Maximize} \quad \frac{1}{2} \alpha^T Q \alpha - e^T \alpha \tag{3}$$

subject to

$$\begin{aligned} 0 \leq \alpha_i \leq C \quad (i = 1, 2, \ldots, M), \\ y^T \alpha = 0, \end{aligned} \tag{4}$$

where e^T is a vector of all ones, Q is a positive semidefinite matrix, and α_i is Lagrange multiplier. The element Q_{ij} of Q is equal to $y_i y_j K(x_i, x_j)$, where $K(x_i, x_j) = \Phi(x_i)^T \Phi(x_j)$ is called kernel function. There are different kernel functions, such as linear, polynomial, and radial basis functions (RBF). Based on the above analysis, we can obtain the ultimate classifier function of SVM as follows:

$$f(x) = \text{sign}\left(\sum_{i=1}^{M} \alpha_i y_i K(x, x_i) + b \right). \tag{5}$$

3. Prediction Model for Banking Systemic Risk

In this section, we take the Chinese banking system as an example to construct an SVM-based prediction model for banking systemic risk and conduct empirical analysis based on the data of the Chinese banking system.

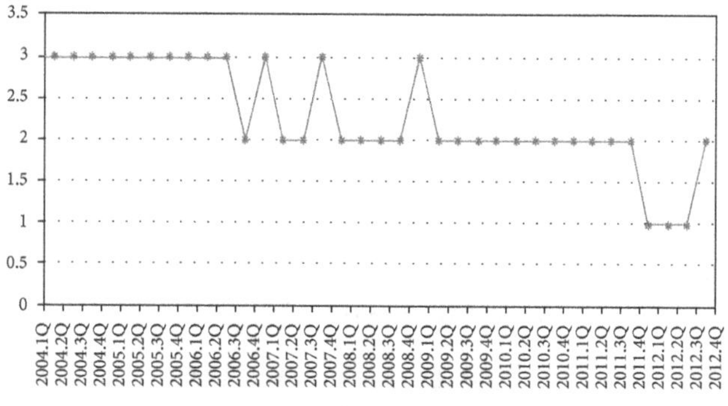

FIGURE 1: Classification result of levels of banking systemic risk from the first quarter of 2004 to the third quarter of 2012. In these Figures 1, 2 and 3 represent respectively, safety, mild safety and mild unsafety.

3.1. Prediction Indicators for Banking Systemic Risk. Micro-prudential indicators cannot provide a systemic perspective on banking systemic risk, while macro-prudential indicators cannot provide distress warnings from individual banks. Recent financial crises show that interbank connections play an important role in propagation mechanisms of banking systemic risk. Thus, the prediction indicators for banking systemic risk should incorporate both Microprudential and macro-prudential perspectives, as well as the characteristics of interbank connections. Based on the relative studies [20–23], prediction indicators and their prudential intervals for banking systemic risk in this paper are given in Table 1, where all indicators are processed chemotactically, X_i ($i = 1, 2, \ldots, 7$) are Microprudential indicators, X_j ($j = 8, 9, 10$) are indicators of interbank connections, and X_k ($k = 11, 12, \ldots, 17$) are macro-prudential indicators.

3.2. Evaluation for Levels of Banking Systemic Risk. In light of the lack of data on levels of banking systemic risk in China, we adopt the principal component analysis to classify levels of banking systemic risk. The basic idea is as follows: constructing three critical value samples based on data of prediction indicators processed chemotactically in Table 1; conducting the principal component analysis on selected samples and the three critical value samples; classifying scores of selected experimental samples according to scores of the three critical value samples. In this paper, we select data of prediction indicators from the first quarter of 2004 to the third quarter of 2012 to collect the experimental data. By conducting the principal component analysis, we can obtain the levels of banking systemic risk, which is presented in Figure 1.

3.3. Empirical Analysis of Prediction Accuracy. In this section, we analyze the prediction accuracy of SVM and also compare its performance with those of back-propagation neural network (BPNN), multiple discriminant analysis (MDA), and logistic regression analysis (Logit). We take the data of prediction indicators from the first quarter of 2004 to the second quarter of 2008 as the training sample and the rest of the data as the testing sample. At the same time, we construct a safety sample based on critical values of the safety condition

TABLE 1: List of prediction indicators for banking systemic risk.

Indicator	Safety	Mild safety	Mild unsafety	Unsafety
Capital adequacy ratio (X_1)	$12\% \le X_1$	$10\% \le X_1 < 12\%$	$8\% \le X_1 < 10\%$	$X_1 < 8\%$
Nonperforming loan ratio (X_2)	$X_2 \le 2\%$	$2\% < X_2 \le 3.5\%$	$3.5\% < X_2 \le 5\%$	$5\% < X_2$
Proportion of a single maximum loan (X_3)	$X_3 \le 6\%$	$6\% < X_3 \le 8\%$	$8\% < X_3 \le 10\%$	$10\% < X_3$
Return on assets (X_4)	$0.25\% \le X_4$	$0.2\% \le X_4 < 0.25\%$	$0.2\% \le X_4 < 0.15\%$	$X_4 < 0.15\%$
Cost-to-income ratio (X_5)	$X_5 \le 20\%$	$20\% < X_5 \le 28\%$	$28\% < X_5 \le 35\%$	$35\% < X_5$
Liquidity ratio (X_6)	$45\% \le X_6$	$35\% \le X_6 < 45\%$	$25\% \le X_6 < 35\%$	$X_6 < 25\%$
Loan-to-deposit ratio (X_7)	$X_7 \le 55\%$	$55\% < X_7 \le 65\%$	$65\% < X_7 \le 75\%$	$75\% < X_7$
Leverage (X_8)	$8\% \le X_8$	$8\% \le X_8 < 6\%$	$6\% \le X_8 < 4\%$	$X_8 < 4\%$
Interdependence (X_9)	$X_9 \le 6\%$	$6\% < X_9 \le 8\%$	$8\% < X_9 \le 10\%$	$10\% < X_9$
External linkages (X_{10})	$X_{10} \le 1.8\%$	$1.8\% < X_{10} \le 2\%$	$2\% < X_{10} \le 2.2\%$	$2.2\% < X_{10}$
Year-on-year growth of GDP (X_{11})	$\lvert X_{11} - 8\%\rvert \le 1.5\%$	$1.5\% < \lvert X_{11} - 8\%\rvert \le 3\%$	$3\% < \lvert X_{11} - 8\%\rvert \le 4.5\%$	$4.5\% < \lvert X_{11} - 8\%\rvert$
Year-on-year growth of CPI (X_{12})	$\lvert X_{12} - 2\%\rvert \le 2\%$	$2\% < \lvert X_{12} - 2\%\rvert \le 4\%$	$4\% < \lvert X_{12} - 2\%\rvert \le 6\%$	$6\% < \lvert X_{12} - 2\%\rvert$
Year-on-year growth of fixed asset investment (X_{13})	$\lvert X_{13} - 16\%\rvert \le 3\%$	$3\% < \lvert X_{13} - 16\%\rvert \le 6\%$	$6\% < \lvert X_{13} - 16\%\rvert \le 9\%$	$9\% < \lvert X_{13} - 16\%\rvert$
Year-on-year growth of national real estate index (X_{14})	$\lvert X_{14}\rvert \le 1.5\%$	$1.5\% < \lvert X_{14}\rvert \le 3\%$	$3\% < \lvert X_{14}\rvert \le 4.5\%$	$4.5\% < \lvert X_{14}\rvert$
Volatility of the Shanghai index (X_{15})	$\lvert X_{15}\rvert \le 10\%$	$10\% < \lvert X_{15}\rvert \le 15\%$	$15\% < \lvert X_{15}\rvert \le 20\%$	$20\% < \lvert X_{15}\rvert$
Ratio of growth of M2 to growth of GDP (X_{16})	$X_{16} \le 1.5$	$1.5 < X_{16} \le 2.2$	$2.2 < X_{16} \le 3$	$3 < X_{16}$
Volatility of benchmark lending rates (X_{17})	$\lvert X_{17}\rvert = 0$	$0 < \lvert X_{17}\rvert \le 0.2\%$	$0.2\% < \lvert X_{17}\rvert \le 0.4\%$	$0.4\% < \lvert X_{17}\rvert$

into the training sample, because of the lack of safety samples in the training sample. Therefore, the size of the training sample is equal to 19, and the sizes of the testing sample is equal to 17. In case of BPNN, the rest data is split into the validation sample and the testing sample, and the size of the validation sample and the testing sample are 4 and 13, respectively.

TABLE 2: The prediction accuracy of SVM.

	Grid Search	Genetic Algorithms	Particle Swarm Optimization
Training sample	100%	100%	100%
Testing sample	94.12%	88.24%	88.24%

3.3.1. Prediction Accuracy of SVM. Construction of prediction models for banking systemic risk based on SVM is to choose the kernel function and the values of model parameters. In this paper, the radial basis function (RBF) is used as the basic kernel function of SVM, because it usually gets better results than other kernel functions [24]. There are two parameters associated with RBF kernels: C and g, playing a crucial role in the performance of SVM. Therefore, improper selection of these two parameters can cause overfitting or underfitting problems. In this paper, we determine the values of parameters C and g by using Grid Search, Genetic Algorithms, and Particle Swarm Optimization.

According to the above method, we first analyze prediction accuracy of prediction models for banking systemic risk based on SVM. Based on the above data, we can obtain values of optimal parameters C and g for SVM model by Grid Search, Genetic Algorithms, and Particle Swarm Optimization. Figures 2, 3, and 4 report, respectively, the results of Grid Search, Genetic Algorithms, and Particle Swarm Optimization. As we observe from Figures 2, 3 and 4, the optimal parameters (C, g) of Grid Search, Genetic Algorithms, and Particle Swarm Optimization are (1.74, 0.57), (1.68, 0.74), and (2.51, 0.43), respectively. After the optimal parametes (C, g) are found, we can obtain that the best prediction accuracy of the testing sample is 94.12%. The performance of SVM under the above mentioned three methods is summarized in Table 2.

3.3.2. Comparison of Prediction Accuracy of SVM, BPNN, MDA, and Logit. Back-propagation neural network (BPNN), multiple discriminant analysis (MDA), and logistic regression analysis (Logit) are widely applied methods for prediction. Therefore, in this paper, we compare the prediction accuracy of SVM with those of BPNN, MDA, and

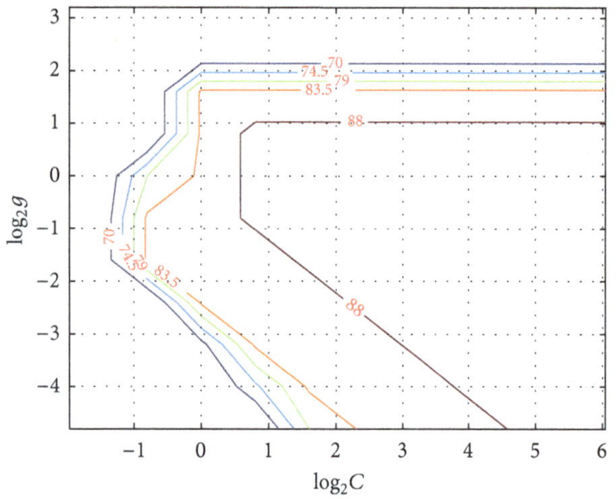

FIGURE 2: Contour lines of Grid Search. Best $C = 1.74$, $g = 0.57$, CV accuracy = 89.47%.

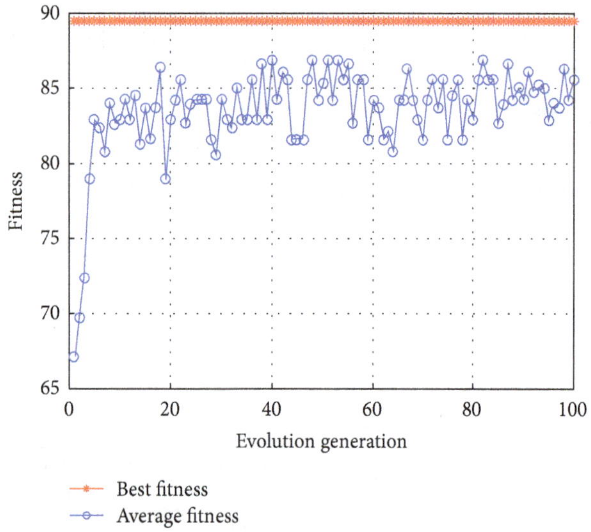

FIGURE 3: Fitness curve of Genetic Algorithms (termination of generation = 100, pop = 20). Best $C = 1.68$, $g = 0.74$, and CV accuracy = 89.47%.

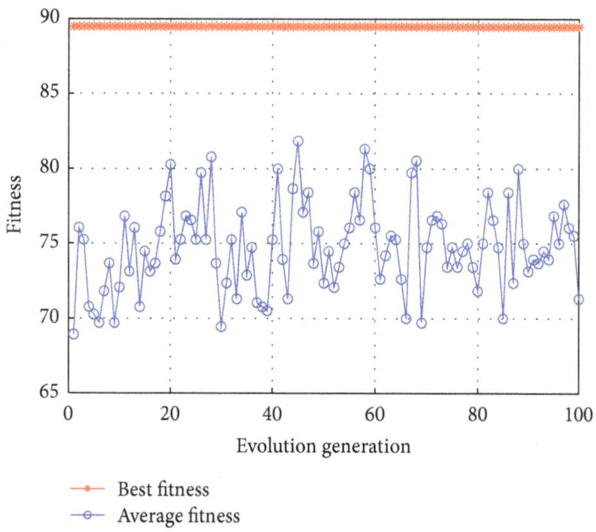

FIGURE 4: Fitness curve of Particle Swarm Optimization (termination of generation = 100, pop = 20, $c_1 = 1.50$, and $c_2 = 1.70$). Best $C = 2.51$, $g = 0.43$, and CV accuracy = 89.47%.

TABLE 3: The best prediction accuracy of SVM, BPNN, MDA, and Logit.

	SVM	BPNN	MDA	Logit
Training sample	100%	100%	94.74%	100%
Testing sample	94.12%	84.62%	76.47%	76.47%

TABLE 4: McNemar values (P values) for the pairwise comparison of performance.

	BPNN	MDA	Logit
SVM	3.37 (0.064)	11.12 (0.001)	10.32 (0.001)
BPNN		1.53 (0.175)	1.29 (0.122)
MDA			0.042 (1.000)

Logit. In case of BPNN, a three-layer fully connected back-propagation neural network is used as a benchmark; 5, 10, 15, 20, 25, and 30 hidden nodes in the hidden layer are analyzed; the maximum number of learning epochs is set to 1000; the learning rate is set to 0.01; the momentum term is set to 0.95; the activation function of the hidden layer and the output layer are, respectively, Tansig and Purelin; the training function is Trainlm. According to parameter adjustment, we can obtain that the best prediction accuracy of the testing sample is found when the number of hidden nodes is 20. The prediction accuracy of the testing sample, the training sample, and the validation sample is 84.62%, 100%, and 100%, respectively.

Table 3 summarizes the best prediction performance of SVM, BPNN, MDA, and Logit in the training sample and the testing sample. From Table 3, we can see that SVM outperforms BPNN, MDA, and Logit: 9.5%, 17.65% and 17.65%, respectively, for the testing sample. In addition, we conduct the McNemar test to examine whether SVM significantly outperforms the other three methods. Table 4 shows the results of the McNemar test. As shown in Table 4, SVM outperforms BPNN at 10% statistical significance level, and SVM outperforms MDA and Logit at 5% statistical significance level. However, Table 4 also shows that the prediction performances of BPNN, MDA, and Logit do not significantly differ from each other. The above result shows that support vector machine has certain advantages for the prediction of banking systemic risk with the complex nonlinear characteristics.

4. Conclusion

In this paper, we apply SVM to predict banking systemic risk. To validate the prediction performance of this approach, we conduct a case study of an SVM-based prediction model for Chinese banking systemic risk. First, we construct the prediction indicators for banking systemic risk from both Microprudential and macro-prudential perspectives, as well as the characteristics of interbank connections. Second, we adopt the principal component analysis to classify levels of banking systemic risk. Based on the above analysis, we conduct empirical analysis of prediction accuracy of the SVM-based prediction model for banking systemic risk. The results of empirical analysis show the capability, accuracy, and high efficiency of the SVM-based prediction model. In addition, compared with the BPNN-based prediction model, multiple discriminant analysis, and logistic regression analysis, the SVM-based prediction model shows superior prediction power. With these results, we claim that SVM can serve as a promising alternative in the prediction of banking systemic risk.

Acknowledgments

This research is supported by NSFC (no. 71071034, no. 71201023), NBRR (no. 2010CB328104-02), and Humanities and Social Science Youth Foundation of the Ministry of Education of China (no. 12YJC630101).

References

[1] J. A. Frankel and A. K. Rose, "Currency crashes in emerging markets: an empirical treatment," *Journal of International Economics*, vol. 41, no. 3-4, pp. 351–366, 1996.

[2] G. L. Kaminsky and C. M. Reinhart, "The twin crises: the causes of banking and balance-of-payments problems," *American Economic Review*, vol. 89, no. 3, pp. 473–500, 1999.

[3] A. Demirgüç-Kunt and E. Detragiache, "Monitoring banking sector fragility: a multivariate logit approach," *World Bank Economic Review*, vol. 14, no. 2, pp. 287–307, 2000.

[4] E. Hanschel and P. Monnin, "Measuring and forecasting stress in the banking sector: evidence from Switzerland," in *Investigating the Relationship between the Financial and Real Economy*, BIS Papers no. 22, 2005.

[5] E. P. Davis and D. Karim, "Comparing early warning systems for banking crises," *Journal of Financial Stability*, vol. 4, no. 2, pp. 89–120, 2008.

[6] M. Oet, R. Eiben, T. Bianco, D. Gramlich, S. Ong, and J. Wang, "SAFE: an early warning system for systemic banking risk," in *Proceedings of the 24th Australasian Finance and Banking Conference*, SSRN, 2011.

[7] C. S. Lin, H. A. Khan, R. Y. Chang, and Y. C. Wang, "A new approach to modeling early warning systems for currency crises: can a machine-learning fuzzy expert system predict the currency crises effectively?" *Journal of International Money and Finance*, vol. 27, no. 7, pp. 1098–1121, 2008.

[8] R. Duttagupta and P. Cashin, "Anatomy of banking crises: a Binary classification tree approach," IMF Working Paper 08/93, 2008.

[9] D. Karim, "The use of Binary recursive trees for banking crisis prediction," Tech. Rep., Brunel University Department of Economics and Finance Working Paper, Brunel University, London, UK, 2008.

[10] L. Laeven and F. Valencia, "Systemic banking crises: a new database," IMF Working Papers, 2008.

[11] M. L. Duca and T. A. Peltonen, "Assessing systemic risks and predicting systemic events," *Journal of Banking Finance*, 2012.

[12] G. Iori, G. de Masi, O. V. Precup, G. Gabbi, and G. Caldarelli, "A network analysis of the Italian overnight money market," *Journal of Economic Dynamics and Control*, vol. 32, no. 1, pp. 259–278, 2008.

[13] D. O. Cajueiro and B. M. Tabak, "The role of banks in the Brazilian interbank market: does bank type matter?" *Physica A: Statistical Mechanics and Its Applications*, vol. 387, no. 27, pp. 6825–6836, 2008.

[14] C. P. Georg, *The Effect of the Interbank Network Structure on Contagion and Common Shocks*, Discussion Paper Series 2: Banking and Financial Studies, 2011.

[15] V. N. Vapnik, *Statistical Learning Theory*, Adaptive and Learning Systems for Signal Processing, Communications, and Control, John Wiley & Sons, New York, NY, USA, 1998.

[16] J. J. Ahn, K. J. Oh, T. Y. Kim, and D. H. Kim, "Usefulness of support vector machine to develop an early warning system for financial crisis," *Expert Systems with Applications*, vol. 38, no. 4, pp. 2966–2973, 2011.

[17] Z. Xie, I. Lou, W. K. Ung, and K. M. Mok, "Freshwater algal bloom prediction by support vector machine in macau storage reservoirs," *Mathematical Problems in Engineering*, vol. 2012, Article ID 397473, 12 pages, 2012.

[18] G. Chen, Y. Zuo, J. Sun, and Y. Li, "Support-vector-machine-based reduced-order model for limit cycle oscillation prediction of nonlinear aeroelastic system," *Mathematical Problems in Engineering*, vol. 2012, Article ID 152123, 12 pages, 2012.

[19] J. Sun and H. Li, "Financial distress prediction using support vector machines: ensemble vs individual," *Applied Soft Computing*, vol. 12, no. 8, pp. 2254–2265, 2012.

[20] M. Arena, "Bank failures and bank fundamentals: a comparative analysis of Latin America and East Asia during the nineties using bank-level data," *Journal of Banking and Finance*, vol. 32, no. 2, pp. 299–310, 2008.

[21] A. Agresti, P. Baudino, and P. Poloni, "The ECB and IMF indicators for the macro-prudential analysis of the banking sector: a comparison of the two approaches," ECB Occasional Paper, 2008.

[22] C. M. Reinhart and K. S. Rogoff, *This Time Is Different: Eight Centuries of Financial Folly*, Princeton University Press, Princeton, NJ, USA, 2009.

[23] International Monetary Fund (IMF), *Global Financial Stability Report*, IMF, Washington, DC, USA.

[24] J. H. Min and Y. C. Lee, "Bankruptcy prediction using support vector machine with optimal choice of kernel function parameters," *Expert Systems with Applications*, vol. 28, no. 4, pp. 603–614, 2005.

Permissions

The contributors of this book come from diverse backgrounds, making this book a truly international effort. This book will bring forth new frontiers with its revolutionizing research information and detailed analysis of the nascent developments around the world.

We would like to thank all the contributing authors for lending their expertise to make the book truly unique. They have played a crucial role in the development of this book. Without their invaluable contributions this book wouldn't have been possible. They have made vital efforts to compile up to date information on the varied aspects of this subject to make this book a valuable addition to the collection of many professionals and students.

This book was conceptualized with the vision of imparting up-to-date information and advanced data in this field. To ensure the same, a matchless editorial board was set up. Every individual on the board went through rigorous rounds of assessment to prove their worth. After which they invested a large part of their time researching and compiling the most relevant data for our readers. Conferences and sessions were held from time to time between the editorial board and the contributing authors to present the data in the most comprehensible form. The editorial team has worked tirelessly to provide valuable and valid information to help people across the globe.

Every chapter published in this book has been scrutinized by our experts. Their significance has been extensively debated. The topics covered herein carry significant findings which will fuel the growth of the discipline. They may even be implemented as practical applications or may be referred to as a beginning point for another development. Chapters in this book were first published by Hindawi Publishing Corporation; hereby published with permission under the Creative Commons Attribution License or equivalent.

The editorial board has been involved in producing this book since its inception. They have spent rigorous hours researching and exploring the diverse topics which have resulted in the successful publishing of this book. They have passed on their knowledge of decades through this book. To expedite this challenging task, the publisher supported the team at every step. A small team of assistant editors was also appointed to further simplify the editing procedure and attain best results for the readers.

Our editorial team has been hand-picked from every corner of the world. Their multi-ethnicity adds dynamic inputs to the discussions which result in innovative outcomes. These outcomes are then further discussed with the researchers and contributors who give their valuable feedback and opinion regarding the same. The feedback is then collaborated with the researches and they are edited in a comprehensive manner to aid the understanding of the subject.

Apart from the editorial board, the designing team has also invested a significant amount of their time in understanding the subject and creating the most relevant covers. They scrutinized every image to scout for the most suitable representation of the subject and create an appropriate cover for the book.

The publishing team has been involved in this book since its early stages. They were actively engaged in every process, be it collecting the data, connecting with the contributors or procuring relevant information. The team has been an ardent support to the editorial, designing and production team. Their endless efforts to recruit the best for this project, has resulted in the accomplishment of this book. They are a veteran in the field of academics and their pool of knowledge is as vast as their experience in printing. Their expertise and guidance has proved useful at every step. Their uncompromising quality standards have made this book an exceptional effort. Their encouragement from time to time has been an inspiration for everyone.

The publisher and the editorial board hope that this book will prove to be a valuable piece of knowledge for researchers, students, practitioners and scholars across the globe.

List of Contributors

Carlos Fernández-Isla, Pedro J. Navarro and Pedro María Alcover
Universidad Polit´ecnica de Cartagena, Campus Muralla del Mar, 30202 Cartagena, Spain

Leopoldo Estrada Vargas and Deni Torres Roman
Department of Electrical Engineering, Center of Research and Advanced Studies (CINVESTAV), 45019 Guadalajara, JAL, Mexico

Homero Toral Cruz
Department of Sciences and Engineering, University of Quintana Roo (UQROO), 77019 Chetumal, QROO, Mexico

Dorin Sendrescu
Department of Automatic Control, University of Craiova, A.I. Cuza 13, 200585 Craiova, Romania

Feng Yu-Qin
School of Automobile and Traffic Engineering, Heilongjiang Institute of Technology, Harbin 150050, China

Leng Jun-Qiang
School of Management, Harbin Institute of Technology, Harbin 150001, China

Leng Jun-Qiang, He Yi and Zhang Gui-e
School of Automobile Engineering, Harbin Institute of Technology, Weihai 264209, China

Wang Peng
Representative Office of PLA in Harbin Railway Administration, Harbin 150001, China

Chunhui Wang, Xiaoyan Li and Xianqing Lv
Laboratory of Physical Oceanography, Ocean University of China, Qingdao 266100, China

Chunhui Wang
Key Laboratory of Marine Spill Oil Identification and Damage Assessment Technology, The Organization of North China Sea Monitoring Center, Qingdao 266033, China

Xiaoyan Li
Institute of Oceanology, Chinese Academy of Science, Qingdao 266071, China

Yinghui He
Department of Mathematics, Honghe University, Mengzi, Yunnan 661100, China

A. Valor, F. Caleyo and J. M. Hallen
Departamento de Ingenier´ıa Metal´urgica, IPN-ESIQIE, UPALM s/n, Edificio 7, Zacatenco, 07738 M´exico, DF, Mexico

L. Alfonso
Universidad Autónoma de la Ciudad de México, 09790 México, DF, Mexico

J. C. Velázquez
Departamento de Ingeniería Química Industrial, ESIQIE-IPN, UPALM Edificio 7, Zacatenco, 07738 México, DF, Mexico

Zhi-Liang Deng
School of Mathematical Sciences, University of Electronic Science and Technology of China, Chengdu 610054, China

Xiao-Mei Yang
School of Mathematics, Southwest Jiaotong University, Chengdu 610031, China

Xiao-Li Feng
Department of Mathematical Sciences, Xidian University, Xi'an 710071, China

Zhijun Zhang, Shiwei Zhang and Tianyi Su
School of Mechanical Engineering and Automation, Northeastern University, Shenyang 110004, China

Shuangshuang Zhao
Shenyang Aircraft Design and Research Institute, Shenyang 110035, China

Zhengchao Xie, Pak Kin Wong, Jing Zhao, Tao Xu, Ka In Wong and Hang Cheong Wong
Department of Electromechanical Engineering, University of Macau, Taipa, Macau, China

Zhipeng Li and Run Zhang
The Key Laboratory of Embedded System and Service Computing of Ministry of Education, Tongji University, Shanghai 201804, China

Hamid Reza Karimi
Department of Engineering, Faculty of Engineering and Science, University of Agder, 4898 Grimstad, Norway

Mohammed Chadli
Laboratory of Modeling, Information and Systems, University of Picardie Jules Verne, 80039 Amiens, France

Peng Shi
College of Engineering and Science, Victoria University, Melbourne, VIC 8001, Australia
School of Electrical and Electronic Engineering, The University of Adelaide, Adelaide, SA 5005, Australia

Yuming Zhai
School of Economics and Management, Shanghai Institute of Technology, Shanghai 201418, China
Antai College of Economics & Management, Shanghai Jiao Tong University, Shanghai 200052, China

Ruixia Yan
School of Management, Shanghai University of Engineering Science, Shanghai 201620, China

Haifeng Liu
College of Humanities, Donghua University, Shanghai 200051, China

Jinliang Liu
Department of Applied Mathematics, Nanjing University of Finance and Economics, Nanjing, Jiangsu 210046, China

Ming Xu, Jinlong Wang and Ang Zhang
Department of Aerospace Engineering, School of Astronautics, Beihang University, Beijing 100191, China

Shengli Liu
DFH Satellite Co., Ltd., Beijing 100094, China

Heli H
Key Laboratory of Manufacturing Industrial Integrated Automation, Shenyang University, Shenyang 110044, China

Dan Zhao
Department of Fundamental Teaching, Shenyang Institute of Engineering, Shenyang 110136, China

Qingling Zhang
Institute of Systems Science, Northeastern University, Shenyang, Liaoning 110004, China

Qing Wu and Wenqing Wang
School of Automation, Xi'an University of Posts and Telecommunications, Xi'an 710121, China

Liang Xu, Junping Du and Qingping Li
Beijing Key Laboratory of Intelligent Telecommunication Software and Multimedia, School of Computer Science, Beijing University of Posts and Telecommunications, Beijing 100876, China

Ying Liu, Zheng Lv and Wei Wang
School of Control Sciences and Engineering, Dalian University of Technology, Dalian 116023, China

Li Li, Shuiyuan Cheng, Jianlei Lang and Dongsheng Chen
College of Environmental & Energy Engineering, Beijing University of Technology, Beijing 100124, China

Li Li
Beijing General Research Institute of Mining & Metallurgy, Beijing 100070, China

Jianbing Li
Environmental Engineering Program, University of Northern British Columbia, Prince George, Canada V2N 4Z9

T. S. Ozsahin and O. Taskıner
Civil Engineering Department, Karadeniz Technical University, 61080 Trabzon, Turkey

Shouwei Li, Mingliang Wang and Jianmin He
School of Economics and Management, Southeast University, Nanjing, Jiangsu 211189, China

www.ingramcontent.com/pod-product-compliance
Lightning Source LLC
Chambersburg PA
CBHW080704200326
41458CB00013B/4964